Methods in Molecular Biology

Series Editor
John M. Walker
School of Life and Medical Sciences
University of Hertfordshire
Hatfield, Hertfordshire, UK

For over 35 years, biological scientists have come to rely on the research protocols and methodologies in the critically acclaimed Methods in Molecular Biology series. The series was the first to introduce the step-by-step protocols approach that has become the standard in all biomedical protocol publishing. Each protocol is provided in readily-reproducible step-by step fashion, opening with an introductory overview, a list of the materials and reagents needed to complete the experiment, and followed by a detailed procedure that is supported with a helpful notes section offering tips and tricks of the trade as well as troubleshooting advice. These hallmark features were introduced by series editor Dr. John Walker and constitute the key ingredient in each and every volume of the Methods in Molecular Biology series. Tested and trusted, comprehensive and reliable, all protocols from the series are indexed in PubMed.

Manuscript Instructions https://drive.google.com/file/d/1K4Az5vpAFVzYadJ3jKx2-QSo7Dqa_SNA/view?usp=sharing

Molecular Markers in Plants

Methods and Protocols

Edited by

Nitish Kumar

Department of Biotechnology, Central University of South Bihar, Gaya, Bihar, India

Humana Press

Editor
Nitish Kumar
Department of Biotechnology
Central University of South Bihar
Gaya, Bihar, India

ISSN 1064-3745 ISSN 1940-6029 (electronic)
Methods in Molecular Biology
ISBN 978-1-0716-5662-4 ISBN 978-1-0716-5663-1 (eBook)
https://doi.org/10.1007/978-1-0716-5663-1

This Humana imprint is published by the registered company Springer Science+Business Media, LLC, part of Springer Nature.
The registered company address is: 1 New York Plaza, New York, NY 10004, U.S.A.

Preface

The present book, titled *Molecular Markers in Plants: Methods and Protocols*, has been developed with the objective of providing a comprehensive yet accessible account of various molecular marker systems and their practical applications. Over the years, a wide range of marker techniques have been developed, each with its own principles, advantages, and limitations. Early methods such as Restriction Fragment Length Polymorphism (RFLP) laid the foundation for DNA-based analysis, while subsequent innovations including Random Amplified Polymorphic DNA (RAPD), Amplified Fragment Length Polymorphism (AFLP), and Simple Sequence Repeats (SSR) expanded the scope and efficiency of genetic studies. More recently, single nucleotide polymorphism (SNP) markers and high-throughput sequencing technologies have further revolutionized the field, enabling large-scale genomic analysis with remarkable accuracy.

This book seeks to bridge the gap between theoretical understanding and practical implementation by presenting detailed methodologies and standardized protocols for commonly used molecular marker techniques. Special emphasis has been placed on clarity, reproducibility, and step-by-step guidance, making the content suitable for students, researchers, and laboratory practitioners alike. Each method is described with careful attention to essential steps such as DNA extraction, quantification, amplification, and analysis, along with important considerations to ensure accuracy and reliability of results. By integrating both fundamental concepts and applied techniques, this work aims to serve as a practical guide for those engaged in plant molecular research.

Despite the remarkable progress achieved, the application of molecular markers requires careful planning, technical expertise, and an understanding of their limitations. Factors such as DNA quality, primer specificity, experimental conditions, and data interpretation play a critical role in determining the success of these techniques. Therefore, this book also addresses common challenges and provides practical insights to help users avoid errors and optimize their experimental outcomes.

It is hoped that this book will serve as a valuable resource for undergraduate and postgraduate students, educators, and researchers in the field of plant sciences and biotechnology. By offering a balanced combination of conceptual knowledge and practical guidance, it aims to inspire further exploration and innovation in the use of molecular markers. Ultimately, a deeper understanding and effective application of these techniques will contribute to sustainable agriculture, biodiversity conservation, and the advancement of scientific knowledge.

Gaya, Bihar, India *Nitish Kumar*

Acknowledgments

First and foremost, I would like to praise and thank God, the Almighty, who has granted countless blessings, knowledge, and opportunity to accomplish the book project work. Thanks to all the authors of the various chapters for their contributions. It has been a bit of a long process from the initial outlines to developing the full chapters and then revising them in the light of reviewer comments. I sincerely acknowledge the authors' willingness to go through this process. I also acknowledge the work and knowledge of the members of our review panels, many of whom had to be done at short notice. Thanks to all the people at Springer Nature, especially Dr. John Walker and Patrick Marton, with whom I corresponded for their advice and facilitation in the production of this book. I am grateful to my family members Mrs. Kiran (Wife), Miss Kartika Sharma and Laavanya Sharma (Daughters), and my parents for their incredible and selfless support all the time.

Gaya, Bihar, India *Nitish Kumar*

Contents

Contributors

ABDUL HASEEB AHMED • *Institute of Horticultural Sciences, University of Agriculture, Faisalabad, Pakistan*

MUHAMMAD TAHIR AKRAM • *Department of Horticulture, PMAS-Arid Agriculture University, Rawalpindi, Pakistan*

ABDULLAH AL-SADI • *Department of Plant Sciences, College of Agricultural and Marine Science, Sultan Qaboos University, Muscat, Oman; College of Agriculture, University of Al Dhaid, Sharjah, UAE*

DEMET ALTINDAL • *Organic Farming Programme, Department of Plant and Animal Production, Fethiye Ali Sıtkı Mefharet Koçman Vocational School, Muğla Sıtkı Koçman University, Muğla, Türkiye*

NÜKET ALTINDAL • *Medical and Aromatic Plants Programme, Department of Plant and Animal Production, Sivaslı Vocational School, Uşak University, Uşak, Türkiye*

NORADDIN HOSSEINPOUR AZAD • *Department of Plant Science and Medicinal Plants, Meshginshahr Faculty of Agriculture, University of Mohaghegh Ardabili, Ardabil, Iran*

K. BASAVARAJ • *ICAR-Indian Institute of Rice Research (IIRR), Hyderabad, Telangana, India*

YUGANTI BEHERA • *Nabapravat Academy for Management Education, Salipur, Cuttack, Odisha, India*

LOUBNA BELQADI • *Department of Plant Production, Protection and Biotechnologies, Institut Agronomique et Vétérinaire Hassan II, Rabat, Morocco*

SNEHA BHUJBAL • *Molecular Biology and Genetic Engineering Laboratory, Vasantdada Sugar Institute, Pune, Maharashtra, India*

SOWPARNIKA BINU • *Department of Botany, Kongunadu Arts and Science College (Autonomous), Bharathiar University, Coimbatore, Tamil Nadu, India*

ETIENNE BUCHER • *Agroscope, Crop Genome Dynamics Group, Nyon, Switzerland*

RIA CAHYANINGSIH • *Research Centre for Biota Systems, National Research and Innovation Agency (BRIN), Jakarta, Indonesia*

BANDELA CHANDANA • *Department of Genetics and Plant Breeding, Bidhan Chandra Krishi Viswavidyalaya, Mohanpur, West Bengal, India*

C.M.N.T. CHANDRASEKARA • *Department of Crop Science, Faculty of Agriculture, University of Ruhuna, Kamburupitiya, Sri Lanka*

OUMAIMA CHETTO • *Institut National de la Recherche Agronomique, Research Unit of Plant Breeding and Plant Genetic Resources Conservation, Rabat, Morocco; Department of Plant Production, Protection and Biotechnologies, Institut Agronomique et Vétérinaire Hassan II, Rabat, Morocco*

SUVARNA RANI CHIMMILI • *ICAR-Indian Institute of Rice Research (IIRR), Hyderabad, Telangana, India*

PATITAPABAN DASH • *Centre for Biotechnology, Siksha 'O' Anusandhan (Deemed to be University), Bhubaneswar, Odisha, India*

RACHAYYA M. DEVARUMATH • *Molecular Biology and Genetic Engineering Laboratory, Vasantdada Sugar Institute, Pune, Maharashtra, India*

MOUNIA ENNAMI • *Department of Plant Production, Protection and Biotechnologies, Institut Agronomique et Vétérinaire Hassan II, Rabat, Morocco*
D.L.C.K. FONSEKA • *Department of Crop Science, Faculty of Agriculture, University of Ruhuna, Kamburupitiya, Sri Lanka*
H.P. GAJERA • *Department of Biotechnology, College of Agriculture, Junagadh Agricultural University, Junagadh, Gujarat, India*
SATHEESH GEORGE • *Department of Botany, St. Joseph's College (Autonomous) Devagiri, Medical College, Devagiri, Kozhikode, Kerala, India*
MADHAVI Y. GHAN • *Molecular Biology and Genetic Engineering Laboratory, Vasantdada Sugar Institute, Pune, Maharashtra, India*
SATYASWARA JASUDASU GOMPA • *ICAR-Indian Institute of Rice Research (IIRR), Hyderabad, Telangana, India*
SHOUVIK GORAI • *Department of Genetics and Plant Breeding, Bidhan Chandra Krishi Viswavidyalaya, Mohanpur, West Bengal, India; Department of Genetics and Plant Breeding, School of Agriculture and Allied Sciences, Seacom Skills University, Bolpur, India*
RAZIEH GOUDARZI • *Department of Tissue and Cell Culture, Agricultural Biotechnology Research Institute of Iran (ABRII), Agricultural Research, Education and Extension Organization (AREEO), Karaj, Iran*
DHANDAPANI GURUSAMY • *Department of Botany, Kongunadu Arts and Science College (Autonomous), Bharathiar University, Coimbatore, Tamil Nadu, India*
SOHAM HAZRA • *Department of Agriculture, Brainware University, Barasat, West Bengal, India*
DARSHNA G. HIRPARA • *Department of Biotechnology, College of Agriculture, Junagadh Agricultural University, Junagadh, Gujarat, India*
MD IMTIYAZ • *Department of Genetics and Plant Breeding, Bidhan Chandra Krishi Viswavidyalaya, Mohanpur, West Bengal, India*
ASHOK JAGTAP • *Molecular Biology and Genetic Engineering Laboratory, Vasantdada Sugar Institute, Pune, Maharashtra, India*
TENNYSON JEBASINGH • *Department of Plant Sciences, School of Biological Sciences, Madurai Kamaraj University, Madurai, Tamil Nadu, India*
AMANPREET KAUR • *Department of Biochemistry, Purdue University, West Lafayette, IN, USA*
MUHAMMAD MUMTAZ KHAN • *Department of Plant Sciences, College of Agricultural and Marine Science, Sultan Qaboos University, Muscat, Oman*
MEHRANA KOOHI-DEHKORDI • *Secondary Metabolites Biotechnology Research Institute, Agricultural Biotechnology Research Institute of Iran (ABRII), Agricultural Research, Education and Extension Organization (AREEO), Isfahan, Iran*
MOHAMED KOUIGHAT • *Research Unit of Plant Breeding and Plant Genetic Resources Conservation, Regional Agricultural Research Center of Meknes, Institut National de la Recherche Agronomique, Rabat, Morocco*
VASANTH KRISHNAN • *Department of Botany, Bharathiar University, Coimbatore, Tamil Nadu, India*
ATUL KUMAR • *Molecular Biology and Genetic Engineering Laboratory, Vasantdada Sugar Institute, Pune, Maharashtra, India*
W.G.I. MADUSHANI • *Department of Crop Science, Faculty of Agriculture, University of Ruhuna, Kamburupitiya, Sri Lanka*

ANIRBAN MAJI • *Department of Genetics and Plant Breeding, Bidhan Chandra Krishi Viswavidyalaya, Mohanpur, West Bengal, India*
V. MANASA • *ICAR-Indian Institute of Rice Research (IIRR), Hyderabad, Telangana, India*
S.K. MANGRAUTHIA • *ICAR-Indian Institute of Rice Research (IIRR), Hyderabad, Telangana, India*
SUDHA MANICKAM • *Department of Plant Biotechnology, Centre for Plant Molecular Biology and Biotechnology, Tamil Nadu Agricultural University, Coimbatore, India*
IRFAN MARTIANSYAH • *Research Centre for Applied Botany, National Research and Innovation Agency (BRIN), Jakarta, Indonesia*
ELAHE MOATAMED • *Department of Genetic Engineering and Biosafety, Agricultural Biotechnology Research Institute of Iran (ABRII), Agricultural Research, Education and Extension Organization (AREEO), Karaj, Iran*
KANAKACHARI MOGILICHERLA • *Department of Biotechnology, ICAR-Indian Institute of Rice Research (IIRR), Hyderabad, Telangana, India*
VIGNESH MOHANAVEL • *Department of Plant Biotechnology, Centre for Plant Molecular Biology and Biotechnology, Tamil Nadu Agricultural University, Coimbatore, India*
MOTAHHAREH MOHSENPOUR • *Department of Genetic Engineering and Biosafety, Agricultural Biotechnology Research Institute of Iran (ABRII), Agricultural Research, Education and Extension Organization (AREEO), Karaj, Iran*
AFRIDI MONDAL • *Department of Genetics and Plant Breeding, Bidhan Chandra Krishi Viswavidyalaya, Mohanpur, West Bengal, India*
SEYYED MOHAMMAD MOUSAVI PAKZAD • *Department of Genetic Engineering and Biosafety, Agricultural Biotechnology Research Institute of Iran (ABRII), Agricultural Research, Education and Extension Organization (AREEO), Karaj, Iran*
GOKUL MURUGESAN • *Department of Botany, Bharathiar University, Coimbatore, Tamil Nadu, India*
RAVEENDRAN MUTHURAJAN • *Department of Plant Biotechnology, Centre for Plant Molecular Biology and Biotechnology, Tamil Nadu Agricultural University, Coimbatore, India*
ABDELGHANI NABLOUSSI • *Research Unit of Plant Breeding and Plant Genetic Resources Conservation, Regional Agricultural Research Center of Meknes, Institut National de la Recherche Agronomique, Rabat, Morocco*
BINDU R. NAIR • *Department of Botany, University of Kerala, Kariavattom, Thiruvananthapuram, Kerala, India*
TANU PRIYA NATARAJAN • *Department of Botany, Bharathiar University, Coimbatore, Tamil Nadu, India*
RAKSHANA PALANISWAMY • *Department of Plant Biotechnology, Centre for Plant Molecular Biology and Biotechnology, Tamil Nadu Agricultural University, Coimbatore, India*
W.M.A.C. PREMASIRI • *Department of Crop Science, Faculty of Agriculture, University of Ruhuna, Kamburupitiya, Sri Lanka*
ARIEF PRIYADI • *Research Centre for Applied Botany, National Research and Innovation Agency (BRIN), Jakarta, Indonesia*
MADHAVI PURANKAR • *Molecular Biology and Genetic Engineering Laboratory, Vasantdada Sugar Institute, Pune, Maharashtra, India*
RASHAD WASEEM QADRI • *Institute of Horticultural Sciences, University of Agriculture, Faisalabad, Pakistan*

SRI RAHAYU • *Research Centre for Applied Botany, National Research and Innovation Agency (BRIN), Jakarta, Indonesia*
VEERA RANJANI RAJAGOPALAN • *Department of Plant Biotechnology, Centre for Plant Molecular Biology and Biotechnology, Tamil Nadu Agricultural University, Coimbatore, India*
RAGHU RAJASEKARAN • *Department of Plant Biotechnology, Centre for Plant Molecular Biology and Biotechnology, Tamil Nadu Agricultural University, Coimbatore, India*
P. RANJITH • *Odisha University of Agriculture & Technology (OUAT), Chiplima, Odisha, India*
R.M.Y.I. RATHNYAKE • *Department of Crop Science, Faculty of Agriculture, University of Ruhuna, Kamburupitiya, Sri Lanka*
SOUMYA S.L. • *Department of Botany, St. Joseph's College (Autonomous) Devagiri, Medical College, Devagiri, Kozhikode, Kerala, India*
RIFKI SADIKIN • *Research Center for Computing, National Research and Innovation Agency (BRIN), Jakarta, Indonesia*
KHUSBOO SAHU • *Centre for Biotechnology, Siksha 'O' Anusandhan (Deemed to be University), Bhubaneswar, Odisha, India*
S.V. SAI PRASAD • *ICAR-Indian Institute of Rice Research (IIRR), Hyderabad, Telangana, India*
SUBHASIS SAMANTA • *Department of Genetics and Plant Breeding, Bidhan Chandra Krishi Viswavidyalaya, Mohanpur, West Bengal, India*
MUKESH SANKAR S. • *ICAR-All India Coordinated Research Project on Spices, Indian Institute of Spices Research, Marikunnu, Kozhikode, Kerala, India*
DIPAK SARODE • *Molecular Biology and Genetic Engineering Laboratory, Vasantdada Sugar Institute, Pune, Maharashtra, India*
DELSE P. SEBASTIAN • *Department of Botany, St. Joseph's College (Autonomous) Devagiri, Medical College, Devagiri, Kozhikode, Kerala, India*
MEHRAN E. SHARIATPANAHI • *Department of Tissue and Cell Culture, Agricultural Biotechnology Research Institute of Iran (ABRII), Agricultural Research, Education and Extension Organization (AREEO), Karaj, Iran*
KANNAN SHRINITHI • *Department of Plant Sciences, School of Biological Sciences, Madurai Kamaraj University, Madurai, Tamil Nadu, India*
GURMUKH SINGH JOHAL • *Department of Botany and Pathology, Purdue University, West Lafayette, IN, USA*
R.C. SIVABHARATHI • *Department of Plant Biotechnology, Centre for Plant Molecular Biology and Biotechnology, Tamil Nadu Agricultural University, Coimbatore, India*
JANANI SREE SIVAKUMAR • *Department of Botany, Bharathiar University, Coimbatore, Tamil Nadu, India*
M. SRINIVAS PRASAD • *ICAR-Indian Institute of Rice Research (IIRR), Hyderabad, Telangana, India*
K. SRUTHI • *ICAR-Indian Institute of Rice Research (IIRR), Hyderabad, Telangana, India*
RAVIMURUGAN SUBASHINI • *Department of Plant Sciences, School of Biological Sciences, Madurai Kamaraj University, Madurai, Tamil Nadu, India*
R.M. SUNDARAM • *ICAR-Indian Institute of Rice Research (IIRR), Hyderabad, Telangana, India*
K. SUNEETHA • *ICAR-Indian Institute of Rice Research (IIRR), Hyderabad, Telangana, India*

KAPIL SUSHIR • *Plant Breeding Division, Vasantdada Sugar Institute, Pune, Maharashtra, India*
SHIVANI TALEKAR • *Molecular Biology and Genetic Engineering Laboratory, Vasantdada Sugar Institute, Pune, Maharashtra, India*
V. UMESH KUMAR • *Department of Genetics and Plant Breeding, Sri Konda Laxman Telangana Horticultural University, Telangana, India*
LANKA NAGA KRISHNA USHASREE • *Department of Biotechnology, Loyola Academy (Autonomous), Secunderabad, Telangana, India*
K. VARADHAN • *Department of Genetics and Plant Breeding, SOAS, Malla Reddy University, Hyderabad, India*
YOGESH VIKAL • *School of Agricultural Biotechnology, Punjab Agricultural University, Ludhiana, India*
VIVI YUSKIANTI • *Research Centre for Biota Systems, National Research and Innovation Agency (BRIN), Jakarta, Indonesia*
RASOOL ASGHARI ZAKARIA • *Department of Plant Production and Genetics, Faculty of Agriculture and Natural Resources, University of Mohaghegh Ardabili, Ardabil, Iran*

Chapter 1

Application of Sequence-Related Amplified Polymorphism (SRAP) Markers for Assessing Molecular Diversity of Trichoderma Associated with Biocontrol Potential Against Phytopathogen

Darshna G. Hirpara and H.P. Gajera

Abstract

The genus *Trichoderma* comprises potent biocontrol fungi known for their antagonistic activity against diverse phytopathogens. Understanding the genetic diversity among *Trichoderma* isolates is essential for identifying strains with enhanced biocontrol potential. Sequence-related amplified polymorphism (SRAP) markers, which target open reading frames (ORFs) and associated regulatory regions, offer a robust and efficient molecular tool for diversity analysis. This chapter describes a detailed protocol for genomic DNA extraction, SRAP-PCR amplification, and data analysis for *Trichoderma* isolates. The approach enables reliable assessment of genetic variation and its correlation with biocontrol traits, such as antagonism against *Sclerotium rolfsii* and production of antifungal metabolites. Representative primer combinations and analytical strategies are provided to facilitate reproducibility and effective strain characterization for biocontrol applications.

Keywords *Trichoderma*, SRAP markers, Molecular diversity, Biocontrol, *Sclerotium rolfsii*, Genetic analysis, PCR, Secondary metabolites

1 Introduction

Biological control using beneficial microorganisms, particularly *Trichoderma* species, has emerged as an eco-friendly and sustainable alternative to chemical fungicides for managing soilborne and foliar plant diseases. *Trichoderma* species are renowned for their diverse mechanisms of antagonism, including mycoparasitism, competition for nutrients and space, and the secretion of antifungal secondary metabolites, which collectively inhibit the growth and spread of plant pathogens [1]. The efficacy of *Trichoderma* as a biocontrol agent (BCA) is highly strain-specific, emphasizing the importance of identifying elite strains with superior antagonistic capabilities for both commercial applications and ecological disease management.

Nitish Kumar (ed.), *Molecular Markers in Plants*, Methods in Molecular Biology, vol. 3097,
https://doi.org/10.1007/978-1-0716-5663-1_1,

Globally, the use of *Trichoderma*-based biocontrol agents has gained significant attention in sustainable agriculture, particularly in regions where chemical fungicides pose environmental and health risks. Several studies have reported the successful application of *Trichoderma* strains to control economically important pathogens such as *Sclerotium rolfsii*, *Rhizoctonia solani*, and *Fusarium* species, demonstrating their potential to reduce yield losses while maintaining soil health. The development and commercialization of elite *Trichoderma* strains require a thorough understanding of their genetic and functional diversity, as these factors directly influence their biocontrol efficacy, adaptability, and metabolite production.

Molecular markers have become indispensable tools for assessing genetic variation and elucidating phylogenetic relationships among microbial isolates. Among these, sequence-related amplified polymorphism (SRAP) markers have emerged as a powerful method for analyzing functional genetic diversity. SRAP is a PCR-based marker system that targets coding regions of the genome, particularly open reading frames (ORFs). It utilizes a pair of primers: a forward primer that anneals to exonic regions and a reverse primer that targets intronic or promoter regions. The resulting polymorphic fragments largely reflect functional variation, making SRAP markers particularly valuable for linking molecular diversity to phenotypic traits such as antagonistic activity, secondary metabolite production, and overall biocontrol potential [1].

Previous studies have successfully employed SRAP markers to characterize genetic diversity in *Trichoderma* isolates. For instance, SRAP profiling has been used to differentiate isolates with varying antifungal activity, to cluster strains based on biocontrol efficacy, and to identify candidate genotypes suitable for large-scale production. These studies demonstrate that SRAP markers provide reliable, reproducible, and functionally informative genetic profiles, making them an ideal choice for selecting elite biocontrol strains.

This chapter presents a comprehensive methodology for applying SRAP markers to assess the molecular diversity of *Trichoderma* isolates, with a focus on their biocontrol potential against the devastating plant pathogen *Sclerotium rolfsii*. By integrating molecular profiling and bioactive constituents with phenotypic evaluation, this approach aims to identify robust, functionally superior strains, thereby providing a foundation for sustainable disease management and the development of commercially viable BCAs.

2 Materials

2.1 Media and Reagents

- Potato dextrose agar (PDA).
- Synthetic medium for bioactive compound extraction.

- Ethyl acetate.
- Chloroform.
- TBA, NaCl.
- GC-MS grade solvents.

2.2 Equipment

- Laminar airflow cabinet.
- Incubator shaker (28°C, 125 rpm).
- Centrifuge (up to 10,000 rpm).
- UV–visible spectrophotometer.
- Gel electrophoresis apparatus.
- Thermal cycler (PCR).
- GC–MS system with RT×5MS capillary column.
- Light microscope with imaging system.
- Rotary vacuum evaporator.

2.3 Fungal Cultures and Growth

- Fungal isolates: Pure cultures of *Trichoderma* strains maintained on Potato Dextrose Agar (PDA) slants.

Isolate no.	*Trichoderma* species	Source	Accession no.
T1	*T. harzianum*	NBAII	Th1
T2	*T. harzianum*	NRRL	13879
T3	*T. harzianum*	NRRL	20565
T4	*T. harzianum*	JAU	Local isolate
T5	*T. viride*	NBAII	Tv23
T6	*T. viride*	NRRL	6418
T7	*T. virens*	NBAII	Tvs12
T8	*T. hamatum*	NBAII	Tha1
T9	*T. koningii*	MTCC	796
T10	*T. pseudokoningii*	MTCC	2048
T11	*T. species*	NRRL	5242
T12	Test pathogen *S. rolfsii*	ITCC	9107.13

- Growth medium: Potato dextrose broth (PDB) for mycelial biomass production.
- Sterile equipment: Petri dishes, conical flasks, surgical blades, pipettes, and sterile water.

2.4 Genomic DNA Extraction

Reagents required for the modified cetyl trimethylammonium bromide (CTAB) method:

- CTAB extraction buffer: Containing CTAB, NaCl, Tris-HCl, EDTA, and 2% (w/v) Polyvinylpyrrolidone (PVP-20).
- RNase A solution.
- Organic solvents.
  1. Chloroform: Isoamyl alcohol (24:1, v/v).
  2. Isopropanol (pre-chilled).
- Washing solution: 70% (v/v) ethanol (pre-chilled).
- DNA elution buffer: TE buffer (10 mM Tris-HCl, 1 mM EDTA, pH 8.0) or sterile deionized water.

2.5 SRAP-PCR Amplification

- Template DNA: High-quality genomic DNA (25–50 ng/μL) from each isolate.
- Primers: Sets of forward (ME) and reverse (EM) SRAP primers (see Table 1).
- PCR reagents:
  - Taq DNA polymerase (1 U per reaction).
  - 10× PCR buffer with $MgCl_2$.
  - dNTPs (0.2 mM each).
  - Sterile deionized water.

2.6 Electrophoresis and Visualization

- Agarose gel: 1.5%–2.0% (w/v) in 1× TBE buffer.
- DNA ladder: 100 bp or 1 kb standard.
- Loading dye and DNA stain: Ethidium bromide or a safer alternative.
- Gel documentation system: For visualization and image capture.

Table 1
Representative SRAP primer combinations for *Trichoderma* molecular diversity analysis

Primer combination	Forward primer (ME)	Reverse primer (EM)	Bands scored	% Polymorphism
ME1/EM2	TGAGTCCAAACCGGATA	GACTGCGTACCA ATTCCG	10	90.0
ME2/EM4	TGAGTCCAAACCGGAAG	GACTGCGTACCAA TTCGA	12	100.0
ME3/EM3	TGAGTCCAAACCGGTGA	GACTGCGTACCAA TTCCC	11	90.9

Source: [1]

3 Methods

3.1 Dual Culture Antagonism Assay

1. Place 5 mm mycelial discs of *S. rolfsii* and *Trichoderma* at opposite sides of PDA plates.
2. Incubate at 28°C for 7 days.
3. Measure radial growth of *S. rolfsii* in control and dual plates.
4. Calculate growth inhibition (%).
5. Record number of sclerotia formed per plate at 7 days after inoculation (DAI).
6. Perform three independent replicates per isolate.

3.2 Microscopic Examination of Antagonism

1. Prepare a thin PDA layer on a sterile glass slide.
2. Inoculate *Trichoderma* and *S. rolfsii* at opposite ends.
3. Incubate in a moist chamber at 28°C for 7 days.
4. Observe interaction using a light microscope at 10× and 100× magnification.
5. Document hyphal coiling, hook-like structures, and lysis as indicators of mycoparasitism (Fig. 1).

3.3 Extraction and GC–MS Profiling of Bioactive Compounds

1. Grow *Trichoderma* isolates in 250 mL synthetic medium (SM) containing 0.1% glucose or *S. rolfsii* cell wall as a carbon source.
2. Incubate for 7 days at 28°C and 125 rpm.
3. Centrifuge at 8000 rpm for 15 min and filter supernatant.
4. Extract with an equal volume of ethyl acetate.
5. Dry over anhydrous sodium sulfate and evaporate to dryness.
6. Reconstitute the extract in ethyl acetate for GC–MS.
7. Perform GC–MS analysis using RT×5MS column; injector temperature 280°C; helium flow 1 mL min^{-1}; column temp 80–290°C at 6°C min^{-1} ramp.
8. Identify compounds using the NIST library; record molecular weight and biological activity (Fig. 2).

3.4 Genomic DNA Isolation from* Trichoderma *[3]

1. Biomass preparation: Inoculate of PDB with a fungal mycelial plug from a fresh culture and incubate at for 4 days in a shaker incubator (28°C) to obtain sufficient mycelial growth.
2. Mycelia harvesting and grinding: Filter the mycelial mat through sterile muslin cloth, rinse with sterile water, blot dry, and flash-freeze the tissue in liquid nitrogen. Grind approximately half of the frozen mycelia into a fine powder using a sterile mortar and pestle.
3. Lysis: Transfer the powder to a microcentrifuge tube. Add of pre-heated 2× CTAB extraction buffer (*see* **Note 1**). Mix

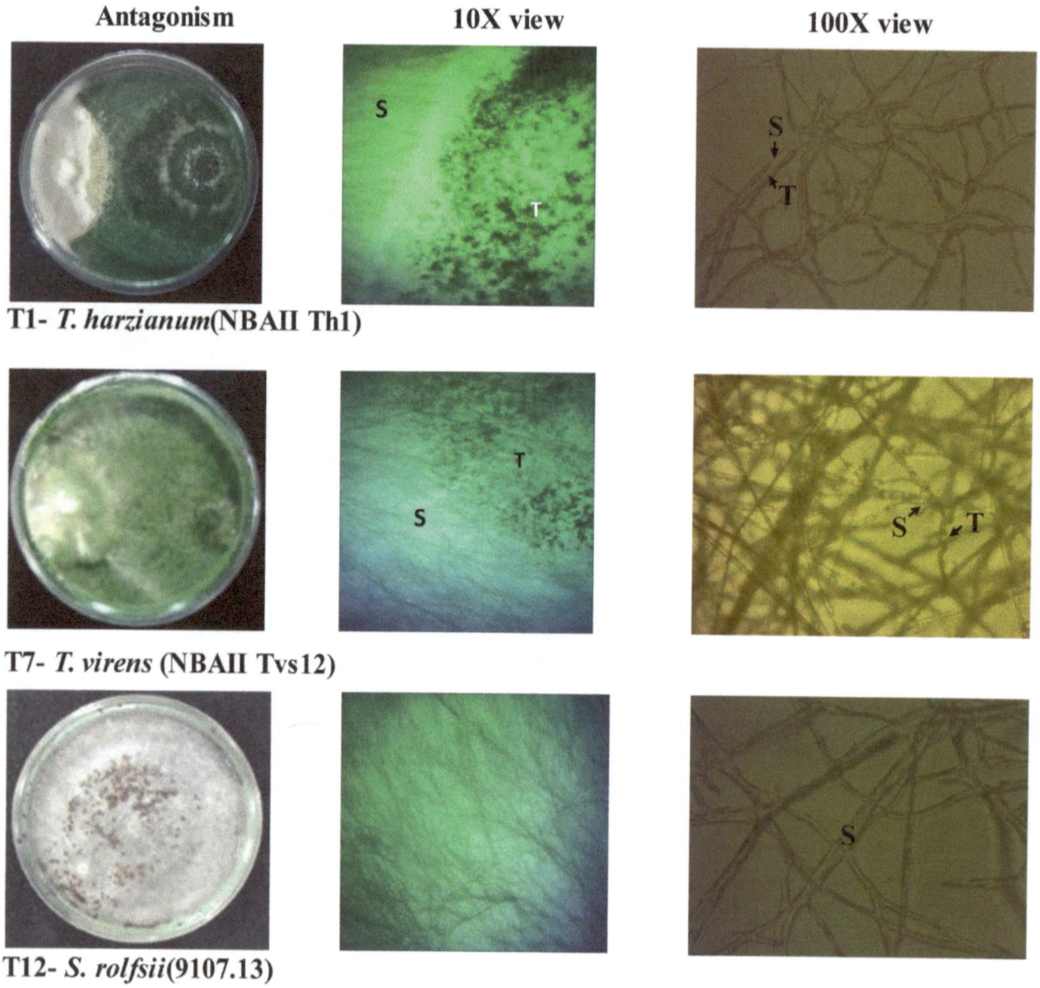

Fig. 1 In vitro antagonism and microscopic characterization of *Trichoderma* inhibiting *Sclerotium rolfsii* at 7 DAI under 10× and 100×(Source: [1])

gently and incubate at 65°C for 45 min, in a water bath with gentle swirling.

4. RNase treatment: Add RNase A (50 μg/mL) and incubate for about 30 min at 37°C.

5. Organic extraction: Add of Chloroform:Isoamyl alcohol (24:1) and mix thoroughly by gentle inversion for 15 times to ensure emulsification of the phase for minutes. Centrifuge 11,000 rpm for 10 min at room temperature. Carefully transfer the upper aqueous layer to a new microcentrifuge tube. Repeat the chloroform extraction step once. Transferred aqueous phase to another fresh tube.

6. DNA precipitation: Add an equal volume of chilled isopropanol and 50 μL of 3 M sodium acetate to the aqueous phase.

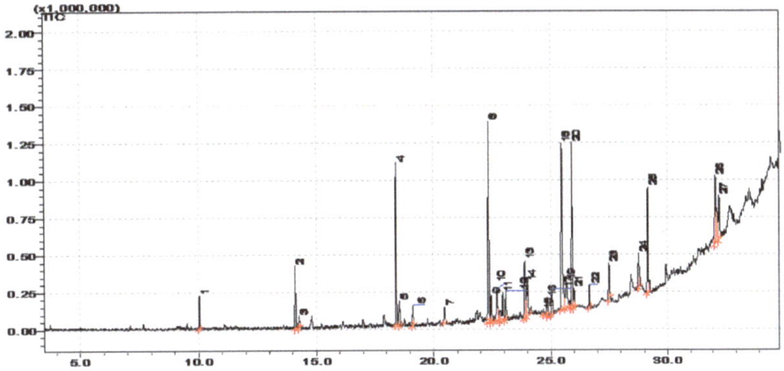

Fig. 2 GC-MS chromatogram of bioactive compounds from potent *T. virens* NBAII Tvs12 (T7) grown on pathogen cell wall at 7 DAI(Source: [1])

Mix gently and incubate at −20°C for 60 min for precipitation of DNA. The tubes were centrifuged at 15,000 rpm, 4°C for 15 min. to pellet the DNA.

7. Washing and drying: Wash the DNA pellet twice with chilled ethanol. Air-dry the pellet briefly (about minutes) until all ethanol evaporates.
8. Elution and quantitation: Dissolve the DNA pellet in TE buffer or sterile water. Quantify the DNA concentration and purity using a NanoDrop spectrophotometer (A260/A280 ratio should be 1.80 to 2.00) and normalize the concentration for PCR.
9. DNA Gel Electrophoresis: The DNA gel was run on 0.8% agarose (Fig. 3).

3.5 SRAP-PCR Amplification [2]

Molecular characterization was performed using SRAP markers following the method of Li and Quiros [2] with minor modifications. The representative list of SRAP primer combinations used in this study is provided in Table 1.

- Utilize a total of 256 SRAP primer combinations in a preliminary test.
- Screen 116 primer combinations (*see* **Note 2**) that produce clear and reproducible amplification profiles for further analysis.
- Use these 116 primer combinations for molecular diversity analysis.
- Set up PCR reactions (Table 2) and perform amplification in a thermal cycler under optimized conditions specified in Table 3 (*see* **Note 3**).
- Run the amplified products on a 1.5% agarose gel by electrophoresis, using a DNA ladder of known molecular weights as a reference.

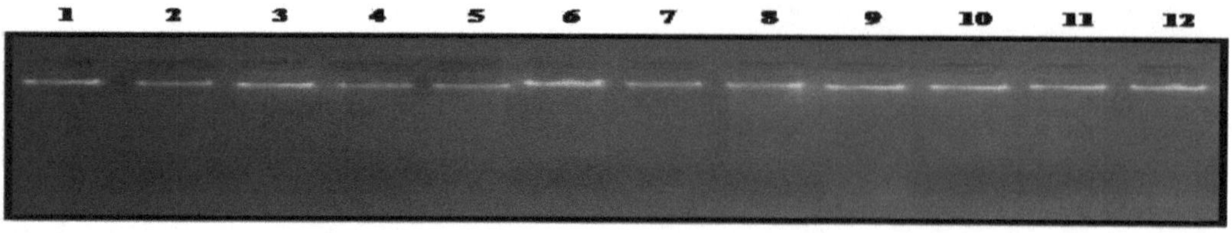

Fig. 3 Electrophoretic banding pattern of genomic DNA from eleven *Trichoderma* strains and *S. rolfsii* on 0.8 % agarose gel(Source: [1])

Table 2
Reaction mixture for PCR amplification

Component	Volume (μL)	Final concentration
Template DNA (25–50 ng/μL)	1.0	25–50 ng
10× PCR buffer	2.5	1×
$MGCL_2$ (25 MM)	1.5	1.5 mM
NTPS (10 MM)	0.5	0.2 mM
Forward primer (10 μM)	1.0	0.4 μM
Reverse primer (10 μM)	1.0	0.4 μM
Taq DNA polymerase (5 U/μL)	0.2	1 U
Sterile DDH_2O	Up to 25 μL	—

Table 3
PCR cycling conditions

Step	Temperature (°C)	Time	Cycles
Initial denaturation	94	5 min	1
Denaturation	94	1 min	5
Annealing (initial)	35	1 min	
Extension	72	2 min	
Denaturation	94	1 min	35
Annealing (final)	50	1 min	
Extension	72	2 min	
Final extension	72	10 min	1
Hold	4	∞	

3.6 Electrophoresis and Data Analysis

1. Load 5 μL of PCR product on a 1.5% agarose gel and run at 100 V for 1 h.

2. Gel electrophoresis: Load the SRAP-PCR product mixed with loading dye onto an agarose gel. Run the gel in buffer until the dye front reaches the end of the gel.
3. Visualization: Stain the gel with a DNA-staining agent, visualize using a Gel Documentation System under UV light, and capture the image (Fig. 4).
4. Data scoring: Score the SRAP bands as binary data (for presence, for absence) for each isolate. Only clear, reproducible polymorphic bands are scored (*see* **Note 4**).
5. Statistical analysis: Use the binary data to construct a genetic distance matrix (e.g., Jaccard's coefficient). Perform cluster analysis, such as unweighted pair group method with arithmetic mean (UPGMA), to generate a dendrogram illustrating the genetic relationships among the isolates (Fig. 5) [1].
6. Correlation with biocontrol data: The genetic data can then be correlated with phenotypic data (antagonism index, secondary metabolite profiles from GC-MS, etc.) to identify molecular markers linked to specific biocontrol traits [1].

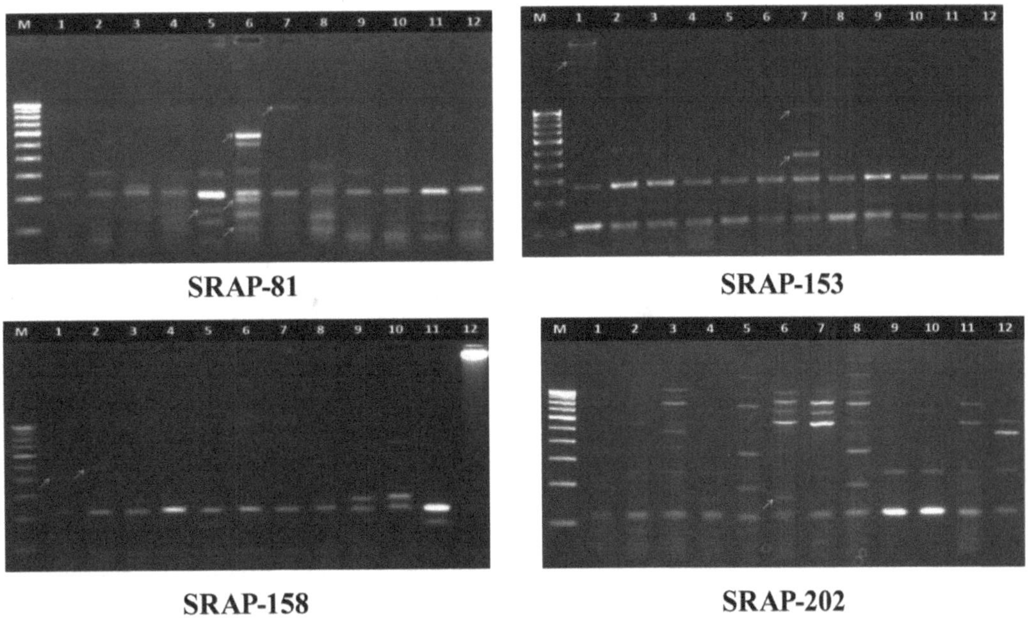

Fig. 4 SRAP profile of antagonist and test pathogen using different SRAP primers [M = DNA Marker (100 bp); 1. *T. harzianum* (NBAII Th1); 2. *T. harzianum* (NRRL 13879); 3. *T. harzianum* (NRRL 20565); 4. *T.harzianum* (Local); 5. *T. viride* (NBAII Tv23); 6. *T. viride* (NRRL 6418); 7. *T. virens* (NBAII Tvs12); 8. *T. hamantum* (NBAII Tha 1); 9. *T. koningii* (MTCC 796); 10. *T. pseudokoningii* (MTCC 2048); 11. *T. species* (NRRL 5242); 12. *S. rolfsii* (9107.13)](Source: [1])

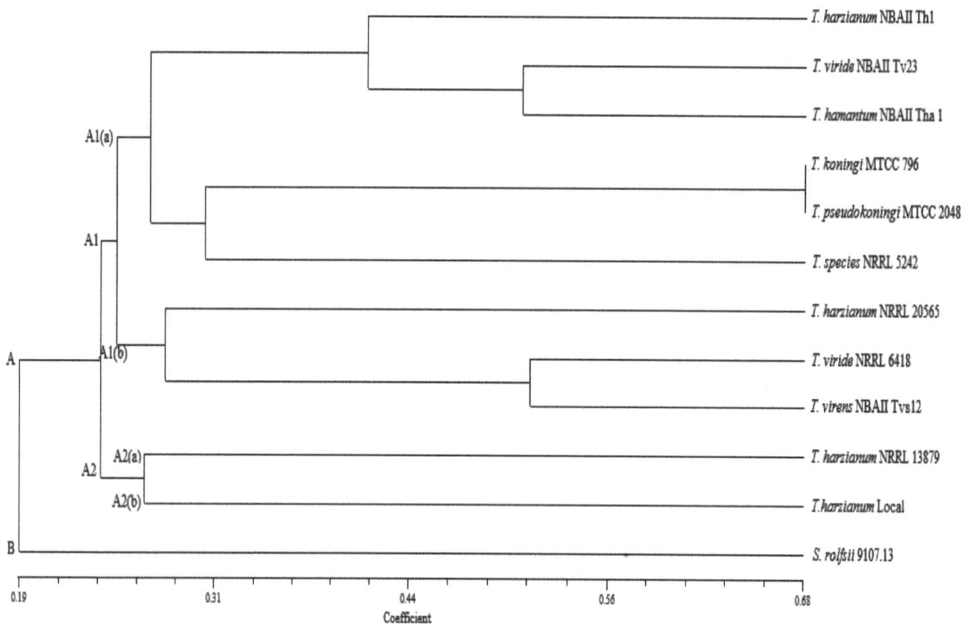

Fig. 5 Dendrogram showing genetic relationships among *Trichoderma* isolates and *S. rolfsii* based on SRAP markers(Source: [1])

4 Notes

1. DNA quality: High-quality, intact genomic DNA is paramount for reproducible SRAP results. Impurities (especially polysaccharides and phenolics) can inhibit *Taq* polymerase. The addition of PVP-20 to the CTAB buffer helps bind phenolics, improving DNA purity for fungal isolates.
2. Primers and polymorphism: SRAP markers are known for high polymorphism. The choice of primer combinations is critical. It is recommended to screen a set of combinations on a small subset of samples to select those yielding the most clear, highly polymorphic, and reproducible banding patterns [1].
3. PCR optimization: The annealing temperature in the SRAP protocol is designed to be touchdown: the initial lower temperature (35°C) allows for less stringent binding to amplify a broader range of fragments, while the subsequent cycles at a higher, more stringent temperature, enhance the specificity of the amplification. Minor adjustments to or primer concentrations may be required for optimal amplification, depending on the thermal cycler and reagents used.
4. Gel interpretation: Scoring SRAP bands can be challenging due to the potential for subtle intensity variations. It is crucial

to run a DNA ladder for accurate fragment size estimation. Bands must be clearly visible and reproducible across multiple runs to be reliably scored.

References

1. Hirpara DG, Gajera HP, Bhimani RD, Golakiya BA (2016) The SRAP-based molecular diversity related to antifungal and antioxidant bioactive constituents for biocontrol potentials of *Trichoderma* against *Sclerotium rolfsii*Scc. Curr Genet 62(3):619–641. https://doi.org/10.1007/s00294-015-0565-x
2. Li G, Quiros F (2001) Sequence-related amplified polymorphism (SRAP), a new marker system based on a simple PCR reaction: its application to mapping and gene tagging in Brassica. Theor Appl Genet 103(2):455–461
3. Narayanasamy PD, Saravana M (2009) RAPD analysis of *Trichoderma* and it's antagonistic affect with *Pseudomonas fluorescens.* J Plant PatholMicrob 20(4):223–233

Check for updates

Chapter 2

Utilizing RNA-Seq-Derived SNPs for KASP Marker Development in Maize

Ashok Jagtap, Amanpreet Kaur, Madhavi Purankar, Dipak Sarode, Rachayya M. Devarumath, Kapil Sushir, Yogesh Vikal and Gurmukh Singh Johal

Abstract

Maize is the third most significant cereal crop, and most of its agronomic traits are quantitative in nature. The fine mapping of genes or quantitative trait loci (QTLs) associated with these traits is essential for marker-assisted selection (MAS) in maize breeding. Although significant progress has been made, there remains a lack of high-density, highly polymorphic single-nucleotide polymorphism (SNP) markers, especially competitive allele-specific PCR (KASP) SNP markers suitable for automated genotyping. The development of next-generation sequencing (NGS) platforms has generated vast amounts of whole-genome and transcriptome data, providing a valuable resource for identifying SNP loci for marker development. KASP markers, being cost-effective and reliable, provide a trustworthy molecular toolbox for the precise genetic analysis of major key agronomic traits and the rapid improvement of maize using MAS.

Keywords KASP, Maize, MAS, RNA-seq, SNP

1 Introduction

Maize is the third most popular staple cereal crop, after wheat and rice [1]. It is one of the most efficient grain crops in terms of yield. Breakthroughs in genomic research have provided efficient tools for identifying genes and quantitative trait loci (QTLs) linked to major agronomic traits, thereby facilitating the development of useful genetic and molecular markers for marker-assisted selection (MAS) to generate climate-resilient varieties [1].

Among the different types of molecular markers, single-nucleotide polymorphisms (SNPs) are the most commonly used in contemporary plant breeding programs owing to their low assay cost, high genomic density, biallelic inheritance, locus specificity, low mutation rate, amenability to high-throughput analysis, and low

Nitish Kumar (ed.), *Molecular Markers in Plants*, Methods in Molecular Biology, vol. 3097,
https://doi.org/10.1007/978-1-0716-5663-1_2,

genotyping error rates [2, 3]. As such, SNPs have become the marker of choice for germplasm analysis, QTL mapping by genetic linkage or association analysis, allele mining, and genomic selection [3, 4].

Most approaches to detecting genetic variation are based on analysis of sequencing data from either whole-genome sequencing (WGS) or whole-exome sequencing (WES) [5, 6]. In recent years, next-generation sequencing (NGS) platforms, especially RNA sequencing (RNA-seq), have been increasingly used to provide a comprehensive understanding of gene expression profiles and to dissect the genetic networks and metabolic pathways that underlie major agronomic traits in maize [7, 8]. In addition to gene expression profiling, RNA-seq can be used as a valuable approach to identify genomic variants in expressed genes [9]. In fact, RNA-seq has been recognized as a cost-effective and efficient alternative to DNA sequencing platforms, such as WGS and WES [9, 10]. Compared with WGS data, RNA-seq data can provide a valuable resource for developing potential functional markers (such as KASP markers) from expressed regions of the genome [11]. Although the detection of variants by RNA-seq may be difficult because of the complexity of the transcriptome and the possibility of false positives [12], there are several advantages of RNA-seq in the discovery of SNPs. RNA-seq enables the correlation of expression profiles (RNA signatures) with phenotypes, thereby facilitating the selection of candidate genes and SNPs for marker development, which in turn can lead to a better genetic understanding of important traits [13, 14]. Additionally, the development of trait-associated molecular markers can improve the efficiency of marker-assisted selection in maize.

The data for SNP genotyping can be generated using different uniplex or multiplex platforms that employ diverse chemistries, detection technologies, and reaction formats [15, 16]. Some of the most commonly used genotyping platforms include TaqMan, Kompetitive allele-specific PCR (KASP), and rhAmp assays; however, the selection of an appropriate platform is dependent on factors such as the size of the study, the number of markers, the availability of the assay platform, cost-effectiveness, and accuracy [17]. Of these, KASP is a next-generation genotyping platform that uses post-PCR fluorescence detection and is known for its cost-effectiveness, high-throughput capacity, and ability to perform well under diverse environmental conditions [18]. The uniplex KASP test has been widely accepted in plant genetics and breeding studies due to its ability to combine PCR amplification with fluorescence-based allele discrimination, offering high scalability and automation potential and making it cost-effective on a per-sample basis [3, 16]. KASP marker systems have already been successfully used in genetic and genotyping studies in different crop species such as pigeon pea [19], chickpea [20], maize [1, 18, 21], wheat [22], and rice [23, 24]. This chapter outlines the development of SNP-based Kompetitive allele-specific PCR (KASP) assay markers from RNA-seq (transcriptomic) data in maize.

2 Materials

2.1 RNA-Seq Data

RNA-seq raw transcriptome sequence read data of the top leaf, pollen, and ovule from maize inbred lines LM11 (HS) and CML25 (HT) generated by using the Illumina HiSeq 2500 sequencing platform (Illumina, San Diego, CA, USA) (*see* **Note 1**)

2.2 DNA Extraction

1. 2× CTAB extraction buffer: 2% of CTAB (cetyl trimethyl-ammonium bromide), 20 mM EDTA (pH 8.0), 1.4 M NaCl, 100 mM Tris base (pH 8.0), 0.2% β-Mercaptoethanol, Distilled water (Table 1).
2. Chloroform/isoamyl alcohol solution in a ratio of 24:1 and stored at 4°C.
3. *RNase* (10 mg/mL).
4. Isopropanol is stored at 4°C.
5. 70% ethanol stored at 4°C.
6. TE buffer: 1 M Tris base (pH 8.0), 0.20 mM EDTA (pH 8.0), Distilled water (Table 1).
7. 6× DNA loading dye (Thermo Scientific™).
8. 0.8% Agarose gel contains 0.1 μg/mL ethidium bromide (final concentration) in Tris Borate EDTA buffer (TBA) (pH 8.8).
9. TBE buffer (Tris-Borate-EDTA), pH 8.0 (10× stock): 45 mM Tris base (pH 8.0), 0.20 mM EDTA (pH 8.0), Distilled water (Table 1).

2.3 Polymerase Chain Reaction

2× KASP master mixture (LGC, Biosearch Technologies), assay mix (LGC, Biosearch Technologies), Nuclease-free water

2.4 Tools and Instruments

Plastic microfuge tubes (1.5, 2.0 mL), Micropipettes, Multichannel-micropipettes, 96-well plates and adhesive seals, 96-well PCR thermocycler or real-time PCR machine, Agarose gel casting tray and gel comb, Agarose gel electrophoresis tank, Power supply, UV transilluminator equipped with a camera, microplate reader.

3 Method

3.1 Identification of SNP and InDels

1. Pre-processing and quality control: The raw RNA-seq read sequences are pre-processed using Trimmomatic v0.36 to remove adapter sequences and low-quality bases. The quality of clean read sequences is evaluated using the FastQC software tool, thereby obtaining high-quality read sequences. The final read sequences are exported in FASTQ format (Figs. 1 and 2).

Table 1
Composition of CTAB, TE, and TBE buffer

Composition of 2× CTAB extraction buffer		
Components	**Working concentration**	**Volume (1 L)**
CTAB (cetyl trimethylammonium bromide)	2%	20 g
EDTA (pH 8.0)	20 mM	7.44 g
NaCl	1.4 M	81.9 g
Tris base (pH 8.0)	100 mM	12.114 g
β-Mercaptoethanol[1]	0.2%	
Distilled water		up to 1 L
Composition of TE buffer (1×)		
Components	**Working concentration**	**Volume (200 mL)**
Tris base (pH 8.0)	1 M	0.316 g
EDTA (pH 8.0)	0.20 mM	0.074 g
Distilled water		up to 200 mL
Composition of TBE Buffer (Tris-Borate-EDTA), pH 8.0 (10× stock)		
Components	**Working concentration**	**Volume (1 L)**
Tris base (pH 8.0)	45 mM	109.02 g
Boric acid	45 mM	55.6 g
EDTA (pH 8.0)	20 mM	7.44 g
Distilled water		up to 1 L

[1]β-mercaptoethanol was added just prior to use

2. Read alignment: High-quality reads are mapped against the maize reference genome (B73 RefGen_v4, MaizeGDB database), which is downloaded from the MaizeGDB database, using the splice-aware aligner tool, STAR, v2.5.2b, using the 2-pass mapping strategy with default parameters to align the sequences [25].
3. Conversion and sorting of alignment files: The generated SAM files have been converted to BAM format using the SAMtools package version 1.4.1 [26]. Next, the BAM files have been sorted, read groups added, and duplicates marked using the Picard tools package v2.13.2.
4. Variant calling: Prediction of variants (SNPs and InDels) is carried out using the Genome Analysis Toolkit (GATK) v4.0.4.0 [27] (Figs. 1 and 2). The steps include:
   - Splitting "N" cigar reads (splice junction reads).

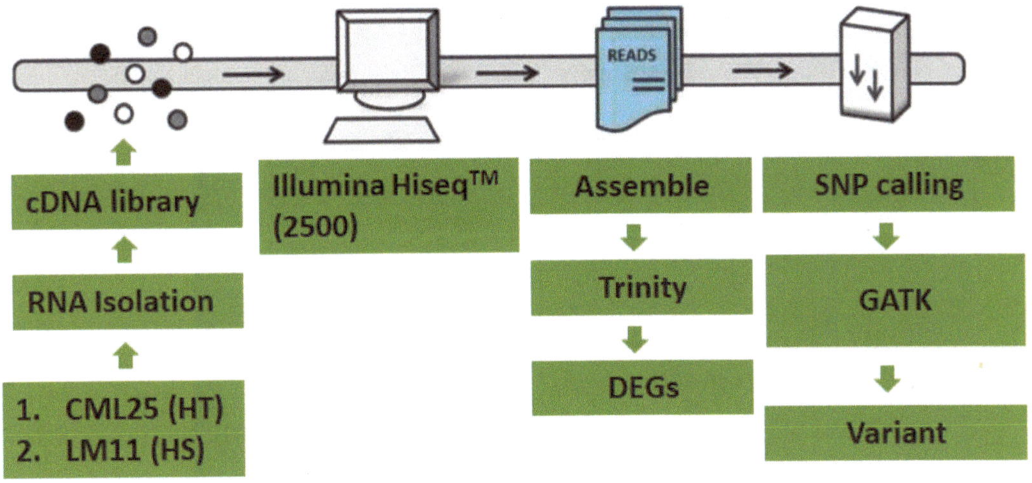

Fig. 1 A simplified workflow of analysis strategies for RNA-seq and SNP discovery

- Base Quality Score Recalibration (BQSR).
- Variant detection using GATK HaplotypeCaller.

5. Variant filtration: Identified variants are filtered to reduce false positive variants, using the GATK Variant Filtration tool. The variants that are filtered only include those that meet the quality criteria specified in Table 2 for the priority variants (Fig. 3).
6. Post-filtration processing: Further filtration and variant statistics analysis are carried out by VCFtools v0.1.13. This software package includes several functionalities such as variant summarization, format conversion, file validation, merging, and subset generation.
7. Comparative SNP analysis and marker development: The final variant call files (VCFs) for two maize inbred lines, LM11 (HS) and CML25 (HT), are compared using VCFtools. The polymorphic SNPs identified between the two lines are used for Kompetitive allele-specific PCR (KASP) marker development.

3.2 Variant Filtering for Suitability of Use as a KASP Marker

1. Retrieval of flanking sequences: The variant call format (VCF) files are analyzed using VCFtools v0.1.13 to retrieve 50 bp flanking sequences on either side of the variants. For that, an in-house customized Perl script is used to automate sequence data retrieval for KASP marker design.
2. Stepwise variant identification: Variants are systematically identified following a stepwise filtering/selection approach as shown in Figs. 2 and 3. This ensured that high-quality variants were identified for use in genotyping applications.

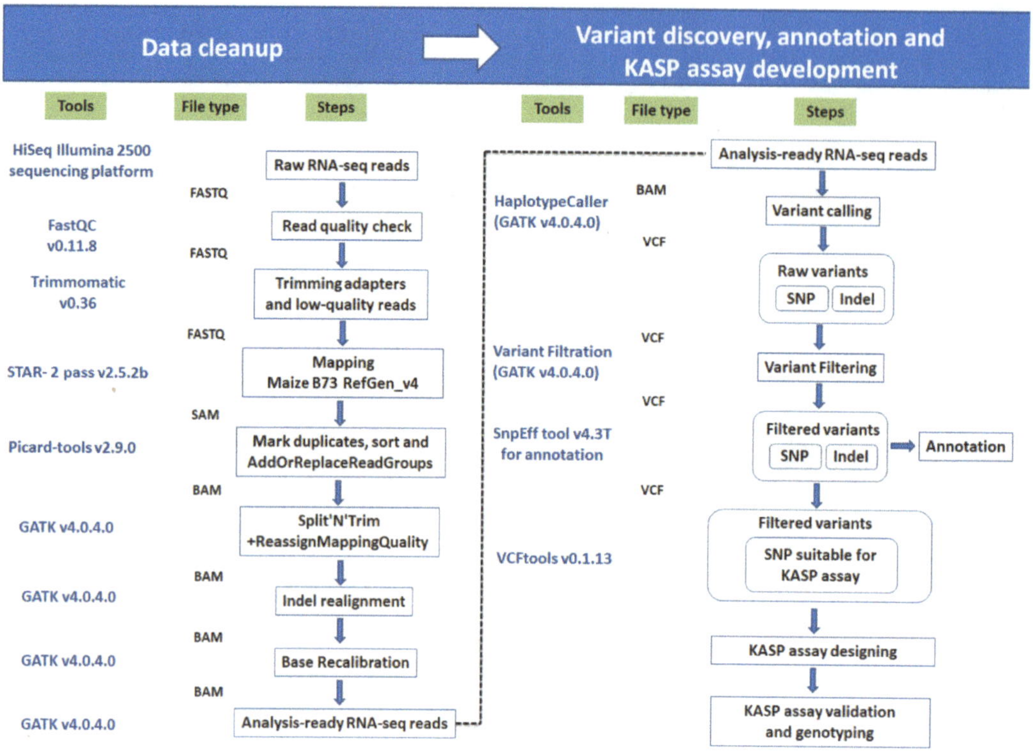

Fig. 2 Variant calling and genetic markers development pipeline (reprinted with permission from (*1*)). Schematic representation of the bioinformatics pipeline for variant calling, annotation, KASP assay development and genotyping from RNA sequencing reads

Table 2
Criteria used in the SNPs filtering

Criteria	Threshold
ReadRankPosSum (RRPS)	RRPS <−8
Quality by depth (QD)	QD <5
Read depth (DP)	DP <10
Fisher's exact test *p*-value (FS)	FS >30
Mapping quality (MQ)	MQ <40
SnpCluster	3 SNPs in 35bp
Mann–Whitney rank-sum (MQRankSum)	MQRankSum<−12.5
Alternative allele supporting read depth	ALTreads<5
Alternative allele ratio (*Het*)	$aa/t \leq 0.10$

3. Selection of potential KASP markers: Only those variants that have passed all filtering criteria, i.e., filtration thresholds, are selected as a potential marker (Fig. 3).
4. Genotyping with KASP markers: The shortlisted SNPs are selected for the development of KASP assay markers to genotype the F_2 mapping population from the genetic cross between LM11 (HS) and CML25 (HT).

3.3 Development of KASP Assay

1. Selection of SNPs for marker development: A total of 100 SNP sites, covering all the ten chromosomes of maize are selected from the leaf-derived variant data set to develop potential KASP assay markers.
2. Primer design for KASP assays: For each of the SNP sites, the SNP sequence, along with 50 bp of flanking sequence on both sides of the SNP, is used for primer design. Two forward primers, each being allele-specific and carrying a distinct fluorescent tail, are designed:
FAM tail: 5′-GAAGGTGACCAAGTTCATGCT-3′
HEX tail: 5′-GAAGGTCGGAGTCAACGGATT-3′

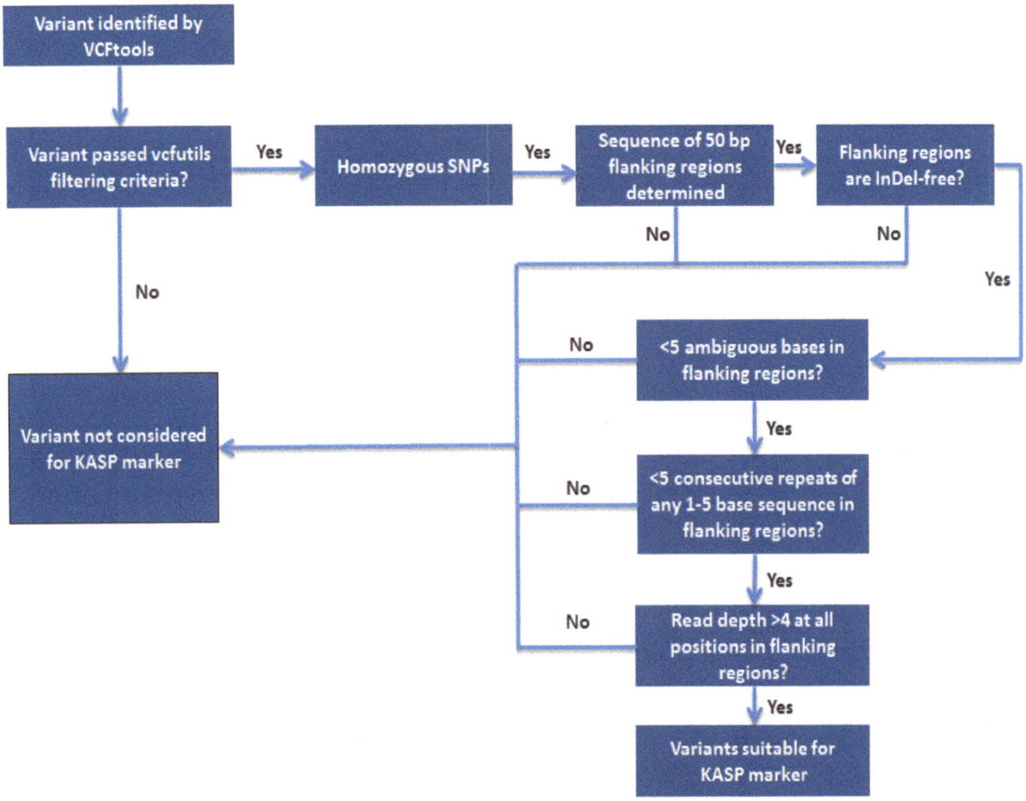

Fig. 3 The criteria used for the identification of potential KASP assay markers from variants identified using SAMtools, GATK, and VCFtools

Each forward primer is designed to have the target SNP at its 3′ end, but a common reverse primer is used for amplification.

3. Synthesis of KASP primers: The design and synthesis of KASP marker primers were performed by LGC Biosearch Technologies (Beverly, MA, USA), and the KASP primer oligos were purchased from the same source.

3.4 Extraction of DNA

Genomic DNA from the collected maize leaf tissue samples of the F_2 population and their parents is extracted using a modified cetyl trimethylammonium bromide (CTAB) method described by Saghai-Maroof et al. [28] (*see* **Note 2**). The extraction process is carried out as follows:

1. Frozen leaf tissues stored at −80°C are removed from the freezer and immediately ground into a fine powder in liquid nitrogen using a pre-chilled mortar and pestle.
2. The powdered tissue is transferred into sterile 2.0 mL microcentrifuge tubes labeled according to sample identification numbers.
3. About 800 μL of pre-warmed CTAB extraction buffer containing 0.2% β-mercaptoethanol is added to each tube. The mixture is incubated at 65°C for 30 min in a water bath, with gentle inversion three to four times to ensure proper mixing. The composition of the CTAB buffer is provided in Table 1.
4. After incubation, the samples are cooled to room temperature for 5 min, and an equal volume (800 μL) of chloroform:isoamyl alcohol (24:1 v/v) is added. The tubes are gently shaken for 40–45 min to allow complete mixing.
5. The samples are centrifuged at 10,000 rpm for 15 min at 4°C using a refrigerated microcentrifuge.
6. The upper aqueous layer, which contains DNA, is carefully transferred into new sterile 1.5 mL microcentrifuge tubes using a micropipette.
7. Approximately two-thirds of the volume of chilled isopropanol is added to each tube to precipitate DNA. The tubes are inverted gently to mix and then incubated at −20°C for 30 min.
8. The precipitated DNA is collected by centrifuging tubes at 10,000 rpm for 15 min at 4°C.
9. The supernatant is discarded carefully, and the DNA pellet is washed with 300 μL of 70% ethanol followed by centrifugation at 10,000 rpm for 5 min. Ethanol is decanted completely.
10. The ethanol wash is repeated once more to remove any remaining impurities.

11. The DNA pellet is air-dried at room temperature for about 30 min and then dissolved in 100 µL of sterile Nanopore water or TE buffer. The composition of the TE buffer is listed in Table 1 (*see* **Note 3**).
12. To remove RNA contamination, 10 µL of RNase (10 mg/mL) is added, and the mixture is incubated at 37°C for 30 min.
13. The DNA samples were kept at −20°C for long-term storage.

3.5 Assessing the Quality of DNA by Electrophoresis

1. Prepare 0.8% agarose by dissolving it in 0.5× TBE electrophoresis buffer (composition given in Table 1).
2. Heat the mixture in a microwave until the solution becomes clear.
3. Allow the agarose solution to cool to approximately 50°C while stirring gently.
4. Add ethidium bromide (EtBr) to achieve a final concentration of 10 µg/mL (5 µL of 10 mg/mL stock per 100 mL) (*see* **Note 4**).
5. Seal the ends of the gel casting tray with tape and place it on a flat surface.
6. Insert the comb on one side of the tray to form wells.
7. Pour the molten agarose carefully into the tray, avoiding air bubbles; if bubbles appear, remove them with a micropipette tip.
8. Leave the gel undisturbed for about 30 min to solidify, then gently remove the comb and tape.
9. Place the gel tray in the electrophoresis tank with the well side facing the cathode (negative terminal).
10. Add 0.5× TBE buffer to the tank until the gel is just covered (*see* **Note 5**).
11. Mix 5 µL of each DNA sample with 1 µL of 6× loading dye.
12. Load the prepared samples into the wells using a micropipette.
13. Load control DNA samples of known concentrations (Invitrogen TrackIt 1 Kb Plus DNA Ladder) in separate wells.
14. Close the lid of the electrophoresis tank and connect the electrodes to the power supply (*see* **Note 6**).
15. Run the gel at a constant voltage of 100 V and a current of 50 A until the tracking dye migrates about two-thirds of the gel length.
16. Turn off the power once migration is complete.

17. Visualize the DNA bands under a UV transilluminator and capture the gel image using a Gel DOC-It™ Imaging System (UVP, USA).
18. Examine the gel profile for band integrity, sharpness, and the absence of RNA or protein contamination.
19. Compare the fluorescence intensity of each DNA sample with the standard DNA concentrations to assess quality.

3.6 Assessing the Quality and Quantity of DNA by the NanoDrop Spectrophotometer Method

1. Turn on the NanoDrop™ ND-1000 spectrophotometer (NanoDrop Technologies Inc., USA).
2. Pipette 1 μL of TE buffer to the pedestal and use it as a blank to calibrate the machine.
3. Pipette 1 μL of each DNA sample into the pedestal and measure the absorbance.
4. Measure the absorbance at 260 and 280 nm wavelengths.
5. Calculate the OD_{260}/OD_{280} ratio to check the purity of DNA, which should be between 1.8 and 2.0 for high-quality DNA.
6. Calculate the DNA concentration using the following formula:
   1 OD at 260 nm = 50 μg double-stranded DNA/mL
   Thus, DNA (μg/mL) = $OD_{260} \times 50$
7. Note down the concentration values for all DNA samples.
8. Normalize the DNA concentration of parental lines LM11 (HS), CML25 (HT), and the F_2 population to 10 ng/μL for genotyping with KASP markers.

3.7 Validation of KASP Markers and Genotyping

1. The genotyping reagents are obtained from LGC, Biosearch Technologies (Beverly, MA, USA).
2. A total of 100 KASP assays are designed (Table 3) and validated using DNA from 90 F_2 individuals, along with parental DNA from LM11 (HS) and CML25 (HT) included in duplicates. Two nontemplate controls are also used to monitor contamination.
3. Each assay is performed in a 96-well plate format with a total reaction volume of 10 μL (*see* **Note 7**).
4. The reaction mixture consists of:
   - 2 μL of template DNA (10 ng/μL final concentration),
   - 5 μL of 2× KASP master mix,
   - 0.14 μL of assay mix, and
   - 2.86 μL of nuclease-free water.
5. The PCR amplification is carried out in a Flex Real-Time PCR System (Applied Biosystems, USA).
6. The thermal cycling conditions include:

Table 3
List of KASP primer sequences (two forward and one reverse) designed for LM11 and CML25 alleles

KASP assay marker name	Primer_AlleleX (5'–3')	Primer_AlleleY (5'–3')	Primer_Common (5'–3')	AlleleX (LM11)	AlleleY (CML25)
MKAM_01_001	TGGGAACGGGACATGTGAGT	CTTGGGAACGGGACA TGTGAGA	GCGCTCTGTGTGTG TGGGTGTA	A	T
MKAM_01_002	GCACATTTCAGCGCTGCAC	GCTGCACATTTCAGC GCTGCAT	ATGAAACTTAATGAC TGCCGAAGCACAAT	G	A
MKAM_01_003	CCCTCGTTCTGCAGGATGTAG	CCCTCGTTCTGC AGGATGTAC	GGCGGACGCGGGG CTGAA	G	C
MKAM_01_004	GTCCCGCAGTAGTCGCAGGT	CCCGCAGTAGTCGCAGGC	CGAGGCGGAGGCCG GCAA	A	G
MKAM_01_005	CGTGGGCAGCCACGGTGAT	GTGGGCAGCCACGGTGAC	CGCCGAGCGGTC CGTCGAA	T	C
MKAM_01_006	CAGGGGAATTCCACTCACTAGATTA	AGGGGAATTCCACT CACTAGATTG	TCAGCCGTTGCCCGT GGCATTT	A	G
MKAM_01_007	GTAGGAAGACTCTGGACACACC	TGTAGGAAGACT CTGGACACACA	CAGGTTGCTGAAG TAGCGGTTGAA	C	A
MKAM_01_008	GAAAGGCGCAGCTTGACGAAG	GGAAAGGCGCAGCT TGACGAAA	CTGTGGTGGGAGGC CGGCAA	C	T
MKAM_01_009	AAGTCGGAGATCTCGAACTCGC	GAAGTCGGAGATCTC GAACTCGT	GCTGGTGGCCTTCTC CCTCAA	G	A
MKAM_01_010	AAGATCTGATTATGCTCTCATGGTCT	AGATCTGATTATGCTC TCATGGTCG	GAATATTGTTGGGT CCAACCATACAACAA	A	C
MKAM_02_011	CGGGCGGTGGGGGATCTA	CGGGCGGTGGGGGATCTG	GCGACCTCCTCCC GCGAGA	T	C
MKAM_02_012	CTACAACATCGGCCATCAGATTAATT	CTACAACATCGGCCAT CAGATTAATG	ACCACCAGTGGAAG CAGCATTACTT	T	G

Table 3 (continued)

KASP assay marker name	Primer_AlleleX (5'–3')	Primer_AlleleY (5'–3')	Primer_Common (5'–3')	AlleleX (LM11)	AlleleY (CML25)
MKAM_02_013	CCGCGCGGCGGGCGAAT	CCGCGCGGCGGGCGAAC	CAGCATCCGCCGC CACCCT	A	G
MKAM_02_014	CCTGTCCAGATATACTGCTAACC	GCCTGTCCAGATATA CTGCTAACT	CAAGCTCGCCATCAA TGGAGTGAAT	C	T
MKAM_02_016	GGCAACAGCCTCATCAGGATTG	GGCAACAGCCTCAT CAGGATTC	TCTCAGATATTCAAT AAGCCACCAAGCAA	G	C
MKAM_02_017	GTGCTTCACCCTGAACTTCCAC	AAGTGCTTCACCCTGA ACTTCCAA	AGTTTGTGTCCTTCG ACGCTGTCAA	G	T
MKAM_02_019	CCCAGGGAACGGAGCAGG	CCCCAGGGAACG GAGCAGA	CGCCGCAGATGCCG ATGCAATT	G	A
MKAM_02_020	CCTCCAAGGCCACCGGC	GTCCTCCAAGGCCA CCGGA	GTGCGGCGCTCCTC GCGTT	G	T
MKAM_03_021	GACAGCGTCCGCAGCGCT	GACAGCGTCCGCAGCGCC	TGGCGCTGCCGGTGG CGAA	T	C
MKAM_03_022	ATGTTGGTGAACACCGTGCCC	CATGTTGGTGAACACC GTGCCT	CGCTGGCCGTGGGGC ACAT	C	T
MKAM_03_023	GACGCCGCGCCACACCC	CGACGCCGCGCCACACCT	GAACCGCTGGTCGT CGAGGTT	C	T
MKAM_03_024	CACCCGCTTGCTTTGAATCCCT	ACCCGCTTGCTTTG AATCCCC	TCGCGATGGCGGAGC AGAAGAA	A	G
MKAM_03_025	CAGGTGACAAATGTGATTTCAAGCC	CAGGTGACAAATGTGA TTTCAAGCT	TCATCCTGTGGCT CTGGAAGGAAA	G	A
MKAM_03_026	CCCAAGAAGCAGTTTCTTGACTCT	CCAAGAAGCAGTTT CTTGACTCG	CCGTTGGCGGCCTTC GCCTT	A	C

Table 3 (continued)

KASP assay marker name	Primer_AlleleX (5'–3')	Primer_AlleleY (5'–3')	Primer_Common (5'–3')	AlleleX (LM11)	AlleleY (CML25)
MKAM_03_027	GACACGGCGGCGCAGCAA	GACACGGCGGCGCAGCAG	CGCACACCTGAAAG CAATGGAGAT	A	G
MKAM_03_028	ACGCATCTGTCTGAGTTGTGCC	CACGCATCTGTCT GAGTTGTGCT	CCTTGGAAGCACAC CCAAGCAATAA	G	A
MKAM_03_029	GCCATGTCGAACCTGACCACA	GCCATGTCGAACCT GACCACG	AGGGAGGCGGAGCG GGGTT	A	G
MKAM_03_030	GGCCGTCAAGTACGTGGACG	AAGGCCGTCAAGTAC GTGGACT	GCCCTGGGGCACG GCGAA	G	T
MKAM_04_031	ACTGACCCCAAGGCCGGC	CACTGACCCCAAGG CCGGT	TCCTGGCGGTAGGGT CGAAGTT	C	T
MKAM_04_032	GTGTCCTTCACCTCGGGACAA	GTGTCCTTCACCTC GGGACAG	TTGGGTGTAAAATATG GGTGTGGTGTTT	A	G
MKAM_04_033	GCCCACGGGGCTCAG	GCTGCCCACGGGGCTCAA	GGGCATGAAGAACA GGAAGGGCTA	C	T
MKAM_04_034	TCTCGACCCTCTGCCG	CTGCTTCTCGACCCT CTGCCA	GGTAGCAGGAAGG CGACGGTAA	G	A
MKAM_04_035	CCCGGGCTTCGCAGAGTCA	CCCGGGCTTCGCA GAGTCT	GTCCGGTGGCAATG CTTATGTGATT	T	A
MKAM_04_036	ACCATCTTGTCTTGCTTGATTGCG	CACCATCTTGTCTTGCT TGATTGCT	TCTTCCGGCCGGCC GCGAA	G	T
MKAM_04_037	GGTTATCCTCCCACAGTACGTG	CGGTTATCCTCCCACA GTACGTT	ATAGGCACCGAACCC TTGAGAACAA	G	T
MKAM_04_038	GTGCGCGACCACGAGCACT	GCGCGACCACGAGCACC	GATTTTGGCGCAC CGGGGCAA	A	G
MKAM_04_039	GTGGGGTCCGCCCGGG	GTGGGGTCCGCCCGGC	CTTCATCAAGCCCC GCTCCACAT	G	C

Table 3 (continued)

KASP assay marker name	Primer_AlleleX (5′–3′)	Primer_AlleleY (5′–3′)	Primer_Common (5′–3′)	AlleleX (LM11)	AlleleY (CML25)
MKAM_04_040	CAGCTGCCAGGATGCTGTGC	GCAGCTGCCAGGATGCTGTGT	GTCCCCTGGGGTTCAATACTATGAT	G	A
MKAM_05_041	GCGGCTGCGTGGACAGGAAA	CGGCTGCGTGGACAGGAAG	AGAAGCTGGACTGGGCCGACAT	A	G
MKAM_05_043	GGTGATGATGGTGAATTTGATGAGC	GGTGATGATGGTGAATTTGATGAGG	CGCTAATGATCCAGGGCTACAAGAA	G	C
MKAM_05_044	ATTAGGTGGGCAACCACGGAGT	AGGTGGGCAACCACGGAGC	ACGGCCCGCTGCTCCCCTT	T	C
MKAM_05_045	ACCGTGCCTATCAGTCGATCTC	CACCGTGCCTATCAGTCGATCTT	GATGAGTAATGTACTTCTTAGCCGCAAT	G	A
MKAM_05_046	ACACATACACCTCTTCTACTGCTCA	ACATACACCTCTTCTACTGCTCG	TTGAGGGCGAGCTCCTCGCATT	A	G
MKAM_05_047	TCATAACGGCATAACCATAGCTAG	CTTCATAACGGCATAACCATAGCTAC	GCTGCTGTAAGCCTGTAATGGTGAA	C	G
MKAM_05_048	CAGCTGCCCCTCCTCCTCTT	AGCTGCCCCTCCTCCTCTC	TCGAATCGTCGGAGGAGCGGAA	T	C
MKAM_05_049	GTGCACGAGATCGGCCACCTA	GCACGAGATCGGCCACCTC	GAGGAGTGGCCGAGCCCCA	T	G
MKAM_05_050	GTGGAGGAGGACAGGGACC	GTGGAGGAGGACAGGGACG	CGAGCTGCTGCTCACGCACAA	C	G
MKAM_06_051	ATCTGGGTCTCAGTCTGCTGC	ATCTGGGTCTCAGTCTGCTGT	GAATAAACGAATACGGGCGGGACAA	G	A
MKAM_06_052	GTTGACTTATACCAGCTCCCTCC	GTTGACTTATACCAGCTCCCTCA	CAACAAAAGAATGATGATGCTGGAAAACAA	C	A
MKAM_06_053	CAGATATCGTGCTATTACTAGTGCG	GCAGATATCGTGCTATTACTAGTGCA	CGCCCCAACAGCACCTCGATAA	G	A

Table 3 (continued)

KASP assay marker name	Primer_AlleleX (5'–3')	Primer_AlleleY (5'–3')	Primer_Common (5'–3')	AlleleX (LM11)	AlleleY (CML25)
MKAM_06_055	CCTCTTTTCTGCATCTGATGTCTTG	GCCTCTTTTCTGCATCTGATGTCTTT	ACGATGCACTGACAACTGGAAGCAT	G	T
MKAM_06_056	GCAGGCTCCGCGCCGTCT	CAGGCTCCGCGCCGTCC	ACGACGGCGCGGAGCCCAT	T	C
MKAM_06_057	GCGGGGCGGACCAGGTC	GCGGGGCGGACCAGGTG	CGCCCGACGGGTCTGGGAT	C	G
MKAM_06_058	ACCAAATAGTTTTGGAACTCTATCCTG	AAAACCAAATAGTTTTGGAACTCTATCCTA	CCATATGGCATCATTTATTTGGGAACGTT	C	T
MKAM_06_059	GCAAAAGCCCAAGACACGCCAT	CAAAAGCCCAAGACACGCCAC	CTGTCCTGAGGGATCCACCTGTT	A	G
MKAM_06_060	GTTGACGGCGCCGGCGTA	GTTGACGGCGCCGGCGTT	GCTGGTGGTGGTGCTCCTTCTA	A	T
MKAM_07_061	CGAGGCGGCAGCCGAGG	GCGAGGCGGCAGCCGAGA	CCACGGCCGGTGCCTCCT	G	A
MKAM_07_062	CCTGCCCAAGAAGACCGCT	CCTGCCCAAGAAGACCGCG	CCTCCTTGGCCGCCGCCTT	T	G
MKAM_07_063	CGTGACACCTCACCCACTCGT	GTGACACCTCACCCACTCGC	ATCCATGAACTGGTGGTCAGGGAA	T	C
MKAM_07_064	CCGTGGCAGCCTGCGGCT	CGTGGCAGCCTGCGGCG	GTTGGCCGCACGCAAGGGAAAA	A	C
MKAM_07_066	GCGGCTGCGTAGGCACA	GCGGCTGCGTAGGCACC	TGCGAGGGCCGCAGCAGGA	T	G
MKAM_07_067	GTAGCCATTCAGGACCAACAA	CTGTAGCCATTCAGGACCAACAG	GCAGGTCCCTGATAGCTGCGTA	T	C
MKAM_07_068	CGGCTCCACGGAGCAGTAC	CCGGCTCCACGGAGCAGTAT	GGCAGGGCTGGCGCTCCAT	C	T

Table 3 (continued)

KASP assay marker name	Primer_AlleleX (5′–3′)	Primer_AlleleY (5′–3′)	Primer_Common (5′–3′)	AlleleX (LM11)	AlleleY (CML25)
MKAM_07_069	CAGTCGACTTGCTCCTGGACATA	AGTCGACTTGCTCCTGGACATG	CAAGCTGACTTCTGCACCTCATCAT	T	C
MKAM_07_070	GACTGGCGCTGCATGTACTTG	CGACTGGCGCTGCATGTACTTA	CAGCGCTCGCTCCGGCTCTT	C	T
MKAM_08_071	ACTAGTGCGCTGGATGCTGAG	GACTAGTGCGCTGGATGCTGAA	TGTCCAGTGCGTCCTGGACGAT	G	A
MKAM_08_072	CGGATAGGTAGGCTGCGGAG	CGGATAGGTAGGCTGCGGAC	AGAGGCAGTGGCTTGGCACCTT	C	G
MKAM_08_073	CATAGTCTCAGTGCATAGAAAAATCAG	AAACATAGTCTCAGTGCATAGAAAAATCAT	GATGATGCTAAACTGAATAAGCACATCAAT	C	A
MKAM_08_074	GGGCAGGCTGGGAATGAGTTT	yGGGCAGGCTGGGAATGAGTTC	TGTGCCGCTGTGGCGCTCATAT	T	C
MKAM_08_075	AACTCGTCCATGATTTCTGTTCATAGT	CTCGTCCATGATTTCTGTTCATAGC	GTTGCTTTTGTTGCACTGAGAAAATCCTT	T	C
MKAM_08_076	GTGCGAATACTTCCTACCATGTGAA	GTGCGAATACTTCCTACCATGTGAT	AAATTGAGTCTTGCCAAATTCACCAGCTA	A	T
MKAM_08_077	ATTTGCCATCTCCACCGTTGAACAA	GCCATCTCCACCGTTGAACAC	TTCTGAACATCAGCCCAGAGGGTAA	T	G
MKAM_08_078	CCGAGCTGCGGGACTCT	CCGAGCTGCGGGACTCC	GGAAGAACGAACCTGAGGGTGAAA	A	G
MKAM_08_079	GTAATGTCGACCGGACCACCA	GTAATGTCGACCGGACCACCG	ACTCCCGATAGTCGCTACCAAGAA	T	C
MKAM_08_080	GCCGGCGGCCGCATCAC	GCCGGCGGCCGCATCAG	CTTCTACGACCTCTTCGCCGCT	C	G

Table 3 (continued)

KASP assay marker name	Primer_AlleleX (5'–3')	Primer_AlleleY (5'–3')	Primer_Common (5'–3')	AlleleX (LM11)	AlleleY (CML25)
MKAM_09_082	GCAGGGGCCGTCCCGTG	GCAGGGGCCGTCCCGTC	GGAACGCCAGGTGCC CCGT	G	C
MKAM_09_083	CCTCGCCAAGGACACCACAA	CCTCGCCAAGGACACC ACAG	CGCACATAGGCGCCG TGGAT	A	G
MKAM_09_084	CGAACGAGATCCACGACCTT	CGAACGAGATCCACGAC CTG	AGCCGCCCTGCAC GGTGGA	T	G
MKAM_09_085	TCCGTCGAGGAGACCGAG	ACTTCCGTCGAGGAG ACCGAT	CGCGCAAACCCTTG CTCCGAT	C	A
MKAM_09_086	CCCTTCCTCCTCGATGACGC	GCCCTTCCTCCTCGA TGACGT	GGCGCAAGGTGCTCA ACGACAA	C	T
MKAM_09_087	CAGAATCGATTGGCACCCAGG	CAGAATCGATTGGC ACCCAGT	TTAATCAAAAGGTATA GGAAGGCGCCA	G	T
MKAM_09_088	TGCATATCCCTGTTACACTGCAG	GTTGCATATCCCTGTT ACACTGCAA	GATCCCTTTCACTGTA CATTGCAAAGAAA	C	T
MKAM_09_089	CATAGCAGATGCCGTGAGGTAG	CATAGCAGATGCCGT GAGGTAC	GTGTGCTGCTGACCC TCGTCAT	G	C
MKAM_09_090	GACCGCGCCAGGGAGCG	GACCGCGCCAGGGAGCC	GGCAAGGTGTCGGCC GAGGA	G	C
MKAM_10_091	GAATGTAGGCGGATTTTAAAGAGGAT	GAATGTAGGCGGATTTTA AAGAGGAA	GACACTGTGATCGGG AGGATATTCAA	T	A
MKAM_10_092	CGGGACTGTCGTGCTCACG	AACGGGACTGTCGT GCTCACA	GAGGAGGCTTCT TGGCCCTGAT	C	T
MKAM_10_093	GCCTGCTGCAGCGTGCAG	GCCTGCTGCAGCGTGCAC	TCTACGGGGACC GCAACCTCAT	C	G
MKAM_10_094	GTTAGTCACAATGTCACCTCTGAGT	GTTAGTCACAATGTCACC TCTGAGA	GGTGTGCTCTGAG TTGAACGTTGAT	T	A

Table 3 (continued)

KASP assay marker name	Primer_AlleleX (5'–3')	Primer_AlleleY (5'–3')	Primer_Common (5'–3')	AlleleX (LM11)	AlleleY (CML25)
MKAM_10_095	GTCCAAGAGGCTATTAAAGAGAC	GCTGTCCAAGAGGCTAT TAAAGAGAA	CTATAAGAGGGAGGC AGCCTGGTT	G	T
MKAM_10_096	GGTTGGCCTGGCCGCCC	GGTTGGCCTGGCCGCCG	GCCGCGGCGCCAGCT ACTA	G	C
MKAM_10_097	ATGAGGCCCCAAATGCAAGTTATAC	AATGAGGCCCCAAATGG AAGTTATAT	CATTCTGAAACTGCTC GGAACTATCTATT	C	T
MKAM_10_ 100	GGGAGGAGGAGGAT GAAGAAGAA	GGAGGAGGAGGATGAAGA AGAC	GTAGTCCAAGGTCAT CTCAGGATCAT	T	G

- An initial hot start at 95°C for 15 min,
- Ten touchdown cycles of 20 s at 95°C and 25 s at 65°C (decreasing by 1°C per cycle), and
- Thirty additional cycles of denaturation at 95°C for 10 s and annealing at 57°C for 60 s.

7. No separate extension step is included since KASP amplicons are generally less than 100 bp in length.
8. After amplification, the plate is read at room temperature using a Tecan Safire plate reader (Tecan Group Ltd., Männedorf, Switzerland) (*see* **Note 8**).
9. The fluorescence signals are analyzed to identify allele-specific clustering and to validate the accuracy of the KASP markers.

3.8 Scoring of KASP Marker Data

1. The genotyping data obtained from KASP assays are analyzed using Klustercaller v2.22.0.5 software (LGC, Biosearch Technologies, Beverly, MA, USA).
2. The fluorescence signals generated during PCR amplification are used to distinguish between different allele types.
3. Each sample's SNP allele call is visually inspected and manually verified within the Klustercaller interface to ensure accurate genotype assignment.
4. The allele discrimination plots generated by the software display three distinct clusters corresponding to homozygous alleles for the two parents and heterozygous alleles for the F_2 individuals.
5. The graphical representation of SNP genotyping results is further visualized (Fig. 4) using SNPviewer software (LGC, Biosearch Technologies, Beverly, MA, USA).
6. The corresponding allele clusters are identified as follows:
   - Allele representing the susceptible parent (LM11),
   - Allele representing the tolerant parent (CML25), and
   - Heterozygous alleles represent the segregating F_2 population.
7. The genotyping accuracy and clustering clarity are verified before downstream data interpretation and marker-trait association analysis.

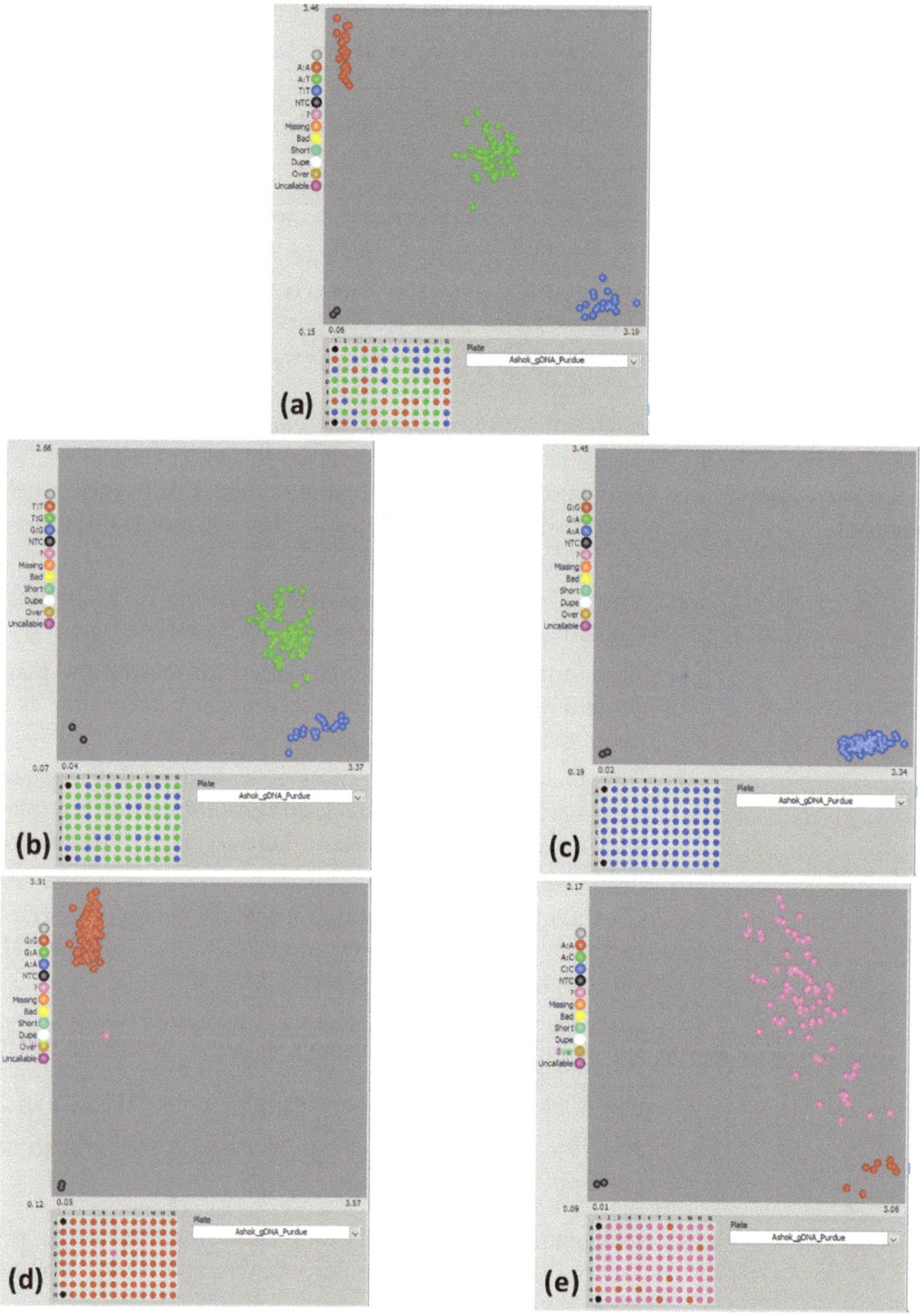

Fig. 4 Snapshots showing SNP genotyping in parental inbred lines LM11, CML25, and their F_2 mapping populations with KASP marker assays. The snapshots show (**a**) polymorphic pattern, occurrence of allele specific to LM11 (blue spots) and CML25 (red spots), and heterozygosity for the corresponding alleles (green spots) in F_2 mapping population. (**b**) occurrence of only one allele (blue spots), and heterozygosity for the corresponding allele (green spots). (**c**) monomorphic pattern, occurrence of allele specific to LM11 (blue spots). (**d**) monomorphic pattern, occurrence of allele specific to CML25 (red spots), and missing data (pink spot). (**e**) The majority of missing data (pink spots), and one allele (red spots), occurrence of failed amplification signals, i.e., invalid markers. In all snapshots, black spots represent no template controls (NTCs)

4 Notes

1. Transcriptomics data for maize inbred LM11 and CML25 are available at NCBI under Bioproject PRJNA656908 (https://www.ncbi.nlm.nih.gov/bioproject/).
2. It is important to always wear laboratory gloves and a lab coat when conducting experiments to ensure occupational health, safety, and welfare. Other protective gear may be needed depending on the experiment, such as when preparing an agarose gel, which may require the use of full-face shields.
3. The DNA pellet can be dissolved in sterile Nanopure water and stored at −20°C (or −80°C for long-term preservation). DNA quantification can be performed with a NanoDrop spectrophotometer, and the concentration adjusted to 10 ng/μL using sterile Nanopure water or TE buffer. Long-term storage of DNA dissolved in sterile Nanopure water at −20°C (for up to 4 years) did not affect the results in the present study.
4. While ethidium bromide is used to visualize PCR double-stranded DNA products, SYBR Safe is recommended, as it is less hazardous. SYBR Safe is light-sensitive and should be protected from light whenever possible. Samples containing SYBR Safe should be covered with aluminum foil, and SYBR Safe electrophoresis tanks should be shielded from direct light to ensure accurate, stable results.
5. For future use, a 10× TBE buffer stock solution can be prepared. Any undissolved materials noticed during the preparation can be dissolved by gently heating the solution in a hot water bath at 60–70°C. It is recommended to change the 0.5× TBE buffer used in the electrophoresis after every two runs. This ensures proper DNA migration through the agarose gel and prevents the accumulation of unwanted byproducts from previous electrophoresis runs.
6. It is important to make sure that the electrodes are correctly connected before starting the electrophoresis process. In this case, the black electrode, which is negative, should be placed on the side of the well, while the red electrode, which is positive, should be placed on the other end of the gel. Since DNA molecules carry a negative charge, they will move towards the positive electrode during electrophoresis. Incorrect electrode positioning will cause leakage of samples from the wells into the buffer tank.
7. It is recommended that, when preparing the reaction mix, a 10% surplus be added to the volumes to ensure enough master mix for all reactions.

8. KASP is an endpoint genotyping assay in which fluorescence is measured after completion of the PCR cycle. The fluorescence should be measured at room temperature, which should be below 40°C for this assay. For this purpose, the Tecan Safire plate reader (Tecan Group Ltd., Männedorf, Switzerland), the PHERAstar plate reader (BMG) or the BMG Omega F plate reader by Biosearch Technologies (Beverly, MA, USA) can be used.

Acknowledgments

This work was supported by the Science and Engineering Research Board (SERB), Department of Science and Technology (DST), Government of India, through the SERB Overseas Visiting Doctoral Fellowship. The research was conducted at the Department of Botany and Plant Pathology, Purdue University, USA.

References

1. Jagtap AB, Vikal Y, Johal GS (2020) Genome-wide development and validation of cost-effective KASP marker assays for genetic dissection of heat stress tolerance in maize. Int J Mol Sci 21:7386
2. Schlötterer C (2004) The evolution of molecular markers—just a matter of fashion?. Nat Rev Genet 5:63–69
3. Kim J, Manivannan A, Kim DS, Lee ES, Lee HE (2019) Transcriptome sequencing assisted discovery and computational analysis of novel SNPs associated with flowering in *Raphanus sativus* inbred lines for marker-assisted backcross breeding. Hortic Res 6:1–12
4. Semagn K, Beyene Y, Babu R, Nair S, Gowda M, Das B, Tarekegne A, Mugo S, Mahuku G, Worku M, Warburton ML, Olsen M, Prasanna BM (2015) Quantitative trait loci mapping and molecular breeding for developing stress resilient maize for Sub-Saharan Africa. Crop Sci 55:1449–1459
5. Adetunji MO, Lamont SJ, Abasht B, Schmidt CJ (2019) Variant analysis pipeline for accurate detection of genomic variants from transcriptome sequencing data. PLoS One 14:e0216838
6. Belkadi A, Bolze A, Itan Y, Cobat A, Vincent QB, Antipenko A, Shang L, Boisson B, Casanova JL (2015) Whole-genome sequencing is more powerful than whole-exome sequencing for detecting exome variants. Proc Natl Acad Sci U S A 112:5473–5478
7. Jagtap AB, Yadav IS, Vikal Y, Praba UP, Kaur N, Gill AS, Johal GS (2023) Transcriptional dynamics of maize leaves, pollens and ovules to gain insights into heat stress-related responses. Front Plant Sci 14:1117136
8. Pronozin AY, Shmakov NA, Afonnikov DA (2026) Diversity of lncRNAs in the pan-transcriptome of maize inbred lines. BMC Genomics 27: 1
9. Oikkonen L, Lise S (2017) Making the most of RNA-Seq: pre-processing sequencing data with Opossum for reliable SNP variant detection. Wellcome Open Res 2:6
10. Al-Ars Z, Wang S, Mushtaq H (2020) SparkRA: enabling big data scalability for the GATK RNA-seq pipeline with Apache Spark. Genes 11:53
11. Chen Z, Tang D, Ni J, Li P, Wang L, Zhou J, Li C, Lan H, Li L, Liu J (2021) Development of genic KASP SNP markers from RNA-Seq data for map-based cloning and marker-assisted selection in maize. BMC Plant Biol 21: 157
12. Guo Y, Zhao S, Sheng Q, Samuels DC, Shyr Y (2017) The discrepancy among single nucleotide variants detected by DNA and RNA high throughput sequencing data. BMC Genom 18:690
13. Yu Y, Wei J, Zhang X, Liu J, Liu C, Li F, Xiang J (2014) SNP discovery in the transcriptome of white Pacific shrimp *Litopenaeus vannamei* by next generation sequencing. PLoS One 9:e87218
14. Piskol R, Ramaswami G, Li JB (2013) Reliable identification of genomic variants from RNA-Seq data. Am J Hum Genet 93:641–651

15. Broccanello C, Chiodi C, Funk A, McGrath JM, Panella L, Stevanato P (2018) Comparison of three PCR-based assays for SNP genotyping in plants. Plant Methods 14:28
16. Ayalew H, Tsang PW, Chu C, Wang J, Liu S, Chen C, Ma XF (2019) Comparison of TaqMan, KASP and rhAmp SNP genotyping platforms in hexaploid wheat. PLoS One 14:e0217222
17. Ahn YK, Manivannan A, Karna S, Jun TH, Yang EY, Choi S, Kim JH, Kim DS, Lee ES (2018) Whole genome resequencing of *Capsicum baccatum* and *Capsicum annuum* to discover single nucleotide polymorphism related to powdery mildew resistance. Sci Rep 8: 5188
18. Shen T, Gao H, Wang C, Zheng Y, Song W, Hou P, Zhu L, Zhao Y, Song W, Guo J (2025) Development and validation of multi-locus GWAS-based KASP markers for maize *Ustilago maydis* resistance. Plants 14:2315
19. Saxena RK, Penmetsa RV, Upadhyaya HD, Kumar A, Carrasquilla-Garcia N, Schlueter JA, Farmer AM, Whaley AM, Sarma BK, May GD, Cook DR, Varshney RK (2012) Large-scale development of cost-effective single-nucleotide polymorphism marker assays for genetic mapping in pigeonpea and comparative mapping in legumes. DNA Res 19:449–461
20. Hiremath PJ, Kumar A, Penmetsa RV, Farmer A, Schlueter JA, Chamarthi SK, Whaley AM, Carrasquilla-Garcia N, Gaur PM, Upadhyaya HD, Varshney RK (2012) Large-scale development of cost-effective SNP marker assays for diversity assessment and genetic mapping in chickpea and comparative mapping in legumes. Plant Biotechnol J 10:716–732
21. Josia C, Mashingaidze K, Amelework AB, Kondwakwenda A, Musvosvi C, Sibiya J (2021) SNP-based assessment of genetic purity and diversity in maize hybrid breeding. PLoS One 16:e0249505
22. Tian Y, Liu P, Zhang X, Liu Y, Kong D, Nie Y, Xu H, Han X, Sang W, Li W (2025) Genome-wide association study and KASP marker development for flour color in winter wheat. Sci Rep 15:39673
23. Yang G, Chen S, Chen L, Sun K, Huang C, Zhou D, Huang Y, Wang J, Liu Y, Wang H, Chen Z (2019) Development of a core SNP array based on the KASP method for molecular breeding of rice. Rice 12:21
24. Ishwarya Lakshmi VG, Sreedhar M, JhansiLakshmi V, Gireesh C, Rathod S, Bohar R, Deshpande S, Laavanya R, Kiranmayee KU, Siddi S, Vanisri S (2022) Development and validation of diagnostic KASP markers for brown planthopper resistance in rice. Front Genet 13:914131
25. Dobin A, Davis CA, Schlesinger F, Drenkow J, Zaleski C, Jha S, Batut P, Chaisson M, Gingeras TR (2013) STAR: ultrafast universal RNA-Seq aligner. Bioinformatics 29:15–21
26. Li H, Handsaker B, Wysoker A, Fennell T, Ruan J, Homer N, Marth G, Abecasis GR, Durbin R (2009) The sequence alignment/map format and SAMtools. Bioinformatics 25:2078–2079
27. McKenna AH, Hanna M, Banks E, Sivachenko A, Cibulskis K, Kernytsky A, Garimella K, Altshuler D, Gabriel S, Daly M, DePristo MA (2010) The Genome Analysis Toolkit: a MapReduce framework for analyzing next-generation DNA sequencing data. Genome Res 20:1297–1303
28. Saghai-Maroof MA, Soliman KM, Jorgensen RA, Allard RWL (1984) Ribosomal DNA spacer-length polymorphism in barley: Mendelian inheritance, chromosomal location, and population dynamics. Proc Natl Acad Sci U S A 81:8014–8018

Chapter 3

Assessment of Clonal Integrity and Virus Indexing in Micro-Propagated Sugarcane Using DNA-Based Markers

Madhavi Y. Ghan, Atul Kumar, Ashok Jagtap, Shivani Talekar, Sneha Bhujbal and Rachayya M. Devarumath

Abstract

Tissue culture-based propagation in sugarcane provides large-scale production of planting material; however, maintaining genetic stability and virus-free status is critical. Genetic fidelity assessment and virus indexing serve as key tools for quality control in micropropagated plants. This chapter discusses available molecular and diagnostic strategies for ensuring clonal uniformity and phytosanitary safety.

Keywords Sugarcane, Tissue culture, Genetic fidelity, Virus indexing, Molecular markers, Molecular diagnosis

1 Introduction

Sugarcane (*Saccharum officinarum* L. and its interspecific hybrids) is a tall, perennial, tropical grass belonging to the family Poaceae, tribe Andropogoneae, and genus *Saccharum*, which comprises six recognized species: *S. officinarum*, *S. spontaneum*, *S. barberi*, *S. sinense*, *S. edule*, and *S. robustum*. Molecular tracing indicates that the original domestication of *S. officinarum* occurred approximately 10,000 years ago in New Guinea and Southeast Asia [1] and that it subsequently spread through Austronesian trade networks for cultivation. Globally, sugarcane is cultivated across more than 100 tropical and subtropical countries on approximately 26.9 million hectares, with a total annual production of approximately 2.03 billion tonnes as of 2023 [2]. It serves as the primary feedstock for approximately 79%–80% of the world's sugar supply and accounts for roughly one-quarter of global ethanol production [3]. Its contribution both as a food-security crop and a renewable energy feedstock positions sugarcane as a crop of the future, especially with reference to addressing the twin global challenges of food demand and fossil fuel dependency [4]. Despite its immense agricultural, industrial, and strategic importance, it faces

Nitish Kumar (ed.), *Molecular Markers in Plants*, Methods in Molecular Biology, vol. 3097,
https://doi.org/10.1007/978-1-0716-5663-1_3,

significant challenges, particularly from persistent, insidious agents such as viruses that cause severe yield losses, reduce the quality of the crop, and thus have profound socioeconomic significance [5].

Sugarcane mosaic disease, Sugarcane streak mosaic disease, Sugarcane yellow leaf disease and leaf fleck disease are major diseases caused by viruses. Sugarcane mosaic disease (SMD) is the oldest, most widespread, and among the most economically consequential viral diseases of sugarcane globally. Under Indian conditions, mosaic disease was found to be caused by both SCMV and SCSMV (alone or in combination), with SCSMV consistently more prevalent than SCMV across all sugarcane-growing zones of the country [6–9]. Yellow leaf disease (YLD), caused by Sugarcane Yellow Leaf Virus (SCYLV) (genus *Polerovirus*, family *Solemoviridae*, formerly *Luteoviridae*), has rapidly emerged as one of the most widespread and damaging viral diseases of sugarcane worldwide, with documented occurrence in more than 25 sugarcane-growing countries across five continents [10,11]. Phylogenetic analyses suggest an ancestral Indian origin for SCYLV global diversity [11,12]. A particularly threatening epidemiological feature of YLD is its frequent occurrence as a mixed infection with Sugarcane Grassy Shoot (SCGS) phytoplasma (16SrXI group). It causes synergistic effects that aggravate yellowing and accelerate the degeneration of infected clones [13]. RT-PCR studies by Viswanathan et al. [8] demonstrated that SCYLV could be readily detected in meristem culture-derived plants and asymptomatic field plants, confirming that routine visual inspection underestimates the true prevalence of the virus in commercial fields. All three viruses mentioned above, SMV, SCSMV and SCYLV, are RNA viruses. Another important viral disease in sugarcane is Leaf Fleck Disease (LFD) caused by *Sugarcane bacilliform virus* (SCBV) (genus *Badnavirus*, family *Caulimoviridae*), which represents a uniquely complex and increasingly significant viral threat to global sugarcane production and germplasm exchange. Unlike the RNA viruses causing mosaic and yellow leaf diseases, SCBV is a para-retrovirus with a double-stranded circular DNA genome [14]. Infected planting material is the primary source of all the above viruses. The transmission and spread of these viruses occur through aphids such as *Aphis gossypii, Longiunguis sacchari*, and *Rhopalosiphum maidis*, and through mealybug species such as *Saccharicoccus sacchari* and *Dysmicoccus boninsis.* Plant hoppers and lacewings are also reported to act as vectors for SCSMV. Phytoplasma is an obligate intracellular parasite of phloem tissues and forms a discrete group within the class Mollicutes. The transmission of phytoplasma occurs mainly through phloem-feeding vectors such as grasshoppers and psyllids [15]. Generally, plants derived through meristem culture are devoid of phytoplasma infection, as these cells are heat-labile. In sugarcane, three different types of phytoplasma-associated diseases exist: Sugarcane grassy shoot disease (SGSD), Sugarcane white leaf disease (SCWL), and Sugarcane yellow leaf disease (SCYL). Detection of phytoplasma is also carried out during disease indexing, as tissue-specific localization and low titers of the organisms make detection difficult with other conventional methods. Detection through nested PCR methods is a specifically effective alternative.

In recent years, sugarcane tissue culture technology has been effectively used to provide high-quality planting material for large-scale propagation. The basic availability of newly released varieties is often limited due to their cultivation in limited areas. Thus, ascertaining the quality of source material is of significant importance as virus-infected material can cause the disease to spread on a larger scale. In national and international programs, stringent policies are in place to ensure that primary source material is certified virus-free before distribution to farmers or participation in varietal exchange programs.

2 Materials

2.1 Sampling and Preparation

Aluminum foil, scissors, sterile test tubes, gloves, ethanol, tissue paper, and marker.

2.2 DNA Isolation

1. Chemical: CTAB, PVP, SDS, Sterile water, EDTA, NaCl, Tris-HCl, Sodium Sulfite, Beta-mercaptoethanol, Phenol, Choloroform, Isoamyl alcohol, Isopropanol, Na- acetate, RNase (10 mg/mL).
2. Glassware and plasticware: Reagent bottles, glass rods, Sterile Eppendorf tubes 2, 1.5 and 0.5 mL.
3. Instrument: Water bath, pH meter, Nanodrop spectophotometer.
4. Miscellaneous: Tissue paper.
5. For the homogenization buffer composition, see Table 1.

2.3 Genetic Fidelity Analysis

A set of ten ISSR primers has been recommended by the National Certification System for Tissue Culture Raised Plants (NCS-TCP), Department of Biotechnology (DBT), Government of India for genetic fidelity analysis of sugarcane. The sequence of ISSR primers used for the analysis is mentioned in Table 2.

2.4 Virus Indexing in Sugarcane Micro-Propagated Plantlets

The set of primers required for all the PCRs of sugarcane virus indexing is mentioned in Table 3.

Table 1
Composition of homogenization buffer used for DNA isolation

Sr. no	Component	Stock concentration	Volume for 25 mL	Final concentration
A	B	C	D	E
1	Tris HCl	2 M	2.5	200 mM
2	EDTA	500 mM	2.5	50 mM
3	NaCl	6 M	7.5	2 M
4	CTAB	20%	2.5	2%
5	Sodium sulfite	0.6%	2.5	0.06%
6	Nuclease-free water		7.5	
	Total volume		25.0	
	Composition of other excipients			
7	SDS	5%		
8	PVP	10%		

All components listed in column B are prepared individually for long-term storage at the concentrations specified in column C. The components can be autoclaved and stored. For DNA isolation, the homogenization buffer can be prepared by adding all components at the volumes listed in column D. The buffer pH should be adjusted to 7.5

Table 2
Sequence of ISSR primers used for the analysis

Sr. No.	Primer	Sequence 5′–3′	Annealing temperature
1	UBC 810	GAGAGAGAGAGAGAGAT	44°C
2	UBC 815	CACACACACACACACAT	50°C
3	UBC 818	CACACACACACACACAG	50°C
4	UBC 826	ACACACACACACACACACC	54°C
5	UBC 827	ACACACACACACACACCA	50°C
6	UBC 834	AGAGAGAGAGAGAGAGYT	52°C
7	UBC 840	GAGAGAGAGAGAGAGAYT	50°C
8	UBC 844	CTCTCTCTCTCTCTCTRA	52°C
9	UBC 847	CACACACACACACACARC	50°C
10	UBC 855	ACACACACACACACACYT	52°C

Other requirements: 2× GT Emerald master mix (TaKaRa- Code RA310), TAE buffer, Agarose, water, ETBr/ other DNA fluorescence dye, Gloves, DNA Ladder, Pipettes Instrument: PCR-Thermocycler, Gel Electrophoresis tank, Power pack, Gel Documentation System with camera

Table 3
Primer sequence used for virus indexing in sugarcane

Sr. no.	Name of the primer	Sequence of the primer 5′ to 3′	Etiological agent detection	Amplicon size	Reference
1.	P1-F	AAGAGTTTGATCCTGGCTCAGGATT	Phytoplasma	~1800 bp	[16]
2.	P7-R	CGTCCTTCATCGGCTCTT			
3.	R16F2n-F	GAAACGACTGCTAAGACTGG		~1200 bp	[17]
4.	R16R2-R	TGACGGGCGGTGTGTACAAACCCG			
5.	SCBVF	ATTGGAAGTAAAAAGAGTGGAT	*Sugarcane bacilliform virus,*	~587 bp	[1][18]
6.	SCBVR	TGTCTRACTCCTTGTTGTA			
7	SCYLVF	ATGAATACGGGCGTAACCGYYCAC	*Sugarcane yellow leaf virus*	~610 bp	[1][18]
8	SCYLVR	GTGTTGGGGRAGCGTCGCYTACC			
9	SCSMVF	CATATGAAGCAGATGATGAGGTACACA	*Sugarcane streak mosaic virus*	~599 bp	[19]
10	SCSMVR	GAGCTCCTACGCAGGTCCGTGTCCT			
11	SCMVF	AGAGGAGTTTGATAGGTGGTA	*Sugarcane mosaic virus*	~465 bp	[1][18]
12	SCMVR	GTCTCGCCGACATTTCCGTA			

Other requirements: RNA isolation kit (Quigen, Cat no: 74904 or Sigma, Cat No-STRN50), cDNA synthesis kit (thermoscientific RevertAid First Strand cDNA Synthesis Kit, K1622), 2× Emerald PCR master mix (TaKaRa, Cat No: RR310), Agarose, TAE buffer

[1]The primers were designed at the referral center for Virus indexing in the National Certification System for tissue-cultured raised plants (NCS-TCP)

3 Methods

In Fig. 1, the flowchart of the procedure for the analysis of genetic fidelity and virus indexing in sugarcane tissue culture plantlets is presented. In this book chapter, we will see step-by-step procedures and specific considerations required during the process. The below flowchart (Fig. 1) outlines the process for sample preparation for testing sugarcane plants. It splits into two main paths: DNA isolation and RNA isolation. The DNA isolation path leads to PCR testing for Genetic fidelity using molecular markers, detection of two pathogens, Phytoplasma and a DNA virus, *Sugarcane bacilliform virus* (SCBV). The RNA isolation path includes c-DNA preparation followed by PCR, which tests for three viruses: *Sugarcane mosaic virus* (SCMV), *Sugarcane streak mosaic virus* (SCSMV) and *Sugarcane yellow leaf virus* (SCYLV). This chart shows the steps and tests used to identify genetic quality and specific diseases in sugarcane samples.

3.1 Sampling

Sampling is the most crucial step for any analysis. The following are the steps for sampling.

1) Sample when received should be first cut into three parts for isolation of RNA, DNA in clean Aluminum foil, which are named as per sample number, and the remaining sample can be saved in −80°C refrigerator (*see* **Notes 1 and 2**)
2) The samples for DNA isolation are cut into smaller sections using clean scissors. About 50–100 mg of the leaf sample is sufficient for DNA isolation.
3) The samples for the RNA are processed immediately or stored in a −80°C refrigerator.

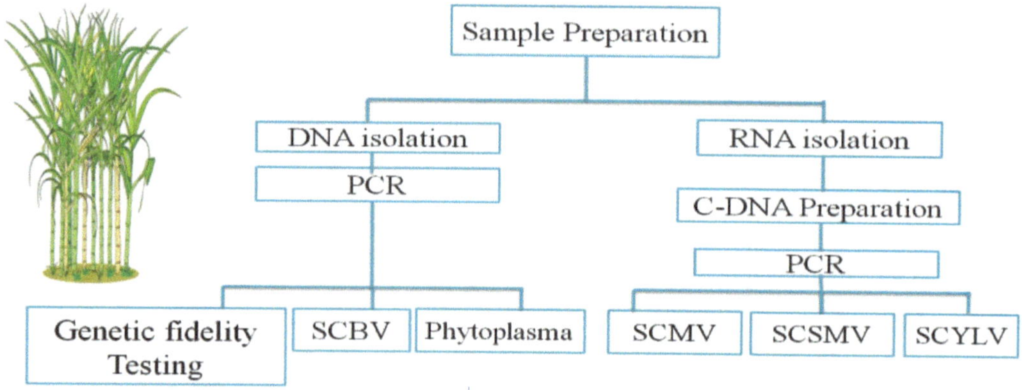

Fig. 1 Schematic representation of workflow for Genetic Fidelity testing and virus indexing in sugarcane micro-propagated plantlets

4) This primary preparation helps avoid repeated freeze-thawing of samples, which is especially important if RNA isolation has to be repeated from the remaining sample.

3.2 DNA Isolation

DNA isolation in sugarcane is carried out by the modified CTAB method [20] (*see* **Notes 3 and 4**).

1) About 900μL of homogenization buffer is added to 2 mL Eppendorf tubes, and 10 μL of β-mercaptoethanol is added to each tube. The tubes are pre-heated to 65°C in a water bath.
2) About 100 mg of the leaf sample is crushed using a mortar and pestle in liquid nitrogen, and the finely crushed powder is directly added to the tubes containing the buffer. 100 μL of both the excipients (5% SDS, 10% PVP) are added to the tube, followed by the addition of 100 μL CTAB. The tubes are vortexed vigorously till light green mixture is obtained. This is followed by incubation at 65°C for 1 h with intermittent mixing.
3) The tubes are cooled to room temperature, and 24:1: Chloroform:Isoamyl alcohol is added to it. Then the tubes are inverted, mixed for about 5–10 min and centrifuged at 8000 rpm for 10 min.
4) The upper aqueous layer is transferred to the fresh sterile 1.5 mL Eppendorf tube. About 10 μL of 10 mg/mL RNase is added to it. The tubes are inverted, mixed, and kept at 37°C for about 30 min.
5) About 500 μL of chilled Isopropanol is added to the tubes, inverted, mixed, and incubated at −80°C for 1 h. About 1/10th volume of 3 M Na-Acetate pH 5.6 is also added to the tube.
6) The tubes are centrifuged at maximum speed for 10–15 min. The supernatant is decanted carefully without disturbing the lower DNA pellet.
7) About 500 μL of 70% ethanol in water is added to the pellet. The tubes are inverted and mixed five times, centrifuged at 14,000 rpm, and the supernatant is discarded.
8) The procedure is repeated once, and the tubes are dried at room temperature.
9) 50–60 μL of TE buffer is added to the DNA and the tubes are incubated at 37°C for 1 h till the pellet is dissolved completely (*see* **Note 5**).
10) The DNA concentration and the 260/280 absorbance ratio are measured using a Nanodrop. Usually, 300-400 ng/μL DNA is obtained (*see* **Notes 6 and 7**).

3.3 Genetic Fidelity Testing Using ISSR Molecular Markers

Initially, the DNA isolated from the samples is added to the PCR tubes. PCR reaction mixture is prepared by component volumes mentioned in E column of Table 4. The mixture volume is calculated by multiplying the number of samples to be analyzed by the volume of each sample. About 17 μL of reaction mixture is added to each PCR tube containing individual sample DNA.

3.3.1 Gel Electrophoresis

After PCR, the reaction mixture can be loaded directly onto the 2% agarose gel prepared in 1× TAE buffer. The DNA ladder is loaded at least once in each lane. The electrophoresis is carried out for about 1.5–2.0 h until separation of all bands is complete. The gel is visualized using a UV gel documentation system (Fig. 2). In the following figure (Fig. 2), three horizontal rows of electrophoretic gel bands compare DNA samples labelled 1 to 22, with lane L showing a size marker on the left. Each row represents a result from a different DNA marker named UBC 815, UBC 834, and UBC 840, respectively. The bands vary in position, indicating differences in DNA fragment sizes among samples. Samples 1-5 show distinct band patterns compared to samples 6 to 22, suggesting genetic variation between these groups. The image highlights how these three markers reveal different DNA fragment patterns across the samples, useful for comparing genetic variation.

3.3.2 Statistical Analysis

For each amplicon size, presence/absence is checked across samples. If any sample of the same variety does not contain any particular band, the band is considered polymorphic. In case where multiple varieties are taken under analysis, like in the above figure (Fig. 2), the polymorphism should be checked within samples of one variety. For example, in panel 'a' of the UBC 815 sample 6-22 are of var Co. 86032, of which sample no. 15 is showing additional bands present

Table 4
PCR composition for the genetic fidelity analysis in sugarcane

Sr. no	Component	Stock concentration	Volume for 1 reaction	Final concentration
A	B	C	E	F
1	DNA	30 ng–50 ng/μL	3.0	
2	GTM	2×	12.5	1×
3	ISSR Primer	10 μM	1.5	0.75 μM
4	Nuclease-free H_2O		3.0	
	Total volume		20.0	

The temperature profile for the PCR is carried out as: initial denaturation at 94°C for 5 min followed by 35 cycles of denaturation at 94°C for 1 min, annealing at 44–54°C for 1 min, and extension at 72°C for 2 min. Final extension is carried out at 72°C for 10 min

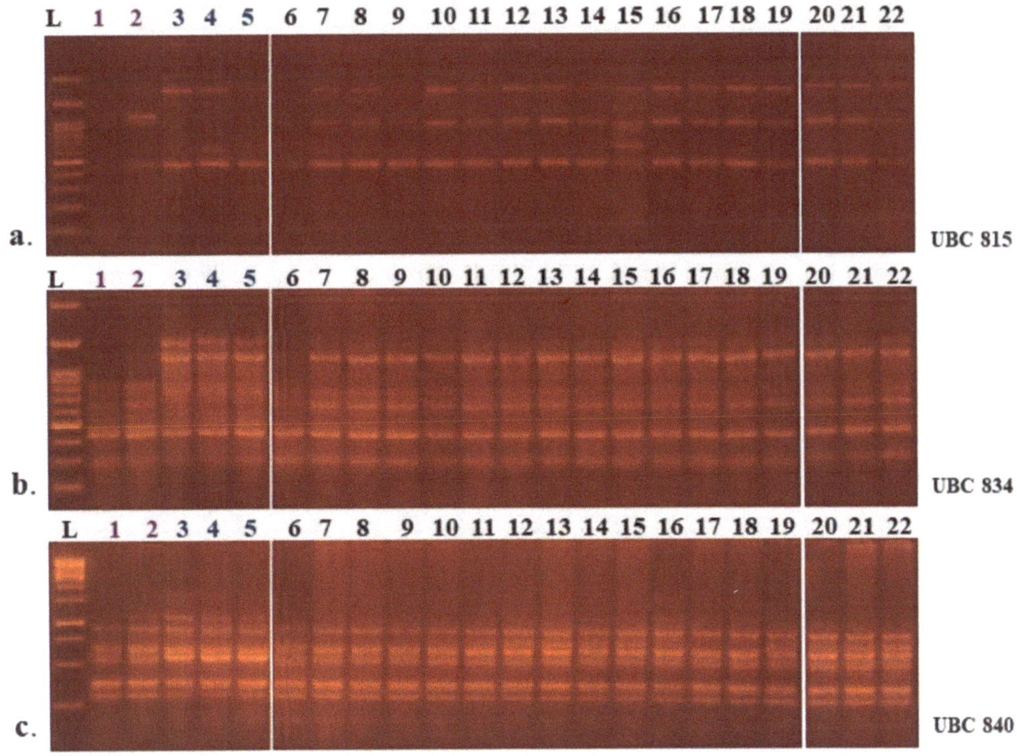

Fig. 2 ISSR PCR gel images of sugarcane micropropagated plantlets of different varieties. 1, 2- Var. MS10001, 3-5- Var. VSI 18121, 6-22: Var. CoM 0265; Panel a. amplification pattern of primer UBC 815, Panel b. amplification pattern of primer UBC 834, and panel c. amplification pattern of primer UBC 840. L-Ladder of 1000 base pairs; the size of the bands in the ladder is as follows: 100, 200,...1000 bp, 1500 bp and 2000 bp (lowest at the bottom)

at about 650 bp, 900 bp and 1250 bp, which are not present in any other sample of the same variety. Such bands are looked for in identifying genomic variation for fidelity testing. In such cases, the variation is reported and the sample (no. 15) is not considered as true to type. In the same panel 'a', samples 1 & 2 are of variety MS 10001, which showed a polymorphic band of about 1100 bp. It is characteristic band for this variety. Although the band is not present in other varieties, but present in both the samples of the same variety, the samples are considered true to type.

3.4 Virus Indexing in Micro Propagated Plantlets

The same DNA used for fidelity analysis can be used for SCBV and phytoplasma detection in micropropagated plantlets.

3.4.1 Detection of Sugarcane bacilliform Virus (SCBV) and Phytoplasma

Sugarcane bacilliform virus causes flex disease in sugarcane. Although it is not a prominent disease, it causes a reduction in photosynthesis. Being an ss DNA virus, it can be easily detected through the PCR technique. The compositions of the PCR mixture are as mentioned in Table 5. The DNA (about 100 ng) of individual samples is added to the pre-labeled PCR tubes, followed by the addition of the 23 μ L of reaction mixture. Tubes are centrifuged for 10–15 s and placed in a thermo-cycler.

3.4.2 Nested PCR for Phytoplasma Detection

Distribution of phytoplasma in leaves is discreet, wherein it might be concentrated in certain parts and not in other parts. Secondly, the phytoplasma titer is very low, and inhibitors such as polysaccharides, polyphenols, and tannins present in plant cells can inhibit phytoplasma amplification. Thus, the nested PCR method is necessary for phytoplasma detection. In this case, the first PCR helps increase the concentration of the primary amplicons. Dilution of the first PCR round reduces inhibitor levels, providing sufficient base amplicons for phytoplasma-specific primers to amplify in the nested PCR, making it a sensitive technique for phytoplasma detection.

The PCR reaction mixture is prepared as mentioned in the Table 6. For the first PCR, DNA isolated from the plant sample is used with PI and P7 primers. The PCR product is diluted 1:20, and the diluted product is used as a template for the nested PCR reaction using R16F2/R2 primers (*see* **Note 8**). The thermal profile for the PCR is also mentioned below.

The temperature profile for PCR is carried out as follows: initial denaturation at 94°C for 5 min, followed by 30 cycles of denaturation at 94°C for 30 s, annealing at 55°C for 1 min, and extension at 72°C for 1 min. The product of the first PCR is diluted to 1:20 in nuclease free water, and 2 μL of PCR product is used for the nested PCR. Also, primers R16F2 and R16R2 are used for the nested reaction. The thermal profile for the nested PCR is similar to that of PCR 1, except that the annealing temperature is 56°C. Presence of ~1800 bp

Table 5
PCR composition for the SCBV detection in sugarcane

Sr. no	Component	Stock concentration	Volume for 1 reaction	Final concentration
A	B	C	E	F
1	DNA	30 ng–50 ng	3 μL	
2	GTM	2×	12.5	1×
3	ScBV Fp	10 μM	1.5	0.6 μM
4	ScBV Rp	10 μM	1.5	0.6 μM
5	Nuclease-free H_2O		6.5	
	Total volume		25.0	

The temperature profile for the PCR is carried out as: initial denaturation at 94°C for 5 min, followed by 35 cycles of denaturation at 94°C for 30 s, annealing at 52°C for 1 min, and extension at 72°C for 1 min. Final extension is carried out at 72°C for 10 min. The PCR product is checked on 1.5% agarose gel

Table 6
PCR reaction mixture composition for the phytoplasma detection in sugarcane

Sr. no	Component	Stock concentration	Volume for 1 reaction	Final concentration
A	B	C	E	F
1	DNA/ Diluted PCR product (*see* footnote 1)	30–50 ng/μL	1 μL	30–50 ng
2	GTM	2×	12.5	1×
3	P1 (First PCR)/R16F2 Fp (Nested PCR)	10 μM	1.5	
4	P7 (First PCR)/R16R2 Rp (Nested PCR)	10 μM	1.5	
5	Nuclease-free H_2O		6.5	
	Total volume		25.0	

[1]About 2 μL of 1:20 diluted PCR1 product used for nested PCR

band in PCR reaction and ~1200 band in nested reaction confirms phytoplasma infection.

3.4.3 Reverse Transcription PCR Analysis for the Detection of RNA Viruses

RNA Isolation

In sugarcane virus indexing, three (*i.e.* SCMV, SCSMV and SCYLV) out of four are RNA viruses. Hence, RNA isolation is a very important step. Following is the method of RNA isolation using the Plant Spectrum Total RNA kit (*see* **Notes 9 to 14** before proceeding with RNA isolation).

1) About 50 mg of plant sample is crushed in a pre-chilled sterile mortar and pestle, the finely chilled powder is added into the pre-chilled 2 mL Eppendorf tubes, and 500 μL of Lysis solution and 5 μL of β-mercaptoethanol are added to each sample. The tubes are vortexed immediately for at least 30 s (*see* **Note 9**).
2) The samples are incubated at 56°C for 3–4 min followed by centrifugation of the tubes at 14,000 rpm for 3 min.
3) The supernatant is carefully transferred to the blue filtration column by pipetting from the bottom of the tube, avoiding disturbance of the debris.
4) To the clear filter, 500 μL of binding solution is added and mixed with pipetting 5–6 times, and 700 μL is transferred to the binding column kept on a 2 mL collection tube and

centrifuged at 14,500 rpm/1 min, and the flow-through is decanted. The process is repeated by transferring the previous filtrate to the binding column.

5) 500 µL of Wash solution 1 is added to the column, the cap is closed and centrifuged at 14,000 rpm for 1 min. The flow through is discarded.
6) 500 µL of wash solution 2 is added to the column, caps are closed, and centrifugation at 14,500 rpm/30 s is carried out. Flow through is discarded. The process with wash buffer-2 is repeated by the addition of another 500 µL of wash solution 2 onto the column.
7) Empty spin is done to dry the columns at 14,500 rpm/30 s and the column is placed in a new collection tube.
8) About 50 µL of elution buffer is added to the center of the tube directly onto the binding column matrix and allowed to sit for 1 min.
9) The tubes are centrifuged at maximum speed at 14,500 rpm. The RNA is kept on ice. The quantity of RNA should be measured using a nanodrop (*see* **Note 15**).

cDNA Preparation

After RNA isolation, the RNA was immediately processed for cDNA preparation. Following is the protocol for using thermoscientific® cDNA preparation kit (Table 7).

PCR Using cDNA

Sugarcane yellow leaf virus (SCYLV), *Sugarcane streak mosaic virus* (SCSMV) and *Sugarcane mosaic virus* (SCMV) DNA can be amplified at the same temperature using primers mentioned in Table 8. Thus, multiplexing is possible.

The following reaction mixture is prepared using 2× Emerald Amp GT PCR master mix (TaKaRa, Cat no: RR310)

The PCR can be carried out by multiplexing two primer pairs for the different viruses. We have observed a mixed infection of SMV and SSMV, along with SCYLV, in some samples. The difference in amplicon size of both SCSMV and SCYLV is only 11 nucleotides (~599 bp for SCSMV and ~610 bp for SCYLV). Thus, it becomes difficult to identify exactly which of these two viruses is present on the gel. Thus, it is preferable to amplify SCYLV and SCSMV separately in two reactions, and SCMV (~465 bp) can be amplified by multiplexing with either of them (Fig. 3). For the multiplex reaction, the reaction mixture remains nearly similar as mentioned in Table 3; but primers of SCMV are also added (1.5 µL each of forward and reverse primer) along with either of SCSMV/SCYLV. Also, cDNA of both the viruses SCMV and either SCSMV/SCYLV is added as 2 µL each in each PCR tube. Subsequently, the amount of NFW is reduced to make up the total volume to 25 µL. The temperature profile is as follows: Initial denaturation of 94°C for 4 min followed by 35 cycles

Table 7
PCR composition for cDNA preparation

Sr. No.	Component	Stock concentration	Volume for 1 reaction (μL)	Final concentration
A	B	C	E	F
1	RNA	100–300 ng/μL	5–10	500 ng–1 μg
2	Reverse primer of the virus	10 μM	1.0	0.5 μM
	The above components are added to the tube individually. Incubate at 65°C in PCR for 5 min. Quickly place on ice.			
3	5× buffer	5×	4.0	1×
4	10 mm dNTP	10 mM	2.0	0.2 mM
5	Ribo lock enhancer (Rl)	1.0 U/μL	1.0	1 U
6	Reverse transcriptase (RT)	1.0 U/μL	1.0	1 U
7	NFW[1]		To make up to 20.0 μL	
	Total volume		20.0	

Add the reaction mixture of components in Sr. No. 3–7 to the PCR tube. Temperature profile of 25°C for 5 min, followed by 42°C for 45 min, and 85°C for 5 min is used for cDNA formation (*see* **Note 16**). cDNA is stored at −20°C till its use in PCR. In the paragraphs below, the procedure for the detection of *Sugarcane yellow leaf virus* (SCYLV), *Sugarcane streak mosaic virus* (SCSMV) and *Sugarcane mosaic virus* (SCMV) is mentioned

[1] *NFW* nuclease-free water

Table 8
PCR composition for RNA virus detection in sugarcane

Sr. no	Component	Stock concentration	Volume for 1 reaction	Final concentration
A	B	C	E	F
1	cDNA	100 ng–1 μg	3 μL	
2	2× GTM Master mix	2×	12.5	1×
3	ScYLV/ScSMV/ScMV Fp	10 μM	1.5	0.6 μM
4	ScYLV/ScMV/ScMV Rp	10 μM	1.5	0.6 μM
5	Nuclease-free H_2O		6.5	
	Total volume		25.0	

of denaturation at 94°C for 1 min, annealing at 59°C for 1 min, and extension at 72°C for 1 min, followed by the final extension at 72°C for 10 min. Figure 3 shows a comparison of five sugarcane diseases with their symptoms and molecular detection results. The diseases are sugarcane leaf fleck disease, sugarcane mosaic disease, sugarcane yellow leaf disease and a group including sugarcane yellow leaf disease,

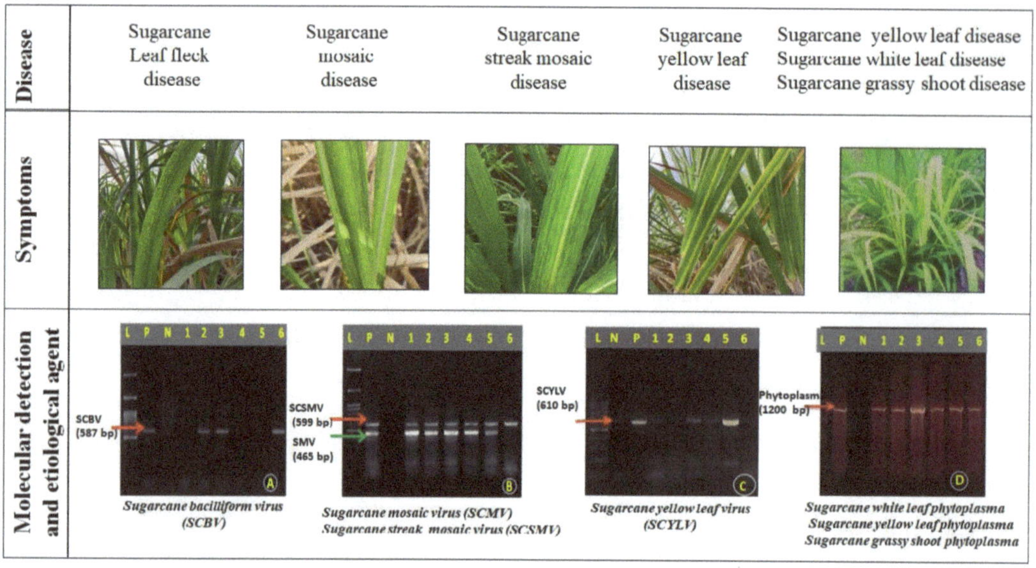

Fig. 3 Disease symptoms and molecular detection of sugarcane viral diseases

sugarcane white leaf disease, and sugarcane grassy shoot disease. Each disease column shows a photo of affected leaves highlighting typical symptoms such as discolouration, streaks or yellowing. Below each symptom, a gel image from molecular testing identifying the specific pathogen by DNA fragment size in base pairs. Red arrows point to the bands indicating the presence of the causative agent: SCBV at 587 base pairs, SSMV at 599 base pairs, SMV at 465 base pairs, SCYLV at 610 base pairs and phytoplasma at 1200 base pairs. This provides help to link visible disease sign with lab confirmation of the exact cause, aiding accurate diagnosis and management of sugarcane disease.

4 Notes

1) It is advisable to clean the leaf surface with tissue paper and ethanol before sampling. The fungal growth often seen in tissue papers wrapped around the sample gives false-positive results in phytoplasma analysis. Also, the midrib sections of the leaf must be included in a sample for phytoplasma detection.
2) For both DNA and RNA isolation, the midrib portion of the sample should be included. It improves SCYLV and phytoplasma detection.
3) For DNA isolation of sugarcane, other protocols of the company's reagents are also available. Commercial kits can reduce

the timing required for the analysis. The protocol for DNA isolation is usually provided by the manufacturer and varies by company.

4) For DNA isolation using commercial kits, a sample quantity of more than 100 mg should be avoided.

5) Using a binding column is also a good alternative to collecting precipitated DNA in 1.5 mL Eppendorf tubes by centrifugation. DNA-binding column available at lower prices than DNA isolation kits. This eliminates the risk of washing off the DNA pellet during decanting and washing.

6) If precipitations of nonsoluble white particles in the TE buffer are observed along with pure DNA, the DNA is allowed to dissolve at 37°C, and the tubes are centrifuged at 6000–8000 rpm, by which the impurities settle to the bottom. The aqueously dissolved DNA should be collected, transferred to a separate tube, and used for PCR.

7) The good quality of DNA shows a 1.7–1.9 absorption ratio at 260/280 nm. A lower ratio often shows protein or carbohydrate contamination. A lower ratio generally suggests carbohydrate contamination. The procedure mentioned in **Note 3** can be followed.

8) The reaction can also be carried out using individual PCR components. The reaction mixture contains 1× buffer, 2 mM $MgCl_2$, 0.2 μM Primer, 1 Unit taq polymerase and nuclease-free water to bring the total volume to 25 μL.

9) We have checked the RNA using both the Trizol method and RNA isolation kits. Although we found more DNA in Trizol, the commercial kit methods are quick and suitable for virus indexing, especially when the quality or quantity of leaf samples is poor.

10) All plasticware, glassware and pipettes used during RNA isolation should be RNase-free. This can be done by maintaining DEPC-treated plasticware separately for RNA isolation procedures, using a separate area with limited movement, surface wiping of the working area with DEPC-treated water, followed by wiping with 70% ethanol, using clean gloves with intermittent ethanol wipes. These are sufficient for the procedure.

11) For DEPC treatment, 1 mL of DEPC should be dissolved in 1 L of sterile distilled water and stirred till oily globules are dissolved. The glass container should be used. The plastic ware can be dipped in this water for about 30 min to 1 h. Taken out and wrapped in autoclave bags. The next day, all the DEPC-treated glassware and plastic wares are autoclaved. Autoclaving of DEPC-treated glass and plasticware is very

important as even trace amounts can cause adenine chemical modification [21].

12) DEPC reacts with polypropylene and polystyrene; glass pipettes should be used for pipetting. The use of DEPC-treated water for making electrophoresis buffers is recommended; DEPC-treated distilled water is boiled to 100°C, then autoclaved, cooled, and kept ready for buffer preparation.

13) Too much of the sample (more than 100 mg) leads to the isolation of poor-quality RNA. Thus, taking too many samples should be avoided.

14) Ethanol addition needs to be done in the wash solution-2 provided with the kit. These preparations must be completed before proceeding with RNA isolation.

15) In spectrophotometric analysis, for pure RNA, the absorption ratio at 260/280 nm calculated as (A260 − A320)/(A280 − A320) should be ideally 2.0–2.3.

16) Some research papers have advised using oligo-DT and random hexamer primers for CDNA preparations. We observed, however, that using a reverse primer enhanced detection efficiency.

References

1. Lebot V (1999) Biomolecular evidence for plant domestication in Sahul. Genet Resour Crop Evol 46(6):619–628
2. FAO (2023) https://doi.org/www.fao.org/faostat/en/#data/QCL. Last accessed 12 Sept, 2026.
3. OECD-FAO (2026) OECD-FAO Agricultural Outlook 2026-2035, OECD Publishing, Paris/FAO, Rome. https://doi.org/10.1787/47874669-en
4. Solomon S (2016) Sugarcane production and development of sugar industry in India. Sugar Tech 18(6):588–602
5. Viswanathan R, Balamuralikrishnan M, Shanmuganathan T (2017) Emerging trends in sugarcane diseases and their management. Sugar Tech 19(1):1–10
6. Rao GP, Chatenet M, Girard JG, Rott P (2006) Distribution of sugarcane mosaic and sugarcane streak mosaic virus in India. Sugar Tech 8(1):79–81
7. Viswanathan R, Balamuralikrishnan M, Karuppaiah R (2008) Identification of three genotypes of sugarcane yellow leaf virus causing yellow leaf disease from India and their molecular characterization. Virus Genes 37(3):368–379
8. Viswanathan R, Karuppaiah R, Balamuralikrishnan M (2009) Identification of new variants of SCMV causing sugarcane mosaic in India and assessing their genetic diversity in relation to SCMV type strains. Virus Genes 39(3):375–386
9. Bagyalakshmi K, Viswanathan R (2021) Development of a scoring system for sugarcane mosaic disease and genotyping of sugarcane germplasm for mosaic viruses. Sugar Tech 23(5):1105–1117
10. ElSayed AI, Komor E, Boulila M, Viswanathan R, Odero DC (2015) Biology and management of sugarcane yellow leaf virus: an historical overview. Arch Virol 160(12):2921–2934
11. Holkar SK, Balasubramaniam P, Kumar A, Kadirvel N, Shingote PR, Chhabra ML, Kumar S, Kumar P, Viswanathan R, Jain RK, Pathak AD (2020) Present status and future management strategies for Sugarcane yellow leaf virus: a major constraint to the global sugarcane production. Plant Pathol J 36(6):536
12. Chinnaraja C, Viswanathan R, Karuppaiah R, Bagyalakshmi K, Malathi P, Parameswari B (2013) Complete genome characterization of Sugarcane yellow leaf virus from India: evidence for RNA recombination. Eur J Plant Pathol 135(2):335–349

13. Nithya K, Parameswari B, Viswanathan (2020) Mixed infection of sugarcane yellow leaf virus and grassy shoot phytoplasma in yellow leaf affected Indian sugarcane cultivars. Plant Pathol J 36(4):364
14. Balan S, Nithya K, Cherian KA, Viswanathan R (2022) True seed transmission of Sugarcane bacilliform virus (SCBV) in sugarcane. Sugar Tech 24(2):513–521
15. McCoy RE, Caudwell A, Chang C, Chen TA, Chiykowski LN, Cousin MT, Dale JL, Leeuw GD, Golino DA, Hackett KJ, Kirkpatrick BC, Marwitz R, Petzold H, Sinha RC, Sugiura M, Whitcomb RF, Yang IL, Zhu B, Seemüller E (1989) Plant diseases associated with mycoplasma like organism. In: Whitcomb T (ed) The mycoplasmas. Academic Press, pp 545–640
16. Deng S, Hiruki C (1991) Amplification of 16S rRNA genes from curable and non-culturable mollicutes. J Microbiol Methods 14:53–61
17. Gundersen DE, Lee IM (1996) Ultrasensitive detection of phytoplasmas by nested PCR assays using two universal primer pairs. Phytopahol Mediterr 90:921–927
18. Kumar A, Ghan MY, Devarumath RM (2025) Comprehensive multiplex RT-PCR and PCR assay for simultaneous detection of mixed infections of major sugarcane viruses. Asian J Microbiol Biotechnol 10(1):65–77
19. Singh D, Tewari AK, Rao GP, Karuppaiah R, Viswanathan R, Arya M, Baranwal VK (2009) RT-PCR/PCR analysis detected mixed infection of DNA and RNA viruses infecting sugarcane crops in different states of India. Sugar Tech 11(4):373–380
20. Purankar MV, Nikam AA, Devarumath RM (2024) Plant regeneration through indirect somatic embryogenesis in sugarcane and evaluation of regenerated plantlets through ISSR molecular markers in genotype CoM 0265. Res J Agric Sci 15(4):1074–1085
21. Henderson RE, Kirkegaard LH, Leonard NJ (1973) Reaction of diethylpyrocarbonate with nucleic acid components. Adenosine-containing nucleotides and dinucleoside phosphates. Biochim Biophys Acta 294:356–364. https://doi.org/10.1016/0005-2787(73)90090-7

Chapter 4

Application of ISSR Molecular Markers in Field Crops

Nüket Altındal and Demet Altındal

Abstract

Assessing genetic variation in field crops is crucial for modern breeding programs and for understanding mechanisms of stress tolerance. The inter simple sequence repeat (ISSR) technique enables amplification of regions between microsatellite loci, offering high reproducibility and discriminatory power. Its applicability across species, cost-effectiveness, and independence from prior DNA sequence information make it a widely adopted method. This chapter systematically outlines the core steps of ISSR analysis, including DNA extraction, PCR optimization, gel electrophoresis, band scoring, and data analysis. Practical recommendations to address technical challenges and enhance experimental efficiency are also provided. Examples from crops such as potato, sage, and Jerusalem artichoke demonstrate the effectiveness of ISSR markers in evaluating genetic diversity and stress adaptation.

Keywords ISSR, Genetic variation, Field crops, Molecular markers, DNA analysis, PCR, Agarose gel electrophoresis

1 Introduction

Comprehensive characterization of genetic diversity in field crops is essential for sustainable agricultural production, breeding efficiency, and adaptation to environmental stresses. Increasing population pressure, climate change-induced abiotic stresses, and food security challenges necessitate the development of high-yielding and stress-resilient genotypes. Molecular-level assessment of germplasm resources has therefore become a central component of modern breeding strategies.

Traditional morphological and agronomic evaluations are often influenced by environmental conditions and may not accurately reflect underlying genetic relationships. DNA-based molecular marker systems provide objective and highly reproducible tools for detecting genetic variation. Among these systems, inter simple sequence repeat (ISSR) markers have gained widespread acceptance due to their high polymorphism, reproducibility, simplicity, and cost-effectiveness [1].

Nitish Kumar (ed.), *Molecular Markers in Plants*, Methods in Molecular Biology, vol. 3097,
https://doi.org/10.1007/978-1-0716-5663-1_4,

ISSR markers amplify genomic regions located between microsatellite loci using a single primer composed of simple sequence repeats. Unlike SSR markers, ISSR does not require prior sequence information, and compared to RAPD, it provides higher reproducibility and specificity [2, 3]. Over the past decades, ISSR markers have been successfully applied in cereals, legumes, oil crops, and tuberous species for genotype identification, diversity assessment, mutation screening, and population structure analysis [3–5].

The method combines the multilocus detection capability of microsatellite-based systems with the technical simplicity of PCR-based dominant markers. ISSR analysis involves high-quality DNA extraction, primer screening, optimization of PCR parameters, agarose gel electrophoresis, and statistical analysis of banding patterns. The resulting profiles are scored as binary matrices (presence/absence), and genetic similarity coefficients are calculated using software such as NTSYS-pc and GenAlEx [6, 7].

ISSR markers have proven particularly valuable in mutation breeding programs, including EMS-induced genetic variation studies, and in stress physiology research aimed at identifying drought-, salinity-, and temperature-tolerant genotypes [1, 3]. In laboratories where advanced sequencing technologies remain cost-prohibitive, ISSR provides a reliable and accessible alternative for large-scale diversity screening. This chapter describes standardized laboratory procedures and practical notes for applying ISSR molecular markers in field crops, following a step-by-step protocol format to ensure reproducibility and analytical reliability.

2 Materials

2.1 General Materials Used in ISSR Analysis

1. CTAB extraction buffer containing:
   a. Tris–HCl.
   b. NaCl.
   c. EDTA (Ethylenediaminetetraacetic acid).
   d. β-mercaptoethanol (*see* **Note 2**).
2. TE (Tris–EDTA) buffer for DNA dissolution and storage (*see* **Note 4**).
3. Template genomic DNA extracted from field-grown plants (*see* **Note 1**).
4. ISSR primers (*see* Table 1; *see* **Note 7**). dNTP mix (200 μM) (*see* **Note 12**).
5. $MgCl_2$ (1.5–2.5 mM) (*see* **Note 13**).
6. Taq DNA polymerase (*Thermus aquaticus*) (*see* **Note 15**).
7. 10× PCR buffer (*see* **Note 16**).

Table 1
ISSR primers used in the study with sequences, lengths, and T_m values

No	Primer set	Sequence (5'→3')	Length (bp)	T_m (°C)	References
1	ISSR4	GAT AGA TAG ATA GAT A	20	38.9	[8]
2	LOL1	(CT)8AC	18	54.0	[9]
3	LOL2	(CT)8GC	18	56.0	
4	LOL3	(CA)6AC	18	41.0	
5	LOL4	(CA)6GT	18	41.0	
6	LOL5	(GA)6GG	18	44.0	
7	LOL6	(GT)6GG	18	44.0	
8	LOL7	(GA)6CC	18	44.0	
9	LOL8	(GT)6CC	18	44.0	
10	LOL9	(CAC)3GC	18	38.0	
11	LOL10	(GAG)3GC	18	38.0	
12	LOL11	(CTC)3GC	18	38.0	
13	PHV3	GCC(CT)8	18	59.0	
14	PHV4	GGC(GT)8	18	59.0	
15	PHV5	ACG(CA)8	18	57.0	
16	PHV6	CCA(CT)8	18	57.0	
17	PHV7	GTG(GT)8	18	57.0	
18	UBC-807	(AG)8T	18	50.0	[10, 11]
19	UBC-810	(GA)8T	18	50.0	
20	UBC-811	(GA)8C	18	52.0	
21	UBC-812	(GA)8A	18	50.0	
22	UBC-814	(CT)8A	18	54.0	
23	UBC-826	(AC)8C	18	52.0	
24	UBC-854	(CT)8RG	18	55.0	
25	UBC-808	AGAGAGAGAGAGAGAGC	18	50.0	
26	UBC-811	GAGAGAGAGAGAGAGAC	18	50.0	
27	UBC-812	GAGAGAGAGAGAGAGAA	18	50.0	
28	UBC-825	ACACACACACACACACT	18	50.0	
29	UBC-834	AGAGAGAGAGAGAGAGYT	18	50.0	
30	UBC-840	GAGAGAGAGAGAGAGAYT	18	50.0	
31	UBC-842	GAGAGAGAGAGAGAGAYG	18	50.0	
32	UBC-880	GGAGAGGAGGAGAGGAGA	18	50.0	

(continued)

Table 1 (continued)

No	Primer set	Sequence (5'→3')	Length (bp)	T_m (°C)	References
33	PCP-1	GACGACGACGACGAC	18	55.0	
34	PCP-2	AGGAGGAGGAGGAGGAGG	18	55.0	[12]
35	PCP-3	GTGCGTGCGTGCGTGC	18	55.0	

8. Agarose powder (1.5%–2.0%) (*see* **Note 17**).
9. 10× TAE buffer (Tris–Acetate–EDTA) (*see* **Note 17**).
10. Ethidium bromide or safer alternative dyes (*see* **Note 22**).
11. Gel loading buffer containing:
    a. Bromophenol blue
    b. Xylene cyanol
    c. Glycerol
12. 100 bp DNA ladder (*see* **Note 13**).
13. Sterile 1.5 and 2 mL microcentrifuge tubes.
14. Pipettes and sterile tips.
15. Microcentrifuge.
16. Programmable thermal cycler.
17. Water bath (65°C).
18. UV transilluminator or gel documentation system (*see* **Note 42**).
19. Ice and ice box for DNA stability.
20. The materials used in ISSR assays are illustrated in Fig. 1.

2.2 Overview of ISSR Workflow

1. Collection of plant material (field-grown plants or callus) (*see* **Note 1**).
2. DNA isolation using the CTAB-based method [13] (*see* **Note 2**).
3. PCR amplification using ISSR primers (*see* **Note 7**).
4. Agarose gel electrophoresis (*see* **Note 17**).
5. Band scoring and binary matrix construction (*see* **Note 26**).
6. Genetic similarity and population structure analysis (*see* **Note 29**).
7. The schematic workflow of ISSR procedures is presented in Fig. 2.

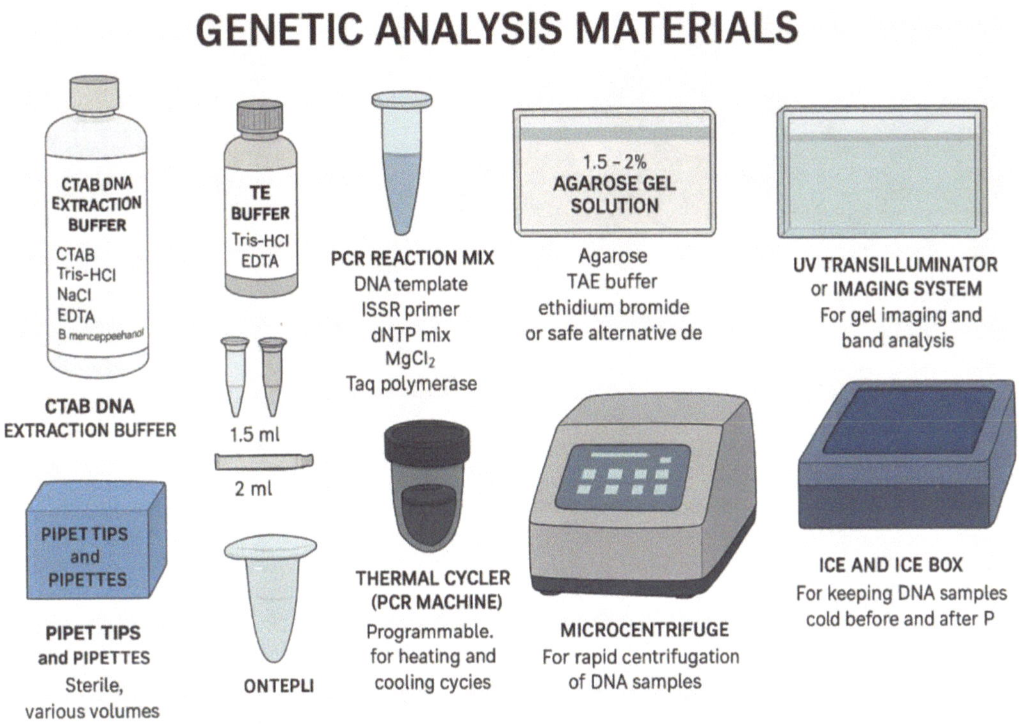

Fig. 1 Genetic analysis materials used in ISSR assays

3 Methods

3.1 DNA Extraction

1. Collect young leaf tissue or callus samples (Fig. 3) (*see* **Note 1**).
2. Store samples at −80°C if not processed immediately (*see* **Note 20**).
3. Grind ~200 mg tissue in liquid nitrogen (*see* **Note 1**).
4. Transfer powder to 2 mL microcentrifuge tube (*see* **Note 1**).
5. Add preheated CTAB extraction buffer (65°C) [13] (*see* **Note 2**).
6. Add 1%–2% PVP and freshly prepared β-mercaptoethanol when working with phenolic-rich plants [14] (*see* **Note 2**).
7. Incubate at 65°C for 45–60 min (*see* **Note 3**).
8. Add equal volume chloroform:isoamyl alcohol (24:1) (*see* **Note 3**).
9. Centrifuge at 10,000 × *g* for 10 min (*see* **Note 3**).
10. Transfer aqueous phase to new tube (*see* **Note 3**).
11. Precipitate DNA with cold isopropanol (*see* **Note 3**).

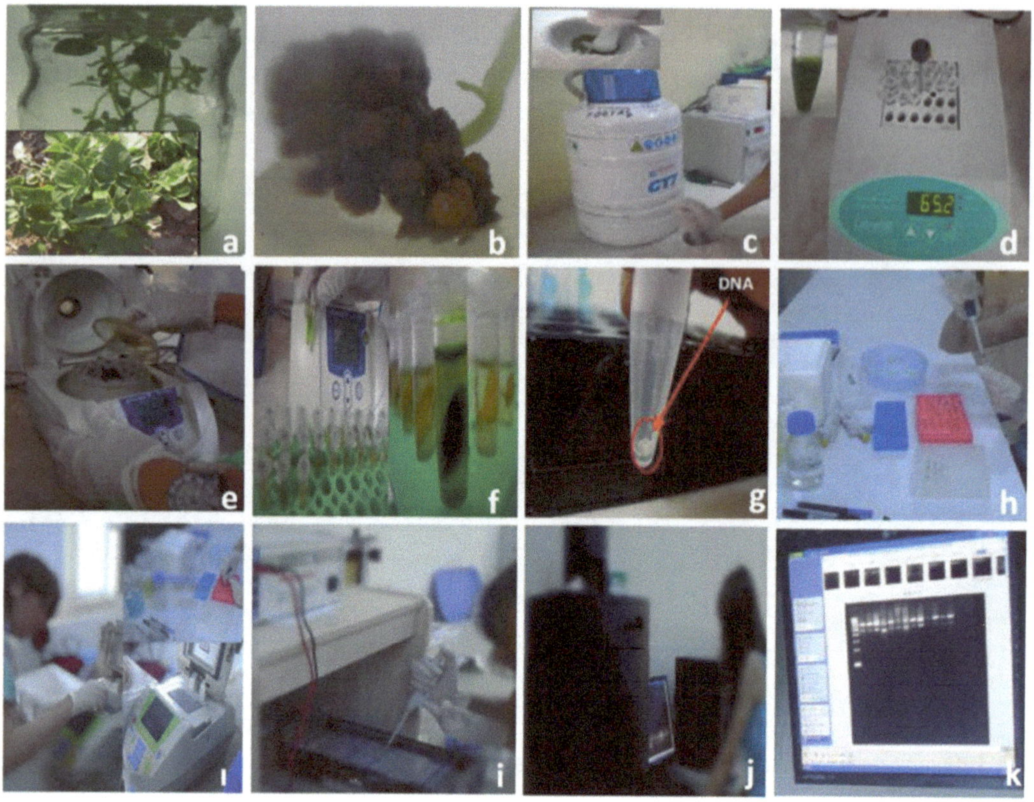

Fig. 2 Schematic workflow of ISSR molecular marker analysis in field crops

12. Incubate at −20°C for 30–60 min (*see* **Note 3**).
13. Centrifuge at 12,000 × *g* for 10 min (*see* **Note 3**).
14. Wash pellet with 70% ethanol (*see* **Note 3**).
15. Air-dry and dissolve in TE buffer (*see* **Note 4**).
16. Measure DNA concentration and purity (A260/280 = 1.8–2.0) [15] (*see* **Note 4**).
17. Re-extract DNA if degraded or impure [15] (*see* **Note 4**).
18. Medicinal and aromatic plants may contain high phenolics and essential oils inhibiting PCR [14] (*see* **Note 3**).
19. Cereals and tuber crops generally yield high-molecular-weight DNA with CTAB [16] (*see* **Note 3**).
20. Callus tissues improve DNA purity due to reduced secondary metabolites [1, 4] (*see* **Note 6**).

3.2 Primer Selection

1. Design single ISSR primers anchored at 3′ or 5′ microsatellite ends [17, 18] (*see* **Note 7**).
2. Select primers with length 18–25 bases (*see* **Note 8**).

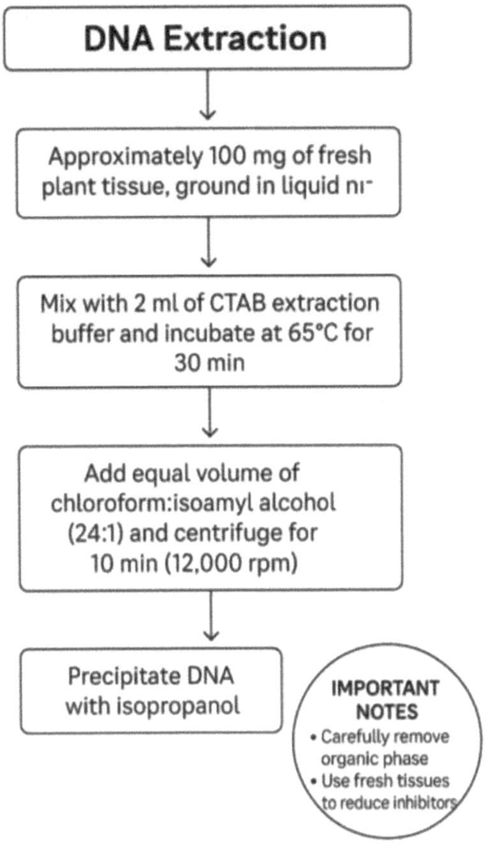

Fig. 3 DNA extraction from plant tissues

3. Maintain GC content at 50%–60% [19] (*see* **Note 8**).
4. Prefer di-, tri-, or tetra-nucleotide motifs (*see* **Note 8**).
5. Screen primers through preliminary amplification tests (*see* **Note 9**).
6. Select primers producing clear and reproducible polymorphic bands [19] (*see* **Note 9**).
7. Use 8–12 primers for reliable diversity analysis in field crops [1] (*see* **Note** 7).
8. Avoid using primers with similar motifs simultaneously (*see* **Note 9**).
9. (ISSR primer details are provided in Table 1 [8–12]) (*see* **Note** 7).
10. The primer selection workflow is shown in Fig. 4.

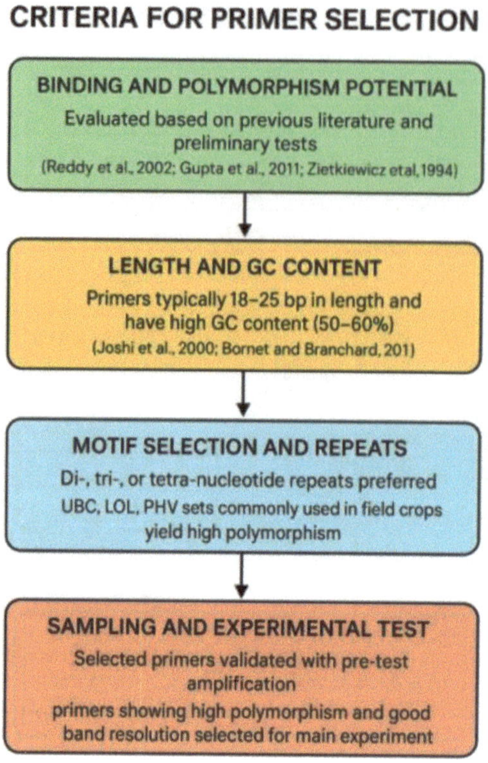

Fig. 4 Primer selection workflow (ISSR)

3.3 PCR Amplification

3.3.1 PCR Reaction Mixture

1. Prepare 25 μL total volume (*see* **Note 12**).
2. DNA template: 20–50 ng (*see* **Note 12**).
3. ISSR primer: 0.2–0.5 μM (*see* **Note 8**).
4. dNTP mix: 200 μM (*see* **Note 12**).
5. $MgCl_2$: 1.5–2.5 mM (*see* **Note 13**).
6. Taq DNA polymerase: 0.5–1 U (*see* **Note 15**).
7. PCR buffer: 1× (*see* **Note 16**).
8. Nuclease-free water to volume (*see* **Note 1**).
9. Role of PCR components described in Table 2 [20] (*see* **Note 12**).
10. Optimization considerations:
    a) Adjust $MgCl_2$ concentration for amplification specificity [21] (*see* **Note 13**).
    b) Perform gradient PCR to optimize annealing temperature [21] (*see* **Note 10**).

Table 2
PCR amplification protocol for ISSR analysis in field crops

Step	Component/procedure	Concentration/duration	Purpose / function
1	DNA template	20–50 ng/μL	Starting material containing the target DNA region for amplification (*see* **Note 12**)
	Primers (forward and reverse)	0.2–0.5 μM	Bind to specific sites on DNA to initiate replication (*see* **Note 8**)
	dNTP mix	200 μM	Provides building blocks for new DNA strand synthesis (*see* **Note 12**)
	Taq DNA polymerase	0.5–1 U	Heat-stable enzyme that synthesizes new DNA strands (*see* **Note 15**)
	PCR buffer	1×	Provides optimal pH and ionic environment for the enzyme (*see* **Note 16**)
	$MgCl_2$	1.5–2.5 mM	Cofactor required for polymerase activity (*see* **Note 13**)
2	Initial denaturation	94°C, 5 min	Separates double-stranded DNA and activates polymerase (*see* **Note 25**)
3	Cycling steps (35–40 cycles)	–	Each cycle includes:
	a) Denaturation	94°C, 30 s	Opens the DNA double helix into single strands (*see* **Note 27**)
	b) Annealing	50–60°C, 45 s	Primers bind to complementary sequences (*see* **Note 28**)
	c) Extension	72°C, 1 min	Taq polymerase synthesizes new DNA (*see* **Note 29**)
4	Final extension	72°C, 10 min	Ensures complete synthesis of PCR products (*see* **Note 30**)
5	PCR product storage	4°C (short-term)/ −20°C (long-term)	Preserves amplified DNA (*see* **Note 32**)
6	Verification (optional)	Agarose gel electrophoresis	Confirms expected fragment size (*see* **Note 41**)

c) Use diluted DNA for starch-rich crops (e.g., potato) (*see* **Note 16**).

d) Add BSA when working with phenolic-rich medicinal plants (*see* **Note 15**).

e) Avoid excessive cycle numbers (>40) to prevent nonspecific bands [21] (*see* **Note 14**).

3.3.2 PCR Cycling Conditions

1. Initial denaturation: 94°C for 5 min (*see* **Note 25**).
2. 35–40 cycles.
3. Denaturation: 94°C for 30 s (*see* **Note 27**).
4. Annealing: 50°C–60°C for 45 s (*see* **Note 28**).
5. Extension: 72°C for 1 min (*see* **Note 29**).
6. Final extension: 72°C for 10 min (*see* **Note 30**).
7. Hold at 4°C (*see* **Note 32**).
8. PCR cycle steps are illustrated in Fig. 5 [20].

3.3.3 Storage of PCR Products

1. Store short-term at 4°C (*see* **Note 32**).
2. Store long-term at −20°C or −80°C (*see* **Note 32**).
3. Avoid repeated freeze–thaw cycles [22] (*see* **Note 32**).

3.4 Gel Electrophoresis and Visualization

1. Prepare 1.5%–2.0% agarose gel in 1× TAE or TBE buffer (*see* **Note 17**).
2. Add ethidium bromide or a safe alternative dye (*see* **Note 22**).
3. Mix PCR products with loading dye (*see* **Note 12**).
4. Load samples with 100 bp DNA ladder (*see* **Note 13**).
5. Run at 80–110 V for 1–2 h (*see* **Note 20**).
6. Separate fragments based on size under an electric field [6] (*see* **Note 20**).
7. Visualize bands under UV transilluminator [7] (*see* **Note 19**).

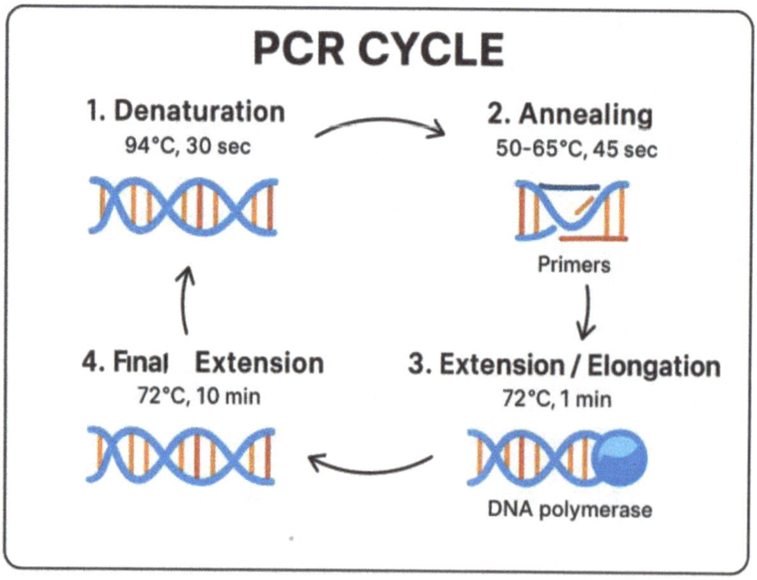

Fig. 5 Polymerase chain reaction (PCR)

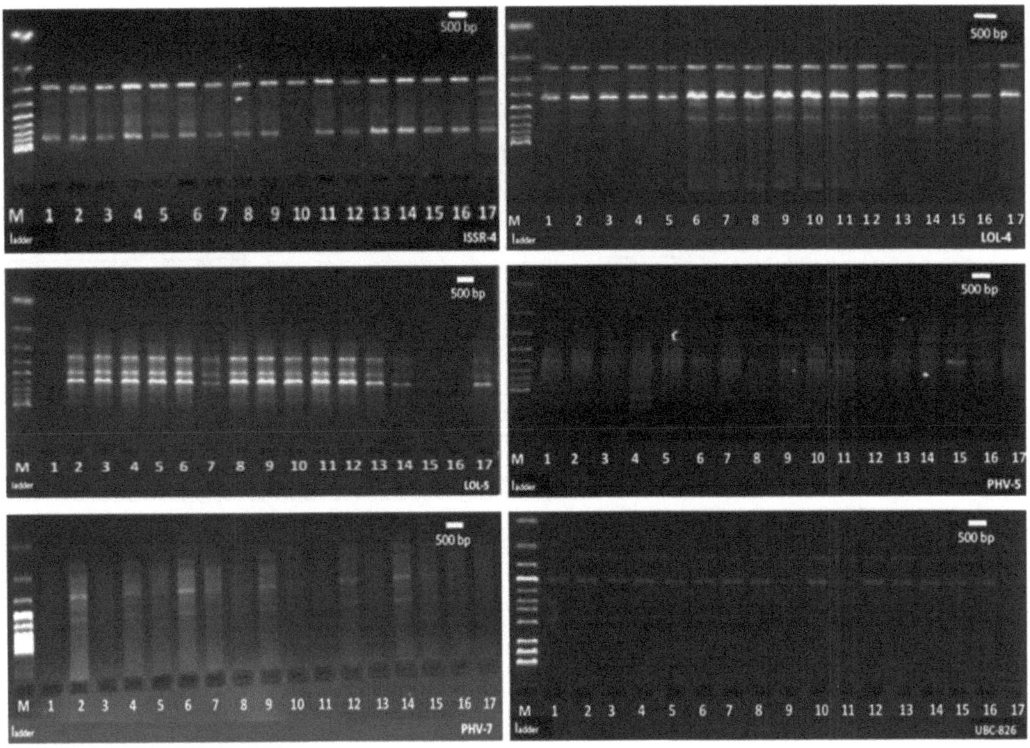

Fig. 6 PCR products amplified from EMS-treated Agria potato callus samples using agarose gel electrophoresis with six selected ISSR primers (>50% polymorphism) Reprinted with permission from [1]

8. Capture gel image using documentation system (Fig. 6) (*see* **Note 19**).
9. Use PAGE for high-resolution separation of small fragments if necessary (*see* **Note 19**).

3.5 Data Analysis

1. Score bands as presence (1) or absence (0) (*see* **Note 23**).
2. Exclude faint or ambiguous bands (*see* **Note 24**).
3. Construct a binary data matrix (*see* **Note 26**).
4. Calculate similarity coefficients (Jaccard or Dice) [7] (*see* **Note 29**).
5. Perform UPGMA clustering to generate a dendrogram (Fig. 7) (*see* **Note 30**).
6. Evaluate cophenetic correlation and Mantel test [23] (*see* **Note 31**).
7. Conduct PCA or PCoA for dimensional reduction (Fig. 8) (*see* **Note 32**).
8. Perform AMOVA to partition genetic variance (*see* **Note 33**).

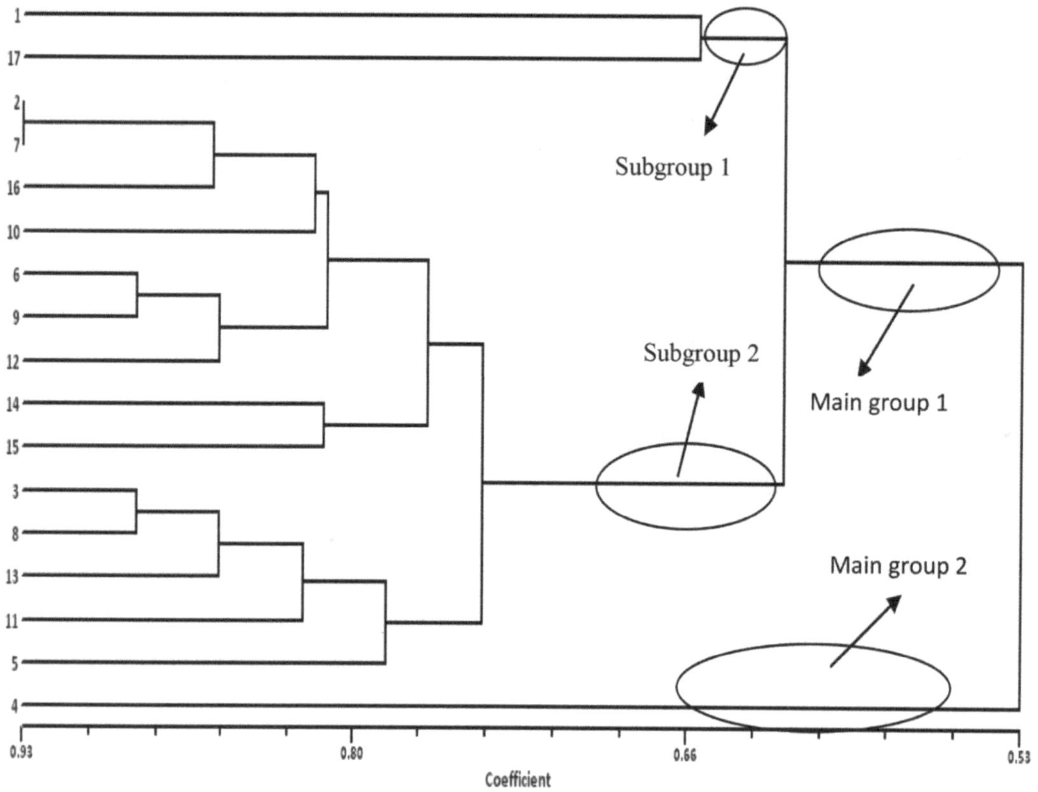

Fig. 7 UPGMA dendrogram showing the genetic relationship among EMS-treated Agria callus Reprinted with permission from [1]

9. Conduct STRUCTURE analysis to determine subpopulation clusters (*see* **Note 32**).
10. Use GenAlEx or NTSYS-pc for statistical analysis [7, 23] (*see* **Note 7**).

4 Notes

1. Use young leaves or callus tissues to obtain high-quality genomic DNA.
2. Add 1%–2% PVP and freshly prepared β-mercaptoethanol to the CTAB buffer to remove polyphenols.
3. Medicinal and aromatic plants containing high phenolic and volatile compounds may inhibit PCR.
4. DNA purity should be within A260/280 = 1.8–2.0.

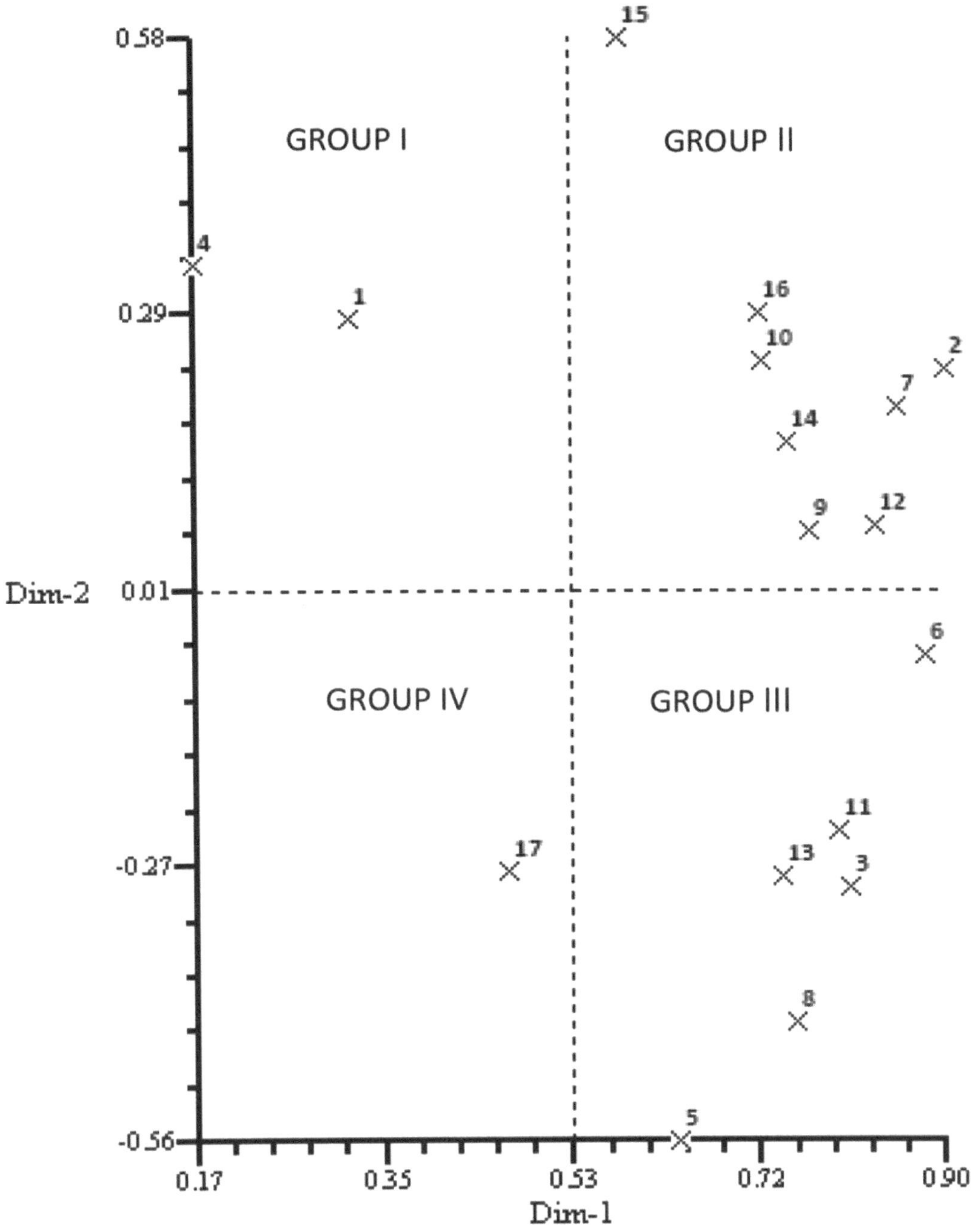

Fig. 8 PCA results of EMS-treated Agria callus Reprinted with permission from [1]

5. Verify DNA quality using spectrophotometry and 1% agarose gel electrophoresis [10, 11].
6. Callus tissues often yield cleaner DNA due to lower phenolic content.

7. Use 8–12 ISSR primers for reliable polymorphism detection.
8. Select primers based on motif type ((GA)n, (CA)n, etc.), GC content (50%–60%), and binding specificity.
9. Avoid using primers with highly similar repeat motifs in the same analysis.
10. Optimize annealing temperature and $MgCl_2$ concentration using gradient PCR.
11. Primers previously validated in field crops (e.g., wheat, maize, and potato) are recommended [1, 12, 14, 17].
12. Optimize template DNA concentration (typically 20–50 ng per reaction).
13. Adjust $MgCl_2$ concentration (1.5–2.5 mM) for improved specificity.
14. Use 30–40 amplification cycles; excessive cycles cause non-specific bands.
15. Add BSA when working with phenolic-rich medicinal species.
16. Dilute DNA templates for starch-rich crops, such as potato, to improve band clarity.
17. Use 1.5%–2.0% agarose for small fragments.
18. Use 0.8%–1.2% agarose for larger fragments [6].
19. Use 6% PAGE for high-resolution separation of 200–800 bp fragments.
20. Run gels at lower voltage for medicinal-aromatic plants to prevent band distortion.
21. Use 1× TBE for sharper band resolution during long runs [6].
22. Prefer safe staining dyes (e.g., SYBR Safe and GelRed) instead of ethidium bromide.
23. Score clear and reproducible bands only.
24. Do not score weak or ambiguous bands.
25. Use image analysis software (ImageJ, GelAnalyzer, Quantity One) to reduce subjectivity.
26. Convert banding profiles into a binary matrix (1 = presence, 0 = absence).
27. Repeat electrophoresis when scoring inconsistencies are observed.
28. Construct a binary (0/1) data matrix.
29. Use Jaccard or Dice coefficients for dominant marker data.
30. Perform UPGMA clustering and generate dendrograms.
31. Validate clustering using cophenetic correlation and the Mantel test.

32. Apply PCA or PCoA for multivariate visualization.
33. Conduct AMOVA to partition genetic variance.
34. Provide raw binary matrices as supplementary material [7, 23].
35. Include negative controls (no DNA) to detect contamination [23].
36. Include positive controls (validated DNA sample) to confirm PCR functionality [23].
37. Repeat experiments if amplification is observed in negative controls [23].
38. Controls are particularly important for starch- or oil-rich species [10, 11].
39. Modify the CTAB protocol based on tissue biochemical properties [10, 11].
40. Use commercial DNA kits for polysaccharide- or polyphenol-rich species [10, 11, 22].
41. Replace ethidium bromide with safer staining alternatives [23].
42. Adjust electrophoresis conditions depending on fragment size distribution.
43. PCR requires approximately 2–3 h [21, 23].
44. Agarose gel electrophoresis and imaging require 1–2 h [23].
45. Maintain consistent thermal cycling parameters across experiments [23].
46. Include one control per 20–30 samples in large-scale screening [23].
47. Standardize gel concentration and running conditions for reproducibility [23].

Acknowledgments

This work was carried out in the Plant Biotechnology Laboratory, Department of Agricultural Biotechnology, Isparta University of Applied Sciences (Süleyman Demirel University), Isparta, Türkiye. The authors gratefully acknowledge the laboratory facilities and technical support provided during the experimental work.

References

1. Altındal N, Karadoğan T, Altındal D (2025) Genetic variation in EMS-treated potato calli assessed by ISSR. Plant Cell Tiss Organ Cult 163:79.https://doi.org/10.1007/s11240-025-03265-2
2. Elgohary ME, Salama EA, El-wakil HM, Abdelsalam NR, Atia MA (2025) Integrating SCoT, CBDP, and ISSR molecular markers for genetic diversity assessment and taxonomic authentication of some Asteraceae

species in Egypt. Genet Resour Crop Evol 72:9399–9419. https://doi.org/10.1007/s10722-025-02506-9
3. Altındal D, Altındal N (2025) Advancements in molecular markers and their application in cereal breeding. In: Molecular markers and crop improvement: principle and application. Springer Nature Switzerland, Cham, 155–167pp
4. Altındal D (2019) Determination of genetic diversity of natural sage populations in Muğ la region of Turkey. Int J Environ Sci Technol 16:5219–5226
5. Altındal N (2019) Molecular characterization of *Helianthus tuberosus* L. treated with ethyl methanesulfonate based on inter-simple sequence repeat markers. Int J Environ Sci Technol 16:5311–5318
6. Nei M, Li WH (1979) Mathematical model for studying genetic variation in terms of restriction endonucleases. Proc Natl Acad Sci U S A 76:5269–5273. https://doi.org/10.1073/pnas.76.10.5269
7. Peakall ROD, Smouse PE (2006) GENALEX 6: genetic analysis in Excel. Population genetic software for teaching and research. Mol Ecol Notes 6:288–295. https://doi.org/10.1093/bioinformatics/bts460
8. Ehsanpour AA, Madani S, Hoseini M (2007) Detection of somaclonal variation in potato callus induced by UV-C radiation using RAPD-PCR. Gen Appl Plant Physiol 33:3–11
9. Marotti I, Alessndra B, Mineli M, Catizone P, Dineli G (2007) Characterization of some Italian common bean (*Phaseolus vulgaris* L.) landraces by RAPD, semi-random and ISSR molecular markers. Genet Resour Crop Evol 54:175–188. https://doi.org/10.1007/s10722-005-3133-4
10. Pasqualone A, Lotti C, Bruno A, De Vita P, Di Fonzo N, Blanço A (2000) Use of ISSR markers for cultivar identification in durum wheat. Options Mediterr 40:157–161
11. Pharmawati M, Guijun Y, Patrick MF (2005) Molecular variation and fingerprinting of Leucadendron cultivars (Proteaceae) by ISSR markers. Ann Bot 95:1163–1170. https://doi.org/10.1093/aob/mci127
12. Meena RK, Raj H, Sharma P, Bhandari MS, Ginwal HS (2021) Low diversity and high genetic divergence depicted the threat of genetic erosion in *Calamus nambariensis* and *Calamus tenuis*. Indian J Biotechnol 20:273–283.
13. Doyle JJ, Doyle JL (1990) Isolation of plant DNA from fresh tissue. Focus 12:13–15
14. Rezadoost MH, Kordrostami M, Kumleh HH (2016) An efficient protocol for isolation of inhibitor-free nucleic acids even from recalcitrant plants. 3 Biotech 6:61. https://doi.org/10.1007/s13205-016-0375-0
15. Golenberg EM, Bickel A, Weihs P (1996) Effect of highly fragmented DNA on PCR. Nucleic Acids Res 24:5026–5033. https://doi.org/10.1093/nar/24.24.5026
16. Abdel-Latif A, Osman G (2017) Comparison of three genomic DNA extraction methods to obtain high DNA quality from maize. Plant Methods 13:1. https://doi.org/10.1186/s13007-016-0152-4
17. Joshi K et al. (2000) ISSR markers in crop improvement. Plant Mol Biol Rep 18:89–99
18. Bornet B, Branchard M (2001) Nonanchored inter simple sequence repeat (ISSR) markers: reproducible and specific tools for genome fingerprinting. Plant Mol Biol Rep 19:209–215
19. Reddy MP, Sarla N, Siddiq EA (2002) Inter simple sequence repeat (ISSR) polymorphism and its application in plant breeding. Euphytica 128:9–17
20. Khehra N, Padda IS, Zubair M (2025) Polymerase chain reaction (PCR). StatPearls. StatPearls Publishing, Treasure Island (FL)
21. Munankarmi NN, Rana N, Bhattarai T, Shrestha RL, Chaudhary S, Shrestha S (2022) Optimization of PCR protocol for ISSR marker based genetic diversity assessment of acid lime. Afr J Biotechnol 21:167–179. https://doi.org/10.5897/AJB2021.17427
22. Lorenz TC (2012) Polymerase chain reaction: basic protocol plus troubleshooting and optimization strategies. J Vis Exp 63:3998. https://doi.org/10.3791/3998
23. Rohlf, FJ, 2000. NTSYS-PC numerical taxonomy and multivariate analysis system, version 2.1. Exeter Software. Applied Biostatistics Inc, New York

Check for updates

Chapter 5

RAPD Markers for the Study of Genetic Diversity in *Echium amoenum* Ecotypes

Noraddin Hosseinpour Azad and Rasool Asghari Zakaria

Abstract

This chapter details a comprehensive, step-by-step RAPD protocol for assessing genetic diversity in 16 Iranian *Echium amoenum* ecotypes, a medicinal plant. The method uses STE (Sucrose–Tris–EDTA) buffer to extract DNA from leaf tissue, followed by spectrophotometric and electrophoretic quality control. PCR amplification with 20 decamer primers yields 248 polymorphic bands (80% polymorphism), separated by agarose gel electrophoresis. NTSYS-pc analysis, using the Dice coefficient, UPGMA clustering, and principal component analysis, reveals geographic–climatic groupings. This approach is applicable not only to *E. amoenum* but also to other less-studied medicinal plants.

Keywords Genetic diversity, Iranian borage, Population structure, RAPD markers

1 Introduction

Random amplified polymorphic DNA (RAPD) as a molecular marker is used to detect genetic variability without prior DNA sequence knowledge. It employs short, random primers in PCR to amplify random genomic regions. The presence or absence of amplified fragments (polymorphisms) reveals genetic differences among individuals or populations. Key features include: no prior sequence information requirement, its nature as a dominant marker system, and its low cost and simplicity [1]. However, reproducibility can be challenging due to sensitivity to experimental conditions. The simplified workflow involves: genomic DNA extraction; PCR amplification with arbitrary primers; agarose gel electrophoresis to separate amplified products; scoring bands to create a binary data matrix; and statistical analysis to assess genetic relationships or diversity [2]. RAPD is applied to assess genetic diversity, construct genetic fingerprints, and aid in linkage mapping and phylogenetic studies. Its limitations include: reproducibility issues; the dominant marker nature limiting heterozygosity information; and difficulty in band comparison across labs

Nitish Kumar (ed.), *Molecular Markers in Plants*, Methods in Molecular Biology, vol. 3097,
https://doi.org/10.1007/978-1-0716-5663-1_5,

[3, 4]. Borage (*Echium amoenum* Fisch & Mey), an endemic Iranian medicinal plant, is traditionally used for its anxiolytic, anti-inflammatory, and antioxidant properties [5, 6]. RAPD analysis has proven effective in revealing genetic variation and differentiation in various plant species, including Boraginaceae and Echinacea, and genera like *Paracaryum*, *Rindera*, and *Echium*. For example, RAPD analysis in *Tanacetum parthenium* has clarified genetic diversity among ecotypes, guiding conservation and domestication. Despite limitations like dominance and PCR sensitivity, which can be addressed through optimization, RAPD markers remain valuable for initial diversity analyses, especially in understudied species like *E. amoenum* [7, 8]. The high sensitivity of RAPD markers to small changes in the PCR condition ($MgCl_2$ concentration, annealing temperature, and quality of template DNA) makes them low in reproducibility, with common differences in banding patterns across laboratories. Being the most dominant markers, they do not differentiate the homozygote and the heterozygote states at a locus, which restrains their informative value in genetic analysis. Moreover, the amplification may be inhibited by plant secondary metabolites, leading to false negatives, and the arbitrary primer binding reduces the resolution of phylogenetic patterns [9, 10]. Though alternative markers like ISSRs and SCOTs exist [11–13]. Both ISSR and SCoT generate more polymorphic bands per primer, with superior marker index (MI) and diversity index compared to RAPD's lower efficiency [14]. This chapter describes a detailed and step-by-step RAPD methodology to determine the genetic diversity and population structure of 16 *E. amoenum* ecotypes from Iran. This approach is applicable not only for *E. amoenum* but also for other less-studied medicinal plants, as genetic diversity measurements provide a scientific basis for prioritizing conservation efforts such as gene banks or in situ reserves [15, 16].

2 Materials

2.1 Plant Materials

Fresh, young leaf tissue from the ecotypes in Table 1 was transported to the Laboratory at 4°C. The tissue was washed with distilled water to remove external contaminants, and then used for DNA extraction (*see* **Note 1**).

2.2 Chemicals and Reagents

The RAPD method's sensitivity to DNA quality necessitates high-quality genomic DNA. Since RAPD markers randomly replicate genomic regions, optimization is required to achieve high band resolution and reproducibility. This optimization depends on factors

Table 1
Geographical characteristics of *Echium amoenum* collection regions

No.	Collection site	Province (Iran)	Code	Latitude (N)	Longitude (E)	Altitude (m)
1	Jannat Rudbar	Gilan	G1	36°50′05″	50°34′34″	1639
2	Galugah (Behshahr)	Mazandaran	P2	36°40′44″	53°46′49″	711
3	Gorgan	Gorgan	O3	36°51′36″	54°30′48″	140
4	Neka	Mazandaran	N4	36°39′14″	53°17′28″	42
5	Ashkorat	Gilan	E5	36°54′33″	50°13′26″	634
6	Isfahan	Isfahan	S6	32°40′23″	51°40′50″	1574
7	Chalus (Kelardasht)	Mazandaran	M7	36°35′12″	51°12′11″	1595
8	Behshahr	Mazandaran	B8	36°42′58″	53°32′14″	0
9	Kermanshah	Kermanshah	K9	34°19′09″	47°03′05″	1413
10	Talesh	Gilan	T10	37°47′42″	48°47′42″	769
11	Do Hezar Highlands	Mazandaran	2000	36°38′30″	50°48′00″	990
12	Amlash (Rudsar)	Gilan	R12	37°05′06″	50°10′30″	116
13	Namin (Heyran Pass)	Ardabil	H13	38°23′06″	48°36′36″	920
14	Qazvin (Alamut)	Qazvin	Z14	36°16′28″	50°00′00″	1298
15	Lorestan	Lorestan	L15	33°28′04″	48°21′00″	1174
16	Suchalma (Neka)	Mazandaran	So16	36°44′11″	53°17′28″	1250

such as the GC ratio of primers and the target genome, as well as the quality and quantity of template DNA. Therefore, an optimized STE (Sucrose–Tris–EDTA) method was used to obtain genomic DNA of suitable quality for PCR. The STE buffer contains Sucrose (4 M), Tris–HCl (2 mM, pH 8), EDTA (20 mM, pH 8), which must be autoclaved after preparation. The reagents and chemicals and their concentrations were prepared according to Table 2. Also, the list of RAPD primers used in polymerase chain reactions (PCR) is shown in Table 3.

2.3 PCR Reaction Mixture

dNTP mixture, magnesium chloride, reaction buffer, and Taq DNA polymerase were obtained from CinnaGen Co. (Tehran, Iran), the specifications of which are given in Table 4.

2.4 Gel Electrophoresis and Product Visualization

Agarose gel electrophoresis was performed to separate and visualize DNA fragments. Gels (1%–2% w/v agarose) were prepared in 1× TBE buffer, supplemented with ethidium bromide (0.5 μg/mL) (Table 5).

Table 2
Chemicals and reagents used for genomic DNA extraction by the optimized STE method

No.	Chemicals and reagents	Concentration	Preparation
1	Na_2–EDTA (Merck, 324503)	20 mM, pH 8	Dissolve 7.44 g Na_2–EDTA·2 H_2O in ~900 mL distilled water, adjust pH to 8.0 with NaOH pellets (it will only dissolve at pH ≈ 8), then bring to 1 L. Sterilize by autoclaving
2	Sucrose (Sigma, S7903)	4 M	Dissolve 136.8 g sucrose in distilled water to a final volume of 100 mL. Mix thoroughly until clear; store at room temperature
3	Tris–HCl (Sigma, T3253)	2 mM, pH 8	Dissolve 242 mg Tris base in ~800 mL distilled water, adjust to pH 8.0 with HCl, and bring to 1 L. Store at room temperature or 4°C
4	SDS (Merck, 71725)	20%	Dissolve 20 g SDS in warm distilled water (~60°C) to a final volume of 100 mL. Cool and store at room temperature (may precipitate below 25°C)
5	LiCl (Sigma, L7026)	8 M	Dissolve 33.93 g LiCl in ~80 mL distilled water, bring to 100 mL with distilled water, and filter-sterilize. Store at room temperature
6	PVP (Sigma, P0930)	1.5%	Dissolve 1.5 g PVP (usually PVP-40) in 100 mL distilled water at ~50°C with stirring until fully dissolved. Cool before use
7	β-mercaptoethanol (Merck, 63689)	0.2%	Add 0.2 mL β-mercaptoethanol to 100 mL buffer (use under fume hood). Mix gently; prepare fresh before use or store at 4°C tightly sealed
8	Chloroform–isoamyl alcohol (Merck)	24: 1	Mix 24 mL chloroform with 1 mL isoamyl alcohol in a glass bottle. Store tightly capped at 4°C, protect from light
9	Ethanol (Merck, 32294)	70%	Mix 70 mL absolute ethanol with 30 mL distilled water. Store tightly sealed at room temperature
10	Isopropanol (Merck, 109634)	As required	Use molecular biology grade; typically used neat (100%). Store at room temperature, tightly closed

To prepare the final STE buffer solution at the onset of extraction, equal volumes of the Na_2–EDTA, sucrose, and Tris–HCl stock solutions should be mixed and utilized

3 Methods

3.1 DNA Extraction

Traditional DNA extraction methods involve hazardous materials like phenol–chloroform, temperature-sensitive components, liquid nitrogen, and RNase. An optimized method developed at the Rice and Citrus Research Institute (Sari, Iran), based on an earlier STE buffer protocol [17], eliminates many of these problematic, harmful materials. While the original STE buffer method used BSA, a costly

Table 3
Characteristics of RAPD primers used in polymerase chain reactions (PCR)

Row	Abbreviation	Number of nucleotides	Primer sequence	GC (%)	Tm (°C)
1	OPD-01	10	5′-ACCGCGAGGG-3′	36	80
2	OPD-02	10	5′-GGACCCAACC-3′	34	70
3	OPD-03	10	5′-GTCGCCGTCA-3′	34	70
4	OPD-05	10	5′-TGAGCGGACA-3′	32	60
5	OPB-05	10	5′-CAGGCCCTTC-3′	34	70
6	OPB-13	10	5′-TTCCCCCGCT-3′	34	70
7	OPB-14	10	5′-TCCGCTCTGG-3′	34	70
8	OPB-20	10	5′-GGACCCTTAC-3′	32	60
9	OPA-01	10	5′-CAGGCCCTTC-3′	34	70
10	OPA-10	10	5′-CTGCTGGGAC-3′	34	70
11	OPA-04	10	5′-AATCGGGCTG-3′	32	60
12	OPH-20	10	5′-GGGAGACATC-3′	32	60
13	OPH-04	10	5′-GGAAGTCGCC-3′	34	70
14	OPH-15	10	5′-AATGGCGCAG-3′	32	60
15	OPH-05	10	5′-AGTCGTCCCC-3′	34	70
16	OPH-01	10	5′-GGTCGGAGAA-3′	32	60
17	OPD-15	10	5′-CATCCGTGCT-3′	32	60
18	OPH-08	10	5′-GAAACACCCC-3′	32	60
19	OPD-11	10	5′-AGCGCCATTG-3′	32	60
20	OPC-08	10	5′-TGGACCGGTG-3′	34	70

Table 4
Specifications of PCR reaction mixture prepared from CinnaGen Co. (Tehran, Iran)

Reagent	Concentration	Storage temperature
dNTP	10 mM	−20°C
$MgCl_2$	50 mM	−20°C
Reaction buffer	10×	−20°C
Taq DNA polymerase	5 U/μL	−20°C

and temperature-sensitive component, the improved method replaced bovine serum albumin (BSA) with polyvinylpyrrolidine (PVP). This modification results in a simple, rapid, and cost-effective genomic DNA extraction protocol suitable for diverse agricultural and medicinal plants. The steps were followed for DNA extraction:

Table 5
Agarose gel electrophoresis reagents and preparation

No.	Component	Concentration	Preparation
1	Agarose (Merck, 1012360500)	0.7%–2% (1 g for 1% gel in 100 mL buffer)	Weigh agarose, add to 1× TBE buffer in flask; microwave in bursts until dissolved (avoid boiling). Cool to ~50°C–60°C. Allow it to solidify for 20–30 min at room temperature or for 10–15 min at 4°C
2	Buffer	1× TBE	1× TBE buffer was prepared by diluting 10× TBE stock (108 g Tris base, 55 g boric acid, 40 mL 0.5 M EDTA, pH 8.0 in 1000 mL distilled water), diluted 1:10 with distilled water
3	Ethidium bromide (Merck, 111608)	0.5 μg/mL final (add ~5 μL of 10 mg/mL stock per 100 mL gel)	Add 5 μL of EtBr stock solution to 100 mL of molten agarose gel, mix thoroughly, and pour the mixture into the gel tray
4	Loading dye (Merck, 69046)	6× stock (e.g., with bromophenol blue)	6× DNA loading dye was prepared by mixing 0.25% (w/v) bromophenol blue, 0.25% (w/v) xylene cyanol FF, 40%–60% (v/v) glycerol (or 15% Ficoll), and 10 mM EDTA in nuclease-free water
5	Lambda DNA/EcoRI+ HindIII Marker (Fermentas, SM0191)	1–5 μL per well	Ready to use

1. Cut 500 mg of leaf tissue into small pieces and place in a porcelain mortar.
2. Add 2 mL of STE buffer and immediately add 300 μL (20%) SDS to thc mortar.
3. Homogenize thoroughly by pounding (homogenization time is critical and should be continued until the mixture turns a dark green color).
4. Transfer 500 μL of the homogenate to 1.5 mL tubes.
5. Add 200 μL of lithium chloride 8 M and 0.2% beta-mercaptoethanol to each tube (excess lithium chloride can inhibit PCR).
6. Add PVP (1.5%) to each tube and mix gently (*see* **Note 2**).
7. Incubate tubes in a 60°C water bath for 45 min, inverting gently every few minutes to ensure uniform heating and mixing.
8. Remove tubes and incubate at room temperature for 10 min.

9. Add an equal volume of chloroform–isoamyl alcohol (24:1 v:v).
10. Centrifuge at 13,400 rpm at room temperature for 15 min.
11. Transfer the supernatant to new tubes.
12. If the solution is not clear, add an equal volume of chloroform–isoamyl alcohol (24:1 v:v) and repeat the centrifugation at 13,400 rpm for 10 min.
13. The supernatant was transferred to a new tube, and two-thirds of its volume of isopropanol (precooled to −20°C) was added. DNA strands became visible following precipitation (*see* **Note 3**).
14. Use a Pasteur pipette or other glass device to remove the DNA strand (DNA adheres to glass due to charge interactions). And wash the DNA with 70% ethanol, being careful not to lose the DNA from the pipette tip. Then allow the DNA to air dry completely on the pipette tip in a vertical position until no alcohol odor is detected (*see* **Note 4**).
15. Dissolve the DNA in 100 μL TE buffer (preheating the TE buffer may improve dissolution) (*see* **Note 5**).
16. Store the DNA at 4°C until use.

3.2 Determination of DNA Quality and Quantity

Genomic DNA quality and quantity were assessed via spectrophotometry and 0.8% agarose gel electrophoresis. Following dilution, DNA concentration was determined spectrophotometrically at 260 nm using the formula: DNA concentration (ng/μL) = A260 × dilution facto × 50. Conversion factor: An absorbance value of 1.0 at 260 nm corresponds to a concentration of 50 ng/μL (50 μg/mL) for purified double-stranded DNA (dsDNA). A260/A280 ratios between 1.7 and 1.9 indicated acceptable nucleic acid purity. DNA electrophoresis on 0.8% agarose gel was performed by mixing 4 μL of extracted DNA with 2 μL of loading buffer and running the samples for 1 h at 85 V in TBE buffer. Finally, DNA bands were visualized and photographed under UV light using a UVP GelDoc device. Samples deemed qualitatively and quantitatively suitable were selected for PCR.

3.3 Steps of the Polymerase Chain Reaction (PCR) with RAPD Marker

The method used for PCR-RAPD was based on the method of Szczecińska et al. [18], which was optimized with some modifications to the PCR reaction.

3.3.1 Primers

Twenty RAPD primers, obtained from CinnaGen Co. (Tehran, Iran), were received lyophilized, each with 5–7 OD at 260 nm. The following calculations, based on the primer catalog's formulas, were used to determine the distilled water volume needed for dilution. For instance, the total amount of the synthesized primer OPA-04 was

171,683 pmol. To obtain a working solution with a final concentration of 100 μM, the amount of water needed should be calculated:

- Total primer amount: 171,683 picomoles (pmol) = 0.171683 μ mol.
- Target working concentration: 100 μM (i.e., 100 μmol/L). This is used to determine the final volume needed to dissolve the entire primer amount.
- For the PCR step, a final primer concentration of 1 μM in an 18 μL reaction is achieved by adding a portion of a 100 μM stock, using the dilution equation $N_1V_1 = N_2V_2$.

A. Prepare a working solution at 100 μM from the total primer amount

1. Convert the total primer amount to micromoles
   - 171,683 pmol × (1 μmol/1,000,000 pmol) = 0.171683 μ mol
2. Relate concentration, amount, and volume
   - Target concentration $C = 100\,\mu M = 100\,\mu mol/L$
   - Volume required for the entire amount: $V = n/C$
     - $n = 0.171683\,\mu mol$
     - $C = 100\,\mu mol/L$
     - $V = 0.171683\,\mu mol/(100\,\mu mol/L) = 0.00171683\,L$
3. Convert volume to microliters
   - 0.00171683 L × 1,000,000 μL/L = 1716.83 μL
   - To dissolve the entire 0.171683 μmol of primer to make a 100 μM solution, you would bring the volume up to 1716.83 μL (approximately 1.717 mL) with water.
   - The amount of the 100 μM primer stock solution required for an 18 μL PCR reaction with a final primer concentration of 1 μM was determined using the dilution formula $N_1 V_1 = N_2 V_2$, where N_1 and V_1 are the concentration and volume of the stock primer solution, and N_2 and V_2 are the desired concentration and total reaction volume, respectively.
   - Substituting values: $100 \times V_1 = 1 \times = 18$, resulting in $V_1 = 0.18\,\mu L$.

3.3.2 PCR Setting Up

RAPD marker reproducibility requires meticulous attention to reaction conditions due to primer sensitivity. Critical factors include DNA quality and quantity, absence of Taq DNA polymerase contamination, primer concentration, magnesium chloride concentration, accurate pipetting, and optimized PCR conditions. To minimize contamination, the preparation process for performing the polymerase chain reaction (PCR) is carried out as follows:

- Ensure DNA quality: Confirm that genomic DNA is of high purity and at a concentration of 40–55 ng/μL to ensure consistent amplification.
- Prepare work area: Autoclave all equipment (tips, tubes, and pipettes) at 121°C and 1.5 bar for 15 min to eliminate potential contaminants.
- Reagent handling: Thaw all PCR reagents (buffers, primers, $MgCl_2$, dNTPs, etc.) on ice to maintain enzyme stability (*see* **Note 6**).
- Tube preparation: Arrange numbered 200 μL microtubes in racks filled with ice to keep samples cool throughout setup.
- Add template DNA: Pipette 1 μL of diluted genomic DNA (containing 40–55 ng) into each microtube after brief gentle centrifugation.
- Master mix preparation:
  a. Prepare a master mix (excluding Taq DNA polymerase) in a sterile 1.5 mL microtube according to the composition in Table 5.
  b. Verify that all components are fully thawed and homogenized by gentle mixing.
- Add Taq polymerase: Introduce the required volume of Taq DNA polymerase into the master mix immediately before dispensing.
- Aliquot reaction mixture: Dispense 18 μL of the prepared master mix into each microtube already containing genomic DNA, ensuring uniform mixing.
- Prevent air bubbles: Minimize bubble formation by rapid dispensing and gently tapping the tubes before closing the caps.
- Include negative control: Prepare a negative control containing all reagents except template DNA to monitor for contamination.
- PCR amplification: Proceed with PCR amplification under optimized thermal cycling conditions.

Detailed information on the preparation of PCR reaction components used for assessing the genetic diversity among 16 ecotypes of *Echium amoenum* using RAPD markers is presented in Table 6.

3.3.3 PCR Program

Primers were amplified using a slightly modified version of the Szczecińska et al. [18] method on an Eppendorf PCR device with gradient capability. The PCR program was consistent across all primers, except for the annealing temperature, which was determined based on GC method estimates of primer melting temperature, as provided by primer suppliers. Specific modifications are detailed in Table 7.

3.4 Electrophoresis of PCR Products

PCR products were analyzed by electrophoresis on a 1.5% agarose gel [(1.5 g agarose in 100 mL 1× TBE buffer, microwaved until boiling)], then cooled to about 55°C–60°C before pouring the gel. The gel was poured into a gel tray with a comb to create 24 wells. Following solidification, the comb was removed, and the gel was transferred to an electrophoresis tank containing 1× TBE buffer. In total, 10 μL of each sample (7 μL PCR product + 3 μL loading buffer) was loaded into 17 wells. Lambda DNA/EcoRI + HindIII Marker (Fermentas) was used for band size determination. Electrophoresis was performed

Table 6
PCR reaction components (18 μL final volume)

No.	Reaction components	Volume (μL) per reaction	Total volume (μL) for 18 reactions[1]
1	PCR buffer	1.8	32.4
2	Magnesium chloride ($MgCl_2$)	1.26	22.68
3	2′-Deoxynucleoside triphosphates (dNTPs)	0.36	6.48
4	Primer	1.8	32.4
5	Taq DNA polymerase	0.24	4.32
6	Ampoule water (nuclease-free water)	11.54	207.72
7	Template DNA	1	–
8	Final volume	18	–

[1] 16 for performing the PCR reactions, one as a negative control, and one to compensate for volume errors when pipetting with the sampler

Table 7
Program set for amplification of RAPD molecular marker primers

	Step	Temperature (°C)	Time (min)	Number of cycles
1	Initial DNA denaturation	94	5	1
2	DNA single-strand denaturation	92	1	38
3	Primer annealing to single-stranded DNA	38–42	1	–
4	Primer extension	72	2	–
5	Final primer extension	72	10	1
6	Hold	4	–	–

at 65 V for 3 h. Then, after washing the Gel with deionized water, the gel was visualized and photographed under UV light using a UVP GelDoc (UVP, an Analytik Jena Company, USA). Images were saved for analysis with statistical and genetic software (Fig. 1).

Gel preparation

- Agarose gel: 1.5% w/v
  - *Procedure: Dissolve 1.5g of agarose in 100mL of 1× TBE buffer, microwave until boiling, then cool to about 55°C–60°C before pouring into the gel tray.*
- Gel tray: Equipped with a comb to create 24 wells.
- Solidification: Allow gel to solidify, then remove the comb to form 24 wells.

Set up and running buffer

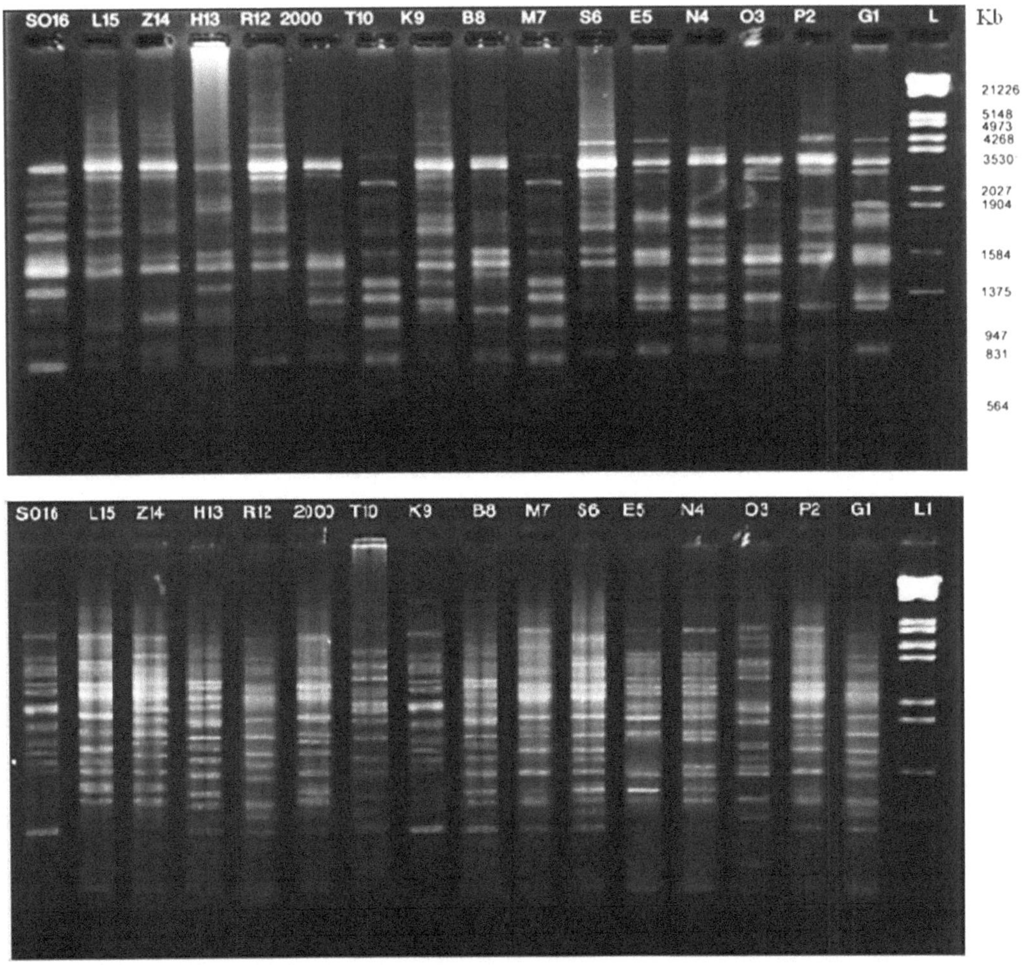

Fig. 1 Polymorphism for OPA-01 (up) and OPD-15 (down) markers in the studied ecotypes

- Buffer: 1× TBE.
- Gel placement: Transfer the solidified gel to an electrophoresis tank containing 1× TBE buffer.

Sample loading

- Sample preparation: Each sample consists of 7 μL PCR product + 3 μL loading buffer (total 10 μL per well).
- Number of wells used: 17 wells loaded with samples.
- Molecular weight marker: Lambda DNA/EcoRI + HindIII Marker (Fermentas) used for band size determination.

Electrophoresis conditions

- Voltage: 65 V.
- Duration: 3 h.
- Post-run processing: Washing the Gel with deionized water.

Visualization and documentation

- Visualization: Gels visualized under UV light. Wear gloves and a UV-protective shield when handling ethidium bromide.
- Imaging: Photographed with a UVP Geldoc system (UVP, a subsidiary of Analytik Jena, USA).
- Data handling: Images saved for subsequent analysis with statistical and genetic software.

3.4.1 Notes and Tips

- Ensure that the ethidium bromide handling and disposal follow your lab's safety protocols. It is worth mentioning that Safe-Stain solution (0.1 μl/mL) can be used as an alternative to ethidium bromide; however, band clarity is superior with ethidium bromide, which is generally preferred for band scoring applications.
- If running at 65 V for 3 h yields faint bands, consider increasing voltage slightly (while monitoring heat) or extending run time.
- For reproducibility, confirm the marker lane placement and include a lane with a ready-to-use DNA ladder on each gel run.
- If you plan to quantify band sizes, consider normalizing imaging exposure and including a calibration step in your analysis workflow.

3.5 Data Analysis

Statistical analysis began by scoring the presence (1) or absence (0) of clear, codable bands for each of the 20 primers. A similarity matrix was then generated in EXCEL for each primer. These matrices were analyzed in NTSYS-pc 2.02e (NT edit section) using a distance/similarity coefficient to produce genetic similarity coefficients (0–1) for all ecotype pairs, which were arranged in a similarity matrix [19].

A dendrogram was constructed from this matrix using SAHN clustering with the UPGMA algorithm.

- Scoring
  - *For each primer, score the presence (1) or absence (0) of each clear, codable band across samples.*
- Data organization
  - *Assemble the scored data into a matrix per primer to create a similarity/dissimilarity context from these scores to compare ecotypes.*
- Similarity calculation
  - *For each pair of ecotypes i and j, calculate the similarity index such as the simple matching coefficient (SMC), Jaccard coefficient, or Nei–Li coefficient (often associated with Dice) (see Sect. 3.5.1).*
  - *This yields a value between 0 and 1, where 1 indicates identical banding patterns.*
- Cluster analysis
  - *Software: NTSYS-pc 2.02e (NT edit section).*
  - *Input: similarity matrix.*
  - *Method: SAHN clustering using the UPGMA or Ward algorithm.*
  - *Output: A dendrogram illustrating genetic relationships among ecotypes (Fig. 2).*
- Cluster validation
  - *Assess the agreement between the observed similarity matrix and the dendrogram using the cophenetic correlation coefficient (MXCOMP model).*
  - *Interpret the coefficient: closer to 1 indicates a better fit between the data and the clustering.*
- Multivariate exploration (PCA)
  - *Tool: EIGEN program (for principal component analysis).*
  - *Objective: Reduce dimensionality and visualize patterns of genetic similarity.*
  - *Visualization: Generate two- or three-dimensional plots to complement the dendrogram and aid interpretation of ecotype relationships (Fig. 3).*
- Optional enhancements
  - *Bootstrapping: If feasible, perform bootstrap analysis to assess cluster stability.*
  - *Permutation tests: Evaluate the significance of groupings under a randomization framework.*

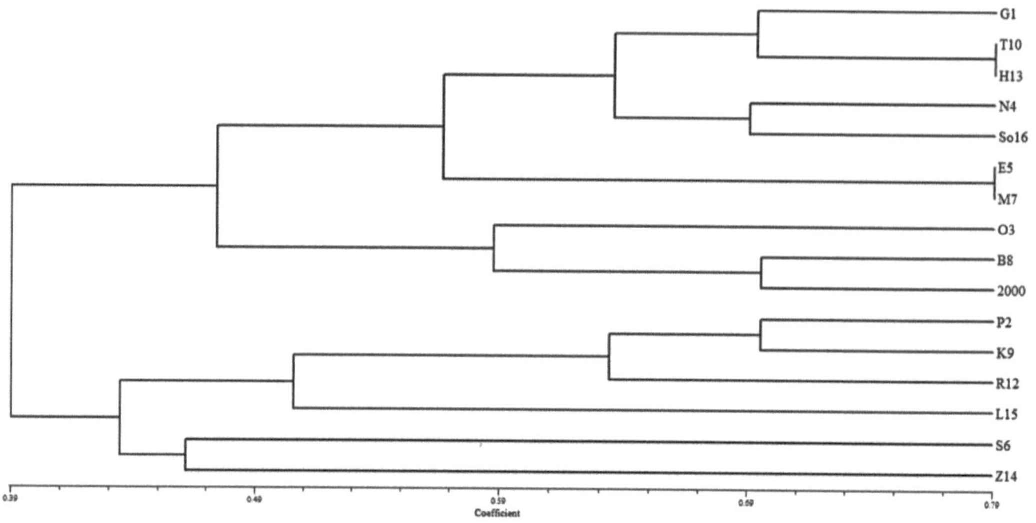

Fig. 2 Dendrogram of data obtained from RAPD molecular marker based on UPGMA (unweighted pair group method with arithmetic mean) algorithm

- *Data visualization: Include color-coded heatmaps of similarity matrices and annotated dendrograms for clarity.*

3.5.1 Similarity Indices

- Simple matching coefficient (SMC)
  - *Formula:* $\mathrm{SMC} = \frac{a+d}{a+b+c+d}$.
  - *What it counts: matches for both presence (a) and absence (d).*
  - *Emphasis: symmetric treatment of presence and absence.*
  - *Suitable for: dominant marker data (e.g., RAPD and AFLP) where absences can be informative (e.g., homozygous recessives).*
  - *Caveat: can overlook common absences; assumes all matches are equally informative.*
- Jaccard coefficient
  - *Formula:* $\mathrm{Jaccard} = \frac{a}{a+b+c}$.
  - *What it counts: shared presences only. Presence-only agreement; double absences are ignored (treated as missing).*
  - *Suitable for: dominant marker data when absences are uninformative or should be discarded.*
  - *Caveat: potential bias if absences carry information.*
- Nei–Li coefficient (dice)
  - *Formula (dice form commonly used):* $\mathrm{Dice/Nei\text{–}Li} = \frac{2a}{2a+b+c}$.

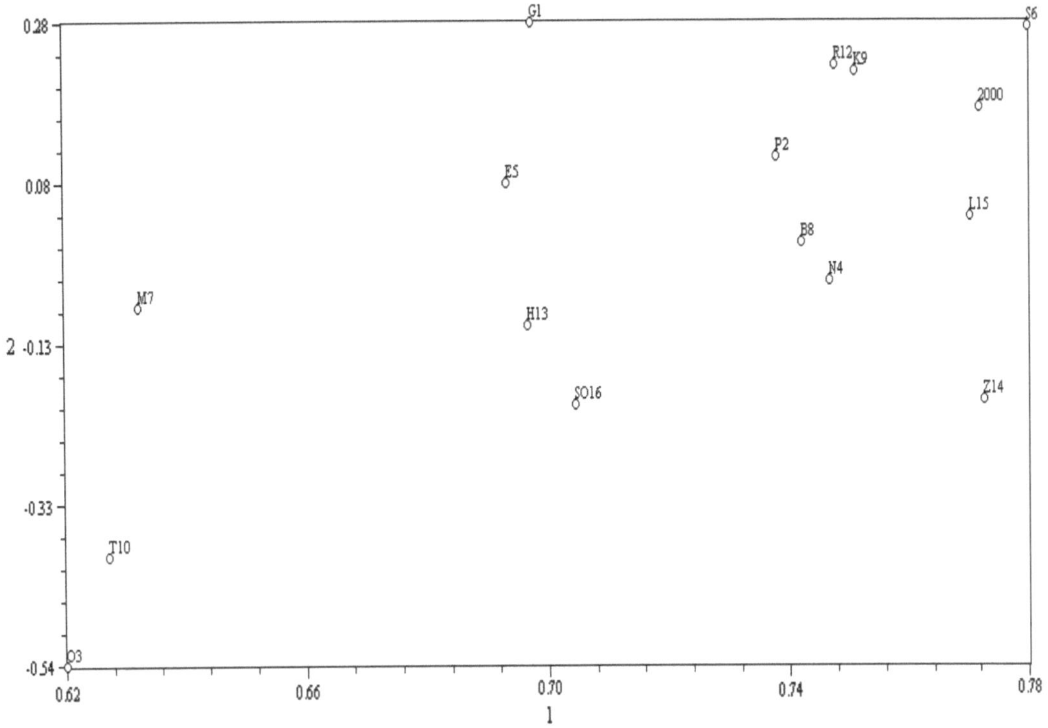

Fig. 3 Two-dimensional plot obtained from principal component analysis

- *What it counts: emphasis on shared presences; absences contribute less.*
- *Emphasis: shared bands are most informative.*
- *Suitable for: codominant data (e.g., RFLP and SSR) or presence/absence data where shared presence is most informative.*
- *Caveat: Absences treated as less informative; may bias if absences are informative.*

- Choose one coefficient and apply it consistently across all pairwise comparisons.
- *a*, *b*, *c*, *d* are clearly defined for your data as:
  - *a = number of bands present in both individuals; b = present in i, absent in j; c = absent in i, present in j; and d = absent in both.*
- Consider your marker type and whether absences carry information to decide between SMC, Jaccard, and Dice/Nei–Li.
- The PCA results largely corroborated the cluster analysis, with minor discrepancies observed in some populations.

3.6 Necessary Considerations and Challenges in Interpreting RAPD Marker Data

1. Marker characteristics and data quality
   - Dominant markers: RAPD produces presence/absence data only, with no reliable information on heterozygosity. This limits estimates of genetic diversity and relatedness.
   - Reproducibility: RAPD can be sensitive to PCR conditions (primer sequence, magnesium concentration, annealing temperature, template quality). Small changes can alter banding patterns. It is recommended to use stringent, consistent protocols; include technical replicates; and report PCR conditions in detail.
   - Band interpretation: Co-migration of fragments can create "false" bands; minor shifts in mobility can misclassify bands. So, use standardized scoring criteria, and consider validating a subset of bands by sequencing or using a second marker type.
   - Scoring error: Scoring mistakes (false positives/negatives) directly affect similarity/distance measures.
   - Recommendation: Blind scoring, multiple scorers with reconciliation, and clear rules for borderline bands.

2. Data handling and preprocessing
   - Binary scoring: Convert gel images to binary matrices (1 = present, 0 = absent). Ensure consistent binarization thresholds.
   - Missing data: RAPD data can have missing bands due to gel quality or low-intensity bands.
   - Recommendation: Define how to handle missing data (ignore, impute, or exclude loci with excessive missingness) and report it.
   - Band matching across samples: Aligning bands across lanes relies on size estimation, which can be imperfect.
   - Recommendation: Calibrate with a standard ladder, set a tolerance window for band matching, and document it.

3. Choice of similarity/dissimilarity measures
   - Impact of measure: Different coefficients weight the presence/absence differently, affecting clustering outcomes.
   - Recommendation: Justify the chosen metric, and consider reporting results with more than one metric to show robustness.
   - Homoplasy risk: Similar band patterns can arise by chance or convergent amplification, not shared ancestry.
   - Recommendation: Acknowledge this limitation and, if possible, corroborate with additional marker types.

4. Clustering and interpretation
   - Algorithm sensitivity: Results can vary with clustering method (UPGMA, NJ, Ward) and distance measure.
   - Recommendation: Test multiple methods or clearly justify the chosen approach; present consensus patterns if they emerge.
   - Cutoff selection: The choice of a threshold to define clusters can be subjective.
   - Recommendation: Explore several cutoffs, report how cluster assignments change, and discuss stability (e.g., bootstrap support).
   - Biological meaning: RAPD clusters may reflect technical variation as well as biology.
   - Recommendation: Integrate with ecological, geographic, or phenotypic data to interpret clusters biologically.
5. Statistical robustness and validation
   - Replicability: Reproducibility across runs and laboratories is a known concern for RAPD.
   - Recommendation: Include inter-run controls, report intra- and inter-lab variability if applicable.
   - Validation with independent markers: Strengthen conclusions by comparing RAPD results with other markers (e.g., SSRs, AFLP, and SNP arrays).
   - Bootstrapping and resampling: Useful to assess cluster stability, though interpretation can be nuanced with dominant markers.
   - Recommendation: If feasible, perform bootstrap analysis and report support values for clusters.
6. Diversity metrics and downstream analyses
   - Diversity estimates: With presence/absence data, calculate genetic diversity indices carefully (e.g., number of polymorphic bands and gene diversity as an approximation).
   - Population structure vs. clustering: Clusters may reflect population structure, but not necessarily discrete populations. Complement with STRUCTURE-like analyses or model-based approaches when possible, even with dominant markers.
   - Comparability: RAPD data can be difficult to compare across studies due to protocol differences; emphasize assay-specific results.
7. Reporting and transparency
   - Documentation: Provide complete methods for primer selection, PCR conditions, gel conditions, scoring rules, and binarization criteria.

- Data availability: Share raw gel images (or band presence/absence matrices) and code for clustering analyses if possible.
- Limitations: Be explicit about the limitations of RAPD data and the inferences you can or cannot support.

Genetic structure has a strong relationship with Geographical distribution, which helps to verify classical isolation-by-distance models [20], and the environmental correlations indicate adaptive divergence has a similar effect on population structure. These results emphasize the importance for conservation practitioners to conserve many populations along the ecological balance of the species since each cluster could be a unique genetic resource with adaptation potentials. RAPD markers were useful in the determination of the overall genetic structure; however, a number of methodological limitations are worth taking note of RAPDs cannot identify heterozygosity and break down fine-scale relationships because of their mostly dominant nature [21]. The intermediate levels of polymorphism (80%), as well as, the small number of ecotypes that were sampled (16 ecotypes) of the extensive environmental gradients in Iran would have underestimated the overall genetic diversity of the species [22]. Such limitations indicate that although RAPDs can offer effective first-line testing, future research is recommended to include codominant markers such as SSRS or SNP arrays to provide more detailed analyses. The ecological relationships observed here, and especially the precipitation-related genetic patterns, indicate that populations of *E. amoenum* could exhibit disparate persistence to climatic change. The ecotypes in highlands that have higher genetic diversity may have alleles that are critical in stress adaptation, and they should be conserved [23]. In a wider view, the findings add to the accumulating evidence that the topographic complexity of Iran has favorable plant diversification via variance and adaptation [24]. Such findings have practical uses not only in conservation but also in generating pharmaceutical products. It is likely that the genetic variation in the ecotypes is associated with the difference in the production of bioactive compounds as observed in other species of the Boraginaceae family [25].

4 Notes

1. Perform a 1-min surface sterilization with bleach to remove surface pollen contaminants, dry leaves in silica gel within 24 h to preserve DNA integrity.
2. Low DNA yield or smeared bands (DNA concentration < 200 ng/μL or A260/280 ratio < 1.7) due to phenolic compounds released during grinding can be mitigated by grinding leaves for less than 90 s in an ice-cold mortar, increasing

PVP to 3% (w/v) if tissue browns heavily, and limiting LiCl to 200 μL to reduce salt carryover that inhibits PCR.

3. Carryover contaminants and PCR inhibition, indicated by a cloudy precipitation pellet during isopropanol washing (Step 13) and faint PCR bands, can be mitigated by an additional chloroform wash (Step 12), RNase treatment before PCR, or, for highland samples (G1 ==1639 m), a CTAB-based protocol with 2% CTAB and 2% PVP.
4. An alternative precipitation step for DNA purification as an alternative to Step 14: first, centrifuge at 3000 rpm for 3 min to pellet the DNA and discard the supernatant; then add 200 μL absolute ethanol and centrifuge at 3500 rpm for 4 min; finally, air-dry the DNA pellet before proceeding to Step 15.
5. If EDTA in TE buffer disrupts enzymatic reactions (Step 16, Sect. 3.1), use nuclease-free water instead of TE buffer.
6. $MgCl_2$ at 1.26 μL (1.8 mM) yields the best results. However, it is recommended to test a gradient of $MgCl_2$ from 1.0 to 2.5 mM to identify the optimal concentration for each primer pair.

References

1. Winget DM, Wommack KE (2008) Randomly amplified polymorphic DNA PCR as a tool for assessment of marine viral richness. Appl Environ Microbiol 74(9):2612–2618. https://doi.org/10.1128/AEM.02829-07
2. Kapteyn J, Goldsbrough P, Simon J (2002) Genetic relationships and diversity of commercially relevant *Echinacea* species. Theor Appl Genet 105(6-7):973–980.
3. Doricchi A, Platnich CM, Gimpel A, Horn F, Earle M, Lanzavecchia G, Cortajarena AL, Liz-Marzán LM, Liu N, Heckel R, Grass RN, Krahne R, Keyser UF, Garoli D (2022) Emerging approaches to DNA data storage: challenges and prospects. ACS Nano 16(11):17552–17571
4. Shvartsman E, Richmond MEI, Schellenberg JJ, Lamont A, Perciani C, Russell JNH, Poliquin V, Burgener A, Jaoko W, Sandstrom P, MacDonald KS (2022) Comparative analysis of DNA extraction and PCR product purification methods for cervicovaginal microbiome analysis using cpn60 microbial profiling. PLoS ONE 17(1):e0262355. https://doi.org/10.1371/journal.pone.0262355
5. Ghasemnezhad A, Aghaei MK, Ghorbanpour M (2019) The importance of Iranian borage (*Echium amoenum* Fisch. Mey): a critical review. In: Swamy MK, Patra JK, Rudramurthy GR (Eds). Medicinal plants: Chemistry, pharmacology, and therapeutic applications. CRC Press, London, pp. 51–57
6. Azadbakht M, Nematzadeh GH, Hosseinpour Azad N, Shokri E (2011) Quantitative and qualitative investigation on pyrrolyzidine alkaloids in roots, leaves, petals and seeds of Iranian *Echium Amoenum* Fisch. & Mey. J Mazandaran Univ Med Sci 20(1):131–137
7. Romeiras MM, Cotrim HC, Duarte MC et al. (2007) Genetic diversity of three endangered species of *Echium* L. (Boraginaceae) endemic to Cape Verde Islands. Biodivers Conserv 16(9):2455–2467.
8. Hosseinpour Azad N, Nematzadeh G, Azadbakht M, Kazemitabar S, Shokri E (2012) Investigation on fatty acids profile in two ecotypes of Iranian *Echium amoenum* Fisch & Mey. Iran J Med Aromat Plants Res 27(4):587–595. https://doi.org/10.22092/ijmapr.2012.4508
9. Amom T, Nongdam P (2017) The use of molecular marker methods in plants: a review. Int J Curr Res Rev 9(17):1–7
10. Hallden C, Nilsson NO, Rading IM, Säll T (1994) Evaluation of RFLP and RAPD markers in a comparison of *Brassica napus* breeding lines. Theor Appl Genet 88(1):123–128
11. Etminan A, Pour-Aboughadareh A, Noori A, Ahmadi-Rad A, Shooshtari L, Mahdavian Z, Yousefiazar-Khanian M (2018) Genetic

relationships and diversity among wild Salvia accessions revealed by ISSR and SCoT markers. Biotechnol Biotechnol Equip 32(3):610–617
12. Nouri A, Golabadi M, Etminan A, Rezaei A, Mehrabi AA (2021) Comparative assessment of SCoT and ISSR markers for analysis of genetic diversity and population structure in some *Aegilops tauschii* Coss. Accessions. Plant Genet Resour 19(5):375–383
13. Jedrzejczyk I (2020) Genome size and SCOT markers as tools for identification and genetic diversity assessment in *Echinacea* genus. Ind Crops Prod 144:112055
14. Gogoi B, Wann SB, Saikia SP (2020) Comparative assessment of ISSR, RAPD, and SCoT markers for genetic diversity in Clerodendrum species of North East India. Mol Biol Rep 47(10):7365–7377
15. Rajeswara Rao BR (2015) Genetic diversity, genetic erosion, conservation of genetic resources, and cultivation of medicinal plants. Genetic diversity and erosion in plants: case histories. Springer International Publishing, Cham, 357–407
16. Niazian M (2019) Application of genetics and biotechnology for improving medicinal plants. Planta 249(5):1285–1298
17. Ahmad SM, Ganaie MM, Qazi PH, Verma V, Basir SF, Qazi GN (2004) Rapid DNA isolation protocol for angiospermic plants. Bulg J Plant Physiol 30(1-2):25–33
18. Meyer AS, Garcia AAF, Souza AP et al. (2004) Comparison of similarity coefficients used for cluster analysis with dominant markers in maize (*Zea mays* L.). Genet Mol Biol 27(3):407–412.
19. Szczecińska M, Sawicki J, Polok K et al. (2006) Comparison of three Polygonatum species from Poland based on DNA markers. Ann Bot Fennici 43:379–388
20. Antunes B, Figueiredo-Vázquez C, Dudek K et al. (2023) Landscape genetics reveals contrasting patterns of connectivity in two newt species (*Lissotriton montandoni* and *L. vulgaris*). Mol Ecol 32(16):4515–4530
21. Nam VT, Hang PLB, Linh NN, Ly LH et al. (2020) Molecular markers for analysis of plant genetic diversity. Vietnam J Biotechnol 18(4):589–608
22. Arjmand A, Ebrahimi M, Moradi N (2023) Genetic diversity of different ecotypes from three *Althaea* species using SCOT phytochemical and molecular markers. Iran J Med Aromat Plants Res 39(4):622–637
23. Cruzan MB, Hendrickson EC (2020) Landscape genetics of plants: challenges and opportunities. Plant Commun 1(4):100–106
24. Shen D (2022) Assessment of genetic variation of genus *Paracaryum* (Boraginaceae) by RAPD markers. Bangladesh J Plant Taxon 29(1):31–41
25. De Lisi A, Montesano V, Negro D, Sarli G, Blanco E, Sonnante G, Laghetti G (2014) Genetic diversity in Borago officinalis germplasm as revealed by seed oils and AFLP polymorphism. Genet Resour Crop Evol 61(4):853–859

Chapter 6

In Silico Pipeline for the Development and Primer Design of Microsatellite Markers in *Mangifera similis* Blume

Arief Priyadi, Sri Rahayu, Irfan Martiansyah, Rifki Sadikin, Vivi Yuskianti and Ria Cahyaningsih

Abstract

Mangifera similis Blume, a wild relative of the mango, is an important genetic resource for improving mango varieties. This chapter outlines an in silico pipeline for the development and primer design of microsatellite (SSR) markers for the chloroplast and the whole genome of *M. similis*, using bioinformatics and next-generation sequencing. The method enables efficient microsatellite identification, saving laboratory time and costs, and is particularly useful for tree species with limited genomic data. The workflow comprises steps such as DNA extraction, sequencing, library construction, in silico SSR mining, primer design, and marker validation. These SSR markers support analyses of genetic diversity and relationships among *M. similis* populations and are applicable across various Mangifera species. Ultimately, this protocol supports effective conservation genetics and sustainable management strategies for *M. similis*, while highlighting the species' significance as a reservoir of traits for mango improvement.

Keywords in silico SSR development, Primer design, Microsatellites, *Mangifera similis*, Sequencing

1 Introduction

Mangifera, a genus comprising over 60 species, including the widely cultivated mango (*Mangifera indica* L.), is distributed throughout tropical and subtropical regions, with its center of diversity in Malesia [1, 2]. Wild relatives such as *Mangifera similis* Blume are valuable genetic resources for breeding programs, offering potential resistance to diseases and pests [3, 4]. This species, native mainly to Borneo and Sumatra, is not only culturally significant and used in local agroforestry for its fruit [5, 6]. Although it was recently assessed as Least Concern [6], ongoing fragmentation continues to threaten its genetic diversity, and ex situ conservation collections may not fully capture its gene pool [7–9].

Despite its importance, there is limited information on the morphology, phenology, cultivation, and stress tolerance of *M. similis*,

Nitish Kumar (ed.), *Molecular Markers in Plants*, Methods in Molecular Biology, vol. 3097,
https://doi.org/10.1007/978-1-0716-5663-1_6,

especially in the context of climate change [10]. The species' long generation time poses challenges for traditional phenotypic studies [2]. Moreover, there is a notable lack of genetic data for *M. similis*, which hampers both effective conservation and its use in mango improvement programs [10, 11]. While SSR (simple sequence repeat) markers have become key tools in plant genetics because of their high levels of polymorphism, co-dominant inheritance, reproducibility, and transferability across species [12–14], only a few genetic studies have focused on wild Mangifera, and none to date have used nuclear SSRs for *M. similis* [11, 15].

SSRs, or microsatellite markers, are short tandem repeats scattered throughout nuclear and organellar genomes [12]. Owing to their high mutation rates, SSR loci often display multiple alleles, making them highly informative for studies of genetic diversity, population structure, and gene flow—information vital for conservation planning and understanding evolutionary processes [16, 17]. In mango, SSRs have been widely used for diversity and population studies [18, 19], but their application in wild relatives remains limited. Developing and applying SSR markers for *M. similis* would enable accurate evaluation of genetic diversity within and among populations, assessment of genetic representation in ex situ collections, and identification of traits valuable for mango improvement [11, 15].

Traditionally, SSR development relied on labor-intensive, costly de novo methods, including library construction, cloning, and Sanger sequencing, which limited their use in nonmodel species such as *M. similis* [20]. Recent advances in next-generation sequencing and bioinformatics have enabled rapid, cost-effective in silico discovery of SSRs, and efficient primer design using existing genomic data, increasing scalability and marker yield [20, 21]. Once developed, SSR markers are readily used in routine PCR-based assays, enabling detailed, reproducible population genetic studies [13]. However, the reliability of in silico approaches depends on the quality and completeness of available sequence data, necessitating thorough locus screening and validation [15].

In summary, the in silico pipeline for SSR development and primer design provides an efficient and scalable approach to generate key molecular tools for population genetics, conservation, and sustainable management of *Mangifera similis* [20, 21].

2 Materials

2.1 Sample Collection, Preservation, DNA Extraction, and Sequencing

1. Plant material: Fully expanded juvenile leaf collected from a living accession of *Mangifera similis* Blume (Coll. no. VI.8a; sourced: Bangka Island, East of Sumatra) cultivated at the Bogor Botanical Garden, Indonesia.

2. Preservation: Leaf tissue preserved using blue silica gel in a teabag, placed in a transparent plastic clip bag.
3. Genomic DNA source: A single, fully expanded leaf (20 mg dried leaf) for DNA extraction.
4. DNA extraction kit: Quick-DNA high molecular weight (HMW) MagBead Kit (D6060) (Zymo Research, CA, USA).
5. DNA quality and quantity assessment: 5400 Fragment Analyzer System (Agilent, CA, USA). For plant genome sequencing with a library type ≤500 bp insert size, the required DNA quantity was ≥700 ng and recommended ≥1400 ng (at least volume 20 mL and concentration 50 ng/mL).
6. Library preparation: Illumina TruSeq DNA Library Preparation Kits (Illumina, CA, USA).
   5′ Adapter:
   5′-AGATCGGAAGAGCGTCGTGTAGGGAAAGAGTG-TAGATCTCGGTGGTCGCCGTATCATT-3′
   3′ Adapter:
   5′-GATCGGAAGAGCACACGTCTGAACTC-CAGTCACGGATGACTATCTCGTATGCCGTCT-TCTGCTTG-3′
7. Sequencing platform: Illumina NovaSeq 6000, service provided by Novogene (Hong Kong, China) via PT. Genetika Science Indonesia (GSI Lab., Banten, Indonesia).

2.2 Raw Reads and Filtered Reads (In Silico)

1. The sequencing output raw reads: Paired-end (PE) 150, insert size 350 bp, total reads: 66.95 M, total bases: 10.04 G (>Q20: 98.0%, >Q30: 93.9%), GC content: 37.26%.
2. Filtered reads: total reads: 66.72 M, total bases: 9.99 G (>Q20: 98.1%, >Q30: 94.1%), GC content: 37.21%.
3. Instrument: Mahameru HPC (high performance computing) by National Research and Innovation Agency (BRIN) 4th generation (https://hpc.brin.go.id/): 92 nodes, 5888 cores, storage capacity 18 PB in total. Up to 128 cores with 256 GB memory for each node CPU: Intel(R) Xeon(R) Gold 6338 CPU @ 2.00 GHz, Arch: X86_64. Operating System (OS): Red Hat Enterprise Linux 8.9 (Ootpa), linux-64. Workload manager: SLURM 22.05.2. Package and Environment manager: Conda 24.7.1, Python 3.11.5. SSH Client: PuTTY 0.83. FTP/SFTP Client: WinSCP 6.5.5.
4. Software for raw reads filtering and quality check: fastp v0.23.4 [22], fastQC v0.11.8 [23].

2.3 Genome Assembly and Annotation

2.3.1 Chloroplast

1. Sequence serves as the seed for de novo Assembly: full sequence of *the rbcL gene from the M. indica* chloroplast genome (NC_035239).
2. Assembler: NOVOPlasty v4.3.1 [24] to be run in Mahameru HPC-BRIN.

2.3.2 Whole Genome

1. Reference Genome: *M. persiciforma* (doi.org/10.6084/m9.figshare.28123553), described in Wu et al. study [25].
2. De novo Assembler: Ray v2.3.1 [26], Scaffolder: RagTag v2.1.0, Assembly completeness benchmarking: BUSCO v6.0.0 [27] with database eudicotyledons_odb12.

2.4 SSR Screening and BLASTn Queries with SSRs

1. The Chloroplast genome assembled by NOVOPlasty v4.3.1, RagTag Scaffolded whole genome assembly contigs, and BUSCOs single-copy sequences in FASTA format.
2. Software: Krait v2.0.6 [28], Notepad++ (https://notepad-plus-plus.org/).

2.5 In Silico Validation of the SSRs Polymorphism

1. Sequence for BLASTn subject of SSRs from *M. similis* chloroplast genome: *Mangifera* spp. Complete chloroplast genome sequences downloaded as a multiple FASTA from the GenBank NCBI.
2. Sequence for BLASTn subject of SSRs from *M. similis* whole genome: Whole genome sequence of *M. persiciforma* in FASTA format (doi.org/10.6084/m9.figshare.28123553), described in Wu et al. study [25].
3. The prepared BLASTn queries containing SSR in FASTA format.
4. Software: BLAST 2.12.0+ [29].

2.6 Bioinformatics Tools

1. Unipro UGENE v53 [30].
2. MEGA v12 [31].
3. SSH Client: PuTTY 0.83.
4. FTP/SFTP Client: WinSCP 6.5.5.

3 Methods

3.1 Sample Collection, Preservation, DNA Extraction, and Sequencing

3.1.1 Plant Material Collection and Preservation

1. Select a healthy, fully expanded leaf,
2. Place the leaf in a teabag.
3. Insert the teabag into a transparent plastic clip bag containing blue silica gel for rapid desiccation (*see* **Note 1**).
4. Store the preserved tissue at room temperature until DNA extraction.

3.1.2 DNA Extraction and Quality Assessment

1. Grind the dried leaf tissue into a fine powder using liquid nitrogen (*see* **Note 2**).
2. Extract genomic DNA using the Quick-DNA HMW MagBead Kit (D6060; Zymo Research), following the manufacturer's protocol.
3. Elute the DNA solution in the supplied buffer in a sterilized 2.5 mL centrifuge tube.
4. Use 2 μL of the gDNA extract for quality and quantity analysis on an Agilent 5400 capillary electrophoresis system (*see* **Note 3**).
5. Submit high-quality DNA samples to a sequencing center for library preparation and sequencing on the Illumina NovaSeq 6000 platform, using paired-end 150 bp reads with a 350 bp insert size (*see* **Note 4**).

3.1.3 High-Performance Computing Environment Preparation

1. Perform all computations on the BRIN 4th-generation HPC system.
2. Connect to the HPC cluster for command-line work via SSH (e.g., using PuTTY).
3. Load and manage all required software tools using Conda 24.7.1 (Python 3.11.5) in your home directory.
4. Use an SFTP client (e.g., WinSCP) for file transfers.
5. Deposit all final results from the genome assembly, annotation, and analysis (L1) into the designated operation/repository storage.

3.1.4 Raw Data Retrieval and Processing

1. After sequencing is completed, access the generated raw reads (fastq.gz files) from the sequencing center's cloud storage (*see* **Note 5**).
2. Use fastp to filter the raw reads and generate filtered reads.
3. Assess the quality of the filtered reads using fastQC.
4. Proceed to the next step, which involves assembling the chloroplast and whole genomes using the filtered reads.

3.2 Genome Assembly

3.2.1 Chloroplast Genome

1. Edit config.txt, especially to add information about the reads to be de novo assembled and the seed sequence on the HPC (*see* **Note 6**).
2. Run NOVOPlasty4.3.1 in the bash terminal, either in interactive or batch mode.
3. Once the run is finished, the output, in FASTA format, can be found in the home directory.
4. Download the FASTA from HPC to a Laptop using WinSCP—the FASTA to be further processed for SSR screening in Krait v2.

3.2.2 Whole Genome

1. Write a Bash script to run Ray in batch mode on the HPC. Perform de novo assembly of the filtered reads using Ray [26] with default parameters.
2. Obtain the main output of the Ray assembly: the contigs in FASTA format.

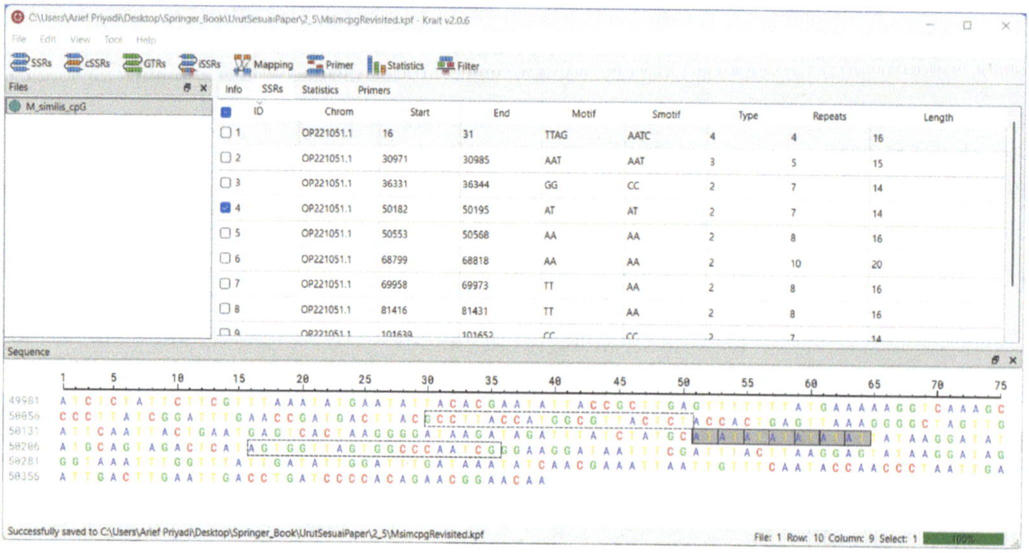

Fig. 1 Display of the Krait2 window after opening a FASTA file and clicking the "SSRs" button

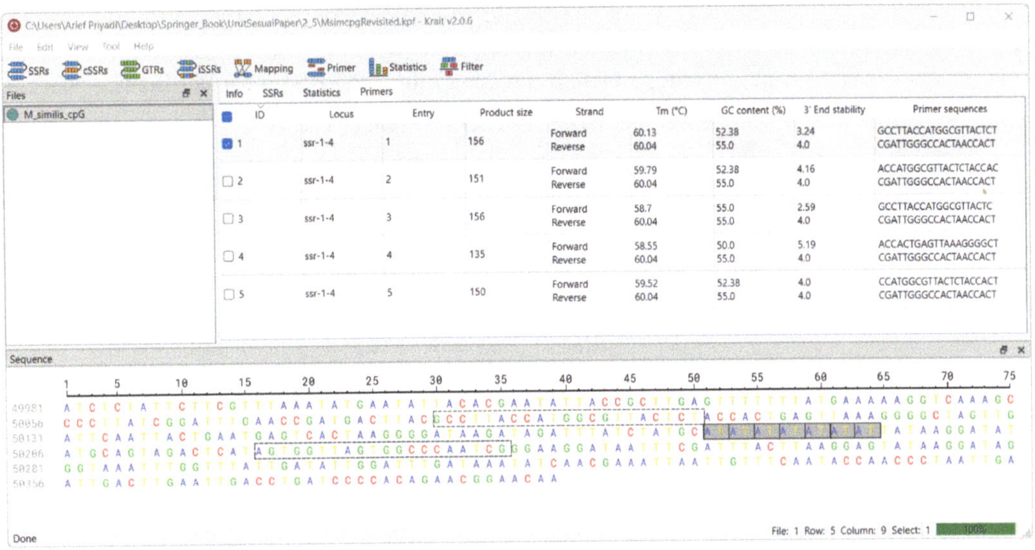

Fig. 2 Display of the Krait2 window after clicking the "Primer" button

3. Scaffold the assembled contigs using RagTag [32], referencing the M. persiciforma genome [25] (doi.org/10.6084/m9.figshare.28123553).
4. Collect the output from the RagTag run, which will be the scaffolds in FASTA format. These can be further processed for SSR scanning in Krait v2.
5. Assess the completeness of the scaffolds using BUSCO v6.0.0 (*see* **Note 7**).

3.3 SSR Loci Screening and Primer Design

3.3.1 Chloroplast Genome SSR Screening

1. Open the FASTA file of the complete chloroplast genome of *M. similis* using Krait v2 [28]. Once the file opens, a tab of "Info" is displayed in the main window of Krait software.
2. Click on the SSRs button to scan the microsatellite loci harbored by the chloroplast genome. The loci are then displayed in the main window, as can be seen in Fig. 1.
3. Click the corresponding radio button in a line, and the sequence with the highlighted SSR locus will be displayed.
4. Click on the "Primer" button to instruct Krait v2 to design the corresponding forward and reverse PCR primers for the corresponding SSR locus. A new tab of "Primers" will be displayed, as can be seen in Fig. 2.

5. Copy and paste the sequences covering the forward-to-reverse PCR Primer region, one by one, into a text editor, such as Notepad++, to create a multiple FASTA format file (*see* **Note 8**).
6. Save this work as a Krait Project (.kpj).

3.3.2 BUSCOs Single-Copy Genes SSR Screening

1. Open the merged single-copy BUSCO sequences in Krait v2 and scan for perfect SSRs.
2. Proceed with the next steps as for the chloroplast genome (Sect. 3.3.1).
3. Save results as a Krait Project (.kpj) and export primer sequences.

3.3.3 RagTag Scaffolds SSR Screening

Follow the same steps as in Sects. 3.3.1 and 3.3.2, but use the FASTA file RagTag generated as input to Krait v2.

3.4 Cross-Species SSR Validation and Polymorphism Analysis

3.4.1 Chloroplast Genome

1. Download complete chloroplast genomes of *Mangifera* spp. from NCBI GenBank as multiple FASTA files to be used as BLASTn subjects.
2. Use the SSR-containing sequences from the *M. similis* generated by Krait v2 for BLASTn Query.
3. Perform local BLASTn to query SSR-containing sequences from the *M. similis* chloroplast genome against the chloroplast genomes of *Mangifera* spp. from NCBI GenBank.
4. Document SSR loci that are transferable and/or polymorphic across species.

3.4.2 BUSCOs Single-Copy Genes and Whole Genome

Follow the procedure described for the chloroplast genome in Sect. 3.4.1. However, use the entire *M. persiciforma* genome [25] as the local BLASTn subject.

4 Notes

1. Collect young, healthy leaves, and immediately preserve them in fresh silica gel. Silica gel must remain active (blue). When it turns pink, moisture absorption decreases, increasing the risk of tissue degradation and fungal and bacterial contamination, which can result in low DNA yield and poor DNA integrity. Replace or reactivate by drying at 60°C for 1 h.

2. All wet lab activities should be performed using sterile gloves and in a clean environment to prevent contamination.
3. DNA samples must meet minimum quantity and purity standards for successful library construction and sequencing. For plant genome sequencing with a library insert size ≤500 bp, the required DNA quantity is ≥700 ng, and the recommended quantity is ≥1400 ng (at least 20 μL and 50 ng/μL).
4. Library preparation and sequencing proceed only after samples pass QC. The manufacturer's protocol includes DNA fragmentation, end repair, adapter ligation, PCR amplification, size selection, purification, and sequencing.
5. Use the sequencing center proceeds for library preparation and sequencing only after the DNA samples pass the quality check. The manufacturer's protocol was followed to shear genomic DNA into short fragments, end-repair the fragments, ligate them with sequencing adapters, amplify by PCR, size-select, purify the DNA library, and finally run the sequencer.
6. Use the provided FTP (File Transfer Protocol) link to download the raw reads to a storage drive on the HPC system. Verify the integrity of the downloaded files by comparing their *md5sum* values with those provided by the sequencing center.
7. Download the Perl script NOVOPlasty4.3.1.pl and the config.txt template from its GitHub Repository (https://github.com/ndierckx/NOVOPlasty) to the home directory on HPC. Also, leave the other settings at their defaults.
8. BUSCO will also generate single-copy BUSCOs as multiple FASTA files. Use the "cat" command in a Bash terminal to merge these FASTA files into a single file. Use the concatenated FASTA file as input for SSR scanning in Krait v2.
9. Use multiple FASTA files as a BLAST query to validate whether the SSR locus is polymorphic [25].

References

1. POWO (2026) Plants of the World Online. Royal Botanic Gardens, Kew. https://powo.science.kew.org
2. Kostermans, AJGH, Bompard, JM, 1993. The mangoes: their botany, nomenclature, horticulture, and utilization. Academic Press
3. Bompard JM (2009) Taxonomy and systematics of the genus Mangifera L. In: Litz RE (ed) The mango: botany, production and uses. 2nd edn. CAB International, pp 19–41
4. Ploetz RC, Benscher D, Vázquez A, Colls A, Nagel J, Schaffer B (2009) Diseases of mango. In: Litz RE (ed) The mango: botany, production and uses. 2nd edn. CAB International, pp 337–362
5. Atmoko T, Gunawan W, Emilia F, Mukhlisi, Prayana A, Arifin Z (2016) Budaya masyarakat Dayak Benuaq dan potensi flora hutan Lembonah. In: Sutedjo (ed) Balai Penelitian Teknologi Konservasi Sumber Daya Alam & PT Borneo Surya Mining Jaya
6. Rahman W, Barstow M (2024) *Mangifera similis*. The IUCN red list of threatened species. https://doi.org/10.2305/IUCN.UK.2024-1.RLTS.T258981334A61525722.en
7. Ariati SR, Astuti RS, Supriyatna I, Yuswandi AY, Setiawan A, Saftaningsih D, Pribadi DO (2019) An alphabetical list of plant species cultivated in the Bogor Botanic Gardens. Center for Plant Conservation Botanic Gardens

8. Narko D, Suprapto A, Lestarini W (2013) An alphabetical list of plant species cultivated in the Purwodadi Botanic Gardens. Purwodadi Botanic Gardens
9. Sujarwo W, Gumilang AR, Hidayat IW (2019) List of living plant collections cultivated in Cibodas Botanic Gardens. Indonesian Institute of Sciences
10. Yuskianti V, Priyadi A, Cahyaningsih R, Matra DD, Sadikin R, Rahayu S (2024) The complete chloroplast genome of *Mangifera similis* Blume (Anacardiaceae) from Bangka Island, Indonesia. J Asia-Pac Biodivers 17(2):351–356
11. Frankham R, Ballou JD, Ralls K, Eldridge M, Dudash MR, Fenster CB, Sunnucks P (2017) Genetic management of fragmented animal and plant populations. Oxford Univ Press
12. Ellegren H (2004) Microsatellites: simple sequences with complex evolution. Nat Rev Genet 5(6):435–445
13. Guichoux E, Lagache L, Wagner S, Chaumeil P, Léger P, Lepais O, Petit RJ (2011) Current trends in microsatellite genotyping. Mol Ecol Resour 11(4):591–611
14. Kalia RK, Rai MK, Kalia S, Singh R, Dhawan AK (2011) Microsatellite markers: an overview of the recent progress in plants. Euphytica 177(3):309–334
15. Hoban S, Bruford M, Jackson JDU, Lopes-Fernandes M, Heuertz M, Hohenlohe PA, Laikre L (2020) Genetic diversity targets and indicators in the CBD post-2020 Global Biodiversity Framework must be improved. Biol Conserv 248:108654
16. Hamrick JL, Godt MW (1996) Effects of life history traits on genetic diversity in plant species. Philos Trans R Soc B Biol Sci 351(1345):1291–1298
17. Selkoe KA, Toonen RJ (2006) Microsatellites for ecologists: a practical guide to using and evaluating microsatellite markers. Ecol Lett 9(5):615–629
18. Ravishankar KV, Mani BHR, Anand L, Dinesh MR (2011) Development of new microsatellite markers from mango (*Mangifera indica*) and cross-species amplification. Am J Bot 98(4):e96–e99
19. Bora L, Singh AK, Kumar A, Metwal M (2018) Morphological and microsatellite marker-based polymorphic assessment of genetic diversity and relationship of mango (*Mangifera indica* L.). Indian J Biotechnol 17:91–99
20. Zalapa JE, Cuevas H, Zhu H, Steffan S, Senalik D, Zeldin E, Simon P (2012) Using next-generation sequencing approaches to isolate simple sequence repeat (SSR) loci in the plant sciences. Am J Bot 99(2):193–208
21. Vieira LDN, Dos Anjos KG, Faoro H, Fraga HPDF, Greco TM, Pedrosa FDO, Guerra MP (2016) Phylogenetic inference and SSR characterization of tropical woody bamboos tribe Bambuseae (Poaceae: Bambusoideae) based on complete plastid genome sequences. Curr Genet 62(2):443–453
22. Chen S, Zhou Y, Chen Y, Gu J (2018) Fastp: An ultra-fast all-in-one FASTQ preprocessor. Bioinformatics 34(17):i884–i890. https://doi.org/10.1093/bioinformatics/bty560
23. Andrews S (2010) FastQC: a quality control tool for high-throughput sequence data. Babraham bioinformatics. Babraham Institute, Cambridge, United Kingdom
24. Dierckxsens N, Mardulyn P, Smits G (2016) NOVOPlasty: de novo assembly of organelle genomes from whole genome data. Nucleic Acids Res 45(4):e18. https://doi.org/10.1093/nar/gkw955
25. Wu J, Bao RX, Liu Y, Long YT, Chen JT, Shi YN, Li BJ (2025) Gap-free telomere-to-telomere assembly of the *Mangifera persiciforma* genome and its evolutionary insights on resistance. Plant Biotechnol J 23(8):3257–3259. https://doi.org/10.1111/pbi.70070
26. Boisvert S, Laviolette F, Corbeil J (2010) Ray: simultaneous assembly of reads from a mix of high-throughput sequencing technologies. J Comput Biol 17(11):1519–1533. https://doi.org/10.1089/cmb.2009.0238
27. Manni M, Berkeley MR, Seppey M, Zdobnov EM (2021) BUSCO: assessing genomic data quality and beyond. Curr Protoc 1(12):e323. https://doi.org/10.1002/cpz1.323
28. Du L, Chen J, Sun D, Zhao K, Zeng Q, Yang N (2025) Krait2: a versatile software for microsatellite investigation, visualization and marker development. BMC Genomics 26(1):72. https://doi.org/10.1186/s12864-025-11252-2
29. Camacho C, Coulouris G, Avagyan V, Ma N, Papadopoulos J, Bealer K, Madden TL (2009) BLAST+: architecture and applications. BMC Bioinformatics 10: 421. https://doi.org/10.1186/1471-2105-10-421
30. Okonechnikov K, Golosova O, Fursov M (2012) Unipro UGENE: a unified bioinformatics toolkit. Bioinformatics 28(8):1166. https://doi.org/10.1093/bioinformatics/bts091
31. Tamura K, Stecher G, Kumar S (2021) MEGA11: molecular evolutionary genetics analysis version 11. Mol Biol Evol 38(7):3022–3027. https://doi.org/10.1093/molbev/msab120
32. Alonge M, Lebeigle L, Kirsche M, Jenike K, Ou S, Aganezov S, Soyk S (2022) Automated assembly scaffolding using RagTag elevates a new tomato system for high-throughput genome editing. Genome Biol 23(1):258. https://doi.org/10.1186/s13059-022-02823-7

Chapter 7

Computational and qPCR-Based Determination of Gene Copy Number as a Molecular Marker for Transgenic Plant Analysis

Motahhareh Mohsenpour, Elahe Moatamed, Seyyed Mohammad Mousavi Pakzad, Razieh Goudarzi and Mehran E. Shariatpanahi

Abstract

The molecular characterization of genetically modified plants is commonly based on the detection of foreign DNA sequences introduced during transformation. While this strategy is effective for conventional transgenic plants, it becomes inadequate for cisgenic and intragenic plants in which the introduced gene originates from the host genome and is regulated by native promoter and terminator elements. Following segregation-based elimination of selectable marker cassettes, such cisgenic plants may become indistinguishable from wild-type plants using standard sequence-dependent PCR assays. This chapter presents a quantitative real-time PCR (qPCR) and computation-based approach for estimating gene copy number as an auxiliary molecular marker for the analysis of cisgenic plant lines. The method relies on quantitative differences in endogenous gene dosage rather than the detection of heterologous DNA. Genomic DNA from putative cisgenic and control plants is analyzed by qPCR using serial template dilutions and primer pairs targeting both a cisgenic candidate gene and an endogenous single-copy reference gene. Linear regression of cycle threshold (Ct) values against $\log_2$-transformed DNA concentrations is performed, and differences in regression intercepts are used to infer relative gene copy number. Because the underlying data processing, regression modeling, and copy number calculations may be technically demanding for routine users, a dedicated standalone computational tool, CisScope, was developed to automate the entire analytical workflow. CisScope integrates data parsing, statistical modeling, and result interpretation into an intuitive graphical interface and is distributed as an executable file, eliminating the need for programming expertise or specialized statistical software. Although not intended for regulatory diagnostics, this integrated experimental, computational framework offers a practical and accessible solution to a major limitation of sequence-based detection strategies in cisgenic plant biotechnology and is well-suited for both research and educational applications.

Keywords Cisgenesis, CisScope, Gene copy number, GMO detection marker, Quantitative real-time PCR (qPCR), Marker-free transgenic plants, Computational analysis, Gene dosage

1 Introduction

Molecular characterization of genetically modified plants has traditionally relied on the detection of foreign DNA elements introduced during transformation, including selectable marker genes,

Nitish Kumar (ed.), *Molecular Markers in Plants*, Methods in Molecular Biology, vol. 3097,
https://doi.org/10.1007/978-1-0716-5663-1_7,

heterologous promoters, or terminator sequences. In conventional transgenic systems, such elements serve as unambiguous molecular signatures and can be readily identified using standard PCR-based assays [1, 2].

However, the increasing adoption of cisgenic and intragenic strategies in plant biotechnology has substantially reduced the effectiveness of these sequence-dependent detection approaches. In cisgenic plants, genes originating from the same species or from sexually compatible species are reintroduced into the host genome under the control of their native regulatory elements. Following genetic segregation, selectable marker cassettes are commonly eliminated, resulting in marker-free cisgenic lines. At the DNA sequence level, these plants may therefore consist exclusively of endogenous genetic elements that are indistinguishable from those present in non-modified plants [3].

As a consequence, conventional PCR assays targeting transgene-associated sequences are no longer sufficient to confirm the presence or molecular configuration of cisgenic insertions. This challenge becomes particularly pronounced when cisgenesis is used to introduce subtle genetic modifications, such as single-nucleotide substitutions or native allele duplications conferring agronomically relevant traits [4]. Under such circumstances, qualitative amplification of the target locus yields identical results in cisgenic and wild-type plants, rendering presence–absence-based molecular assays ineffective.

Moreover, event-specific identification of cisgenic plants, similar to conventional transgenic events, fundamentally requires precise determination of the transgene integration site within the host genome [5]. Such event-specific assays rely on junction sequences spanning the insert-genome border, whose characterization typically necessitates labor-intensive molecular approaches, including genome walking, Southern blotting, or next-generation sequencing. In cisgenic plants, where the inserted sequence is fully endogenous, the absence of unique foreign DNA further complicates insertion site identification and renders the development of event-specific detection methods particularly challenging [3, 6].

These limitations necessitate the consideration of alternative molecular features that extend beyond primary sequence identity. Gene copy number represents one such quantitative molecular parameter. Even when a cisgenic construct is sequence-identical to an endogenous gene, the integration of an additional copy typically alters the genomic dosage of the corresponding locus [7, 8].

Quantitative real-time PCR (qPCR), in combination with appropriate endogenous reference genes, offers a sensitive and widely accessible approach for detecting relative differences in gene copy number [2, 6, 9, 10]. By analyzing the relationship between cycle threshold (Ct) values and template DNA concentration across a defined dilution series, quantitative information reflecting underlying gene dosage can be obtained.

The protocol described in this chapter is based on linear regression analysis of Ct values plotted against log_2-transformed genomic DNA concentrations. For each gene analyzed, a regression model is fitted, and the intercept value, corresponding to the theoretical Ct at unit template concentration, is estimated. Relative gene copy number is inferred from differences in intercepts between a target gene and an endogenous single-copy reference gene. This regression-based strategy is conceptually related to established qPCR quantification models [11, 12] but circumvents direct reliance on ΔΔCt calculations by exploiting regression-derived parameters that are robust to variation across dilution points [9, 13].

The workflow presented here encompasses several key procedural components: (i) extraction and accurate quantification of high-quality genomic DNA [14], (ii) preparation of a defined serial dilution series, (iii) qPCR amplification of target and reference loci using SYBR Green chemistry in compliance with established best-practice guidelines [15–16], (iv) organization of Ct values in a standardized data format, and (v) computational analysis of the resulting data. While routine laboratory procedures such as plant DNA extraction and qPCR setup are not described in exhaustive technical detail, they are briefly outlined to define the experimental context and to specify the type and quality of data required for successful analysis.

The primary emphasis of this protocol lies in the generation of appropriately structured qPCR data suitable for downstream computational evaluation. To facilitate reproducible and user-friendly data analysis, a dedicated computational tool, CisScope, has been developed. CisScope automates data handling, regression analysis, and gene copy number estimation through an intuitive graphical interface and is distributed as a standalone executable file. As a result, users are not required to perform manual calculations or possess specialized expertise in programming or statistical modeling; instead, they need only generate high-quality qPCR data.

Overall, this chapter presents a practical and reproducible qPCR- and computation-based protocol for estimating gene copy number as an auxiliary molecular marker in the analysis of cisgenic and intragenic plant lines. Intended for research and educational applications rather than regulatory diagnostics, the approach provides a structured solution to a central analytical limitation associated with marker-free cisgenic plants and supports standardized interpretation of gene dosage variation in modern plant biotechnology.

2 Materials

Note to the reader: This protocol is designed to be independent of specific DNA extraction kits, qPCR chemistries, or real-time PCR platforms. Users are expected to generate high-quality qPCR Ct data using locally optimized and validated laboratory procedures. The materials listed below define the ***minimum experimental requirements and data structure*** necessary for downstream computational copy-number analysis using the CisScope software.

2.1 Plant Genomic DNA Samples

High-quality genomic DNA was isolated from putative cisgenic plant lines and corresponding non-transgenic (wild-type) control plants. DNA integrity and purity should be suitable for quantitative real-time PCR applications. Genomic DNA samples are adjusted to a uniform working concentration (e.g., 50 ng μL^{-1}) using DNase-free double-distilled water.

2.2 DNA Dilution Series

A serial dilution of genomic DNA prepared from each sample to generate at least three template input levels per reaction. Typical template amounts may include 200, 100, and 50 ng per qPCR reaction, although equivalent dilution schemes may be used, provided that they span a linear dynamic range suitable for regression analysis.

2.3 qPCR Reaction Components

A fluorescence-based real-time PCR reaction system compatible with SYBR Green or equivalent double-stranded DNA binding chemistries, including:

- 2× qPCR master mix.
- Gene-specific forward and reverse primers.
- Genomic DNA template.
- DNase-free double-distilled water.

Reactions are assembled in optically compatible qPCR consumables, such as individual tubes, tube strips, or multiwell plates (e.g., 96-well formats), appropriate for the real-time PCR instrument in use.

Final reaction volume is typically 20 μL; minor variations are acceptable provided that reaction volume, consumable type, and sealing method are applied consistently across all samples and dilution series.

2.4 Primers

- Target gene primers: primers specific to the cisgenic candidate gene, amplifying an amplicon <200 bp, with verified specificity and no predicted homo or heterodimer formation.
- Reference gene primers: primers specific to a confirmed single-copy endogenous reference gene (*see* **Note 3**), amplifying

an amplicon <200 bp, with comparable amplification efficiency and no predicted primer–dimer formation.

Primer validation (specificity, efficiency, and melting curve profile) should be performed according to standard qPCR best practices.

2.5 Real-Time PCR Instrumentation

A real-time PCR thermocycler capable of fluorescence detection and melting curve analysis, compatible with the chosen qPCR chemistry. Instrument model and manufacturer are not restricted, provided consistent performance across dilution series and targets.

2.6 DNA Quantification

A spectrophotometric or fluorometric device suitable for nucleic acid quantification, used to normalize genomic DNA concentrations prior to dilution series preparation.

2.7 Data Recording and File Preparation

Quantification cycle (Ct) values obtained from real-time PCR experiments are organized in a structured data file for downstream computational analysis. The input file is required to include information identifying the sample category (e.g., cisgenic or control), the amplified gene, the corresponding DNA template concentration used in each reaction, and the associated Ct value generated by the qPCR instrument.

2.8 Computational Analysis Tool

CisScope software: a standalone executable application developed for automated parsing of qPCR Ct data, linear regression modeling, and relative gene copy number estimation. CisScope performs all computational steps described in this chapter.

3 Methods

3.1 Genomic DNA Preparation and Quantification

1. Extract genomic DNA from putative cisgenic plants and corresponding non-transgenic control plants using a standard plant DNA extraction protocol (*see* **Note 1**).
2. Quantify genomic DNA using a UV–vis spectrophotometer.
3. Measure each DNA sample three times independently and calculate the mean concentration value.
4. Adjust all DNA samples to a working concentration of 50 ng μL^{-1} using DNase-free double-distilled water.

3.2 Preparation of Serial DNA Dilutions

1. Prepare a 1:2 serial dilution series for each DNA sample using DNase-free double-distilled water.
2. Generate three template concentrations corresponding to 200, 100, and 50 ng of genomic DNA per qPCR reaction.
3. Use a constant template volume of 4 μL for all qPCR reactions.

4. Label dilution levels numerically as Con=4 (200 ng), Con=2 (100 ng), and Con=1 (50 ng) for downstream data organization.

3.3 Quantitative Real-Time PCR Setup

1. Prepare qPCR reactions in a final volume of 20 μL containing SYBR Green master mix, forward and reverse primers, genomic DNA template, and DNase-free double-distilled water (see Sect. 2).
2. Set up three technical replicates for each dilution level of each gene (target and reference) (*see* **Note 8**).
3. Include non-template controls (NTCs) for both primer pairs.
4. Perform qPCR amplification using a real-time PCR instrument compatible with SYBR Green chemistry.

3.4 qPCR Cycling Conditions and Specificity Assessment

1. Program the thermal cycler according to primer characteristics.
2. Use an initial denaturation step followed by 35 amplification cycles consisting of denaturation, annealing, and extension.
3. Perform a final extension step followed by melting curve analysis using the default instrument settings.
4. Confirm that amplification produces a single melting peak and that no primer–dimer formation is detected (*see* **Note 2**).

3.5 Ct Data Collection and Organization

1. Collect cycle threshold (Ct) values automatically generated by the real-time PCR instrument.
2. Organize Ct values into a tabular data file with four columns in the following order: Cat, Gene, Con, Ct.
3. Populate the data file as shown below, using three technical replicates for each dilution level:
   - Cat: sample category (e.g., *Control* or *Cisgenic*).
   - Gene: gene name (reference or target).
   - Con: dilution level (4, 2, or 1).
   - Ct: measured Ct value.
4. Replace *ref* with the name of the endogenous reference gene and *target* with the name of the target gene.
5. Replace *n* with the corresponding Ct values obtained from the qPCR experiment.

3.6 Computational Analysis Using CisScope

1. Launch the CisScope application (Fig. 1).
2. Select **Browse** under *Select CSV/TXT file* and load the prepared data file.
3. Enter the **Reference Gene** and **Target Gene** names exactly as specified in the data file.

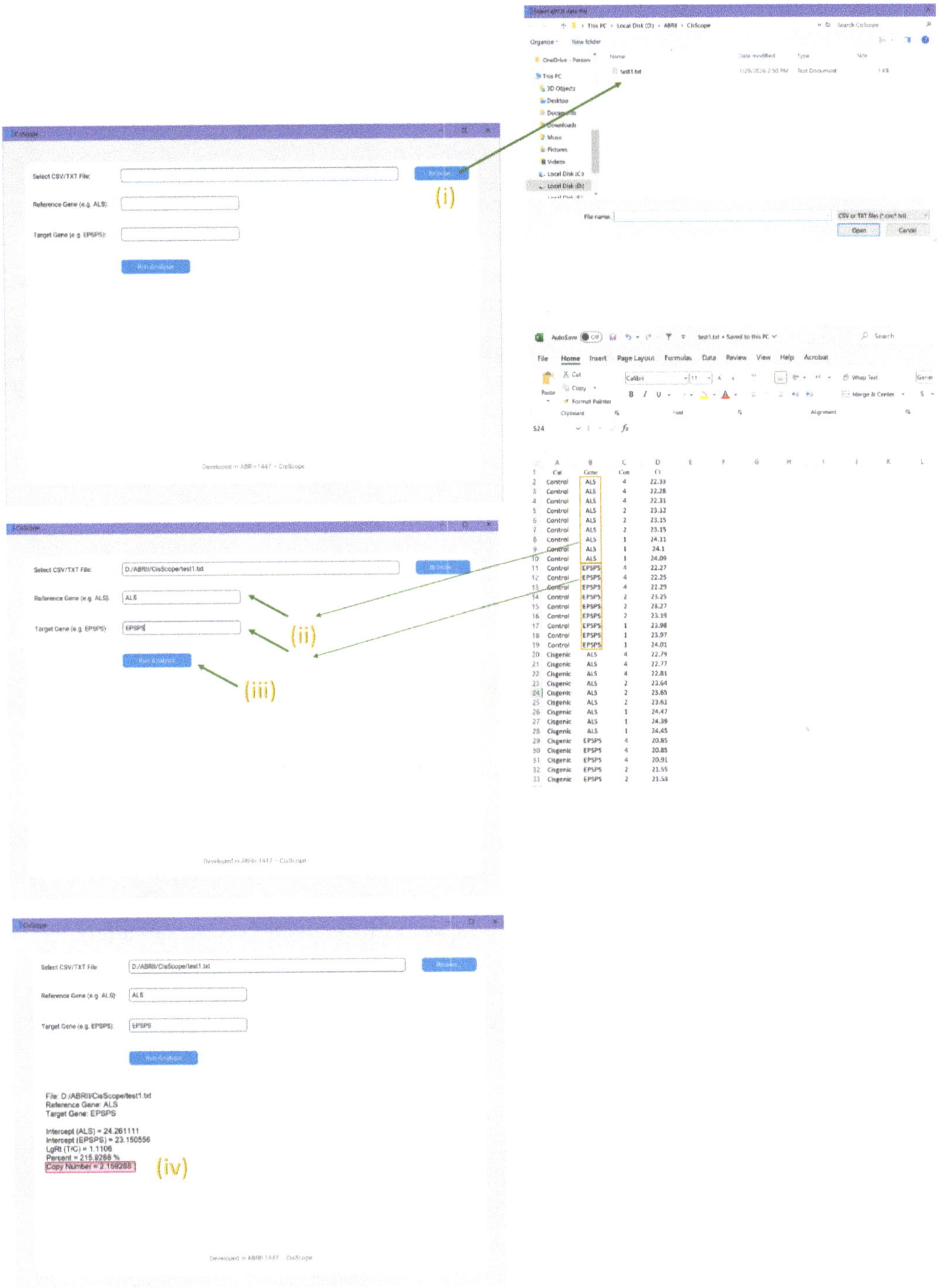

Fig. 1 Step-by-step workflow of CisScope-based data analysis. Graphical overview of the CisScope analytical workflow, including (i) loading of the formatted Ct data file, (ii) specification of reference and target gene identifiers, (iii) execution of automated regression analysis, and (iv) visualization of output parameters, including regression intercepts and estimated gene copy number. The software interface enables users to perform complete data analysis without manual calculations or specialized computational expertise (*see* **Notes 9 and 13**)

4. Click **Run Analysis** to execute automated regression analysis and copy number estimation.
5. Record the output parameters displayed in the results window, including regression intercepts and relative gene copy number values.

3.7 Interpretation of Results

1. Compare the estimated copy number values obtained for putative cisgenic plants with those of non-transgenic control plants.
2. Identify deviations from single-copy expectations as indicative of additional gene insertions (*see* **Notes 11 and 12**).

An example illustrating Ct data organization together with representative analysis output is presented in the "Example Application and Data Interpretation" section.

4 Notes

1. *Quality of genomic DNA*: The accuracy of copy number estimation critically depends on the quality and integrity of genomic DNA. Degraded DNA or the presence of residual polysaccharides or phenolic compounds may affect amplification efficiency and regression linearity. DNA extraction methods should therefore be optimized for the plant species under investigation, and only high-quality DNA preparations should be used (see Sect. 3.1).
2. *DNA quantification and dilution accuracy*: Accurate DNA quantification is essential for reliable regression analysis. Multiple spectrophotometric measurements are recommended to minimize pipetting or instrument-related variability. Errors introduced during preparation of the serial dilution series directly affect slope and intercept estimation and may lead to inaccurate copy number calculations (see Sects. 3.1 and 3.2).
3. *Selection of the reference gene*: The endogenous reference gene must be present as a single copy in the genome and should not be affected by structural variation or copy number polymorphisms. Prior validation of the reference gene in the relevant plant genotype is strongly recommended, particularly in species with large or complex genomes.
4. *Primer design considerations*: Primer pairs should amplify short amplicons (<200 bp) to ensure efficient and uniform amplification across dilution points. The absence of homo and heterodimer formation should be verified in silico and experimentally. Primer pairs producing nonspecific amplification or primer–dimer artifacts may distort melting curve profiles and compromise Ct reliability (see Sect. 3.4).

5. *Amplification efficiency requirements*: Amplification efficiency for both target and reference genes should ideally fall within the range of 90%–100%. Substantial differences in efficiency between primer pairs may bias regression intercept comparisons. Efficiency should be evaluated using preliminary qPCR experiments prior to copy number analysis.
6. *Compatibility of target and reference qPCR programs*: Target and reference gene primer pairs should be compatible with a common thermal cycling program. Using different cycling conditions for target and reference reactions may introduce systematic bias and is therefore discouraged.
7. *Melting curve analysis*: A single, well-defined melting peak is required for each primer pair. Multiple peaks or shoulders indicate nonspecific amplification and invalidate Ct-based quantitative comparisons. Melting curve analysis should be evaluated for every qPCR run (Fig. 2).
8. *Replication strategy*: While three technical replicates per dilution point are generally sufficient, inclusion of biological replicates is strongly recommended when material is available. Biological replication improves robustness and facilitates the detection of sample-specific anomalies.
9. *Organization of Ct data*: Strict adherence to the prescribed data format is required for correct CisScope operation. Column names and gene identifiers must be consistent and case-sensitive. Mislabeling of sample categories, gene names, or dilution levels may lead to erroneous regression grouping or failed analysis (see Sect. 3.5).
10. *Interpretation of regression intercept differences*: The intercept value represents the theoretical Ct at unit template concentration and is sensitive to systematic shifts in amplification behavior. Copy number estimates should therefore be interpreted

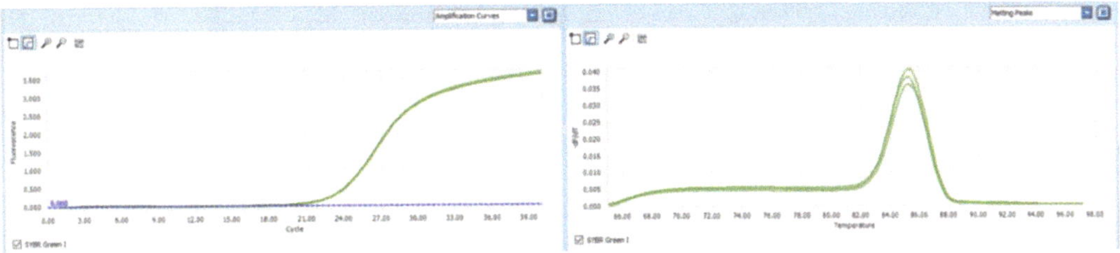

Fig. 2 qPCR amplification and melting curve analysis. Representative amplification plots and corresponding melting curves are shown for a single three-technical-replicate reaction derived from one genomic DNA dilution. For accurate gene copy number estimation, all serial dilutions are required to be analyzed for both the target gene and the endogenous single-copy reference gene. Across all dilutions, technical replicates should exhibit highly consistent amplification behavior, with closely overlapping amplification curves and uniform, single-peak melting profiles, indicating robust amplification efficiency and primer specificity (*see* **Notes 4–7**)

in relative terms and in comparison with an appropriate non-transgenic control rather than as absolute genomic values.

11. *Expected copy number resolution*: This method is best suited for detecting integer or near-integer changes in gene copy number (e.g., one versus two copies). Small deviations may fall within experimental variability and should be interpreted cautiously, particularly in genomes with high ploidy or segmental duplication.
12. *Scope and limitations of the method*: The protocol is intended for research and educational applications and is not designed for regulatory diagnostics. Results should be interpreted as supportive molecular evidence rather than definitive proof of cisgenic status.
13. *Role of CisScope in data analysis*: CisScope automates regression fitting and numerical calculations but does not correct for experimental errors introduced during sample preparation or qPCR execution. The responsibility for generating high-quality input data remains with the user (see Sect. 3.6).

References

1. Schmidt M, Parrott W (2001) Quantitative detection of transgenes in soybean [*Glycine max* (L.) Merrill] and peanut (*Arachis hypogaea* L.) by real-time polymerase chain reaction. Plant Cell Rep 20:422–428
2. Song P, Cai C, Skokut M, Kosegi B, Petolino J (2002) Quantitative real-time PCR as a screening tool for estimating transgene copy number in WHISKERS™-derived transgenic maize. Plant Cell Rep 20:948–954
3. Xu W, Liang J, Wang F, Yang L (2024) Comparative evaluation of gene copy number estimation techniques in genetically modified crops: insights from Southern blotting, qPCR, dPCR and NGS. Plant Biotechnol J 22:3456
4. Chen X, Yu H, Huang C, Hou C, Guan H, Xie J (2025) Methods established for EPSPS gene mutation detection in glyphosate-resistant rice (*Oryza sativa* L.). Plants 14:2256
5. Chamani Mohasses F, Mousavi Pakzad SM, Moatamed E, Entesari M, Bidadi H, Molaahmad Nalousi A et al (2024) Efficient genetic transformation of rice using Agrobacterium with a codon-optimized chromoprotein reporter gene (ChromoP) and introducing an optimized iPCR method for transgene integration site detection. Plant Cell Tissue Organ Cult 156:5
6. De Saeger J, Park J, Thoris K, De Bruyn C, Chung HS, Inzé D et al (2022) IMPLANT: a new technique for transgene copy number estimation in plants using a single end-point PCR reaction. Plant Methods 18:132
7. Mason G, Provero P, Vaira AM, Accotto GP (2002) Estimating the number of integrations in transformed plants by quantitative real-time PCR. BMC Biotechnol 2:20
8. Bubner B, Baldwin IT (2004) Use of real-time PCR for determining copy number and zygosity in transgenic plants. Plant Cell Rep 23:263–271
9. Yuan JS, Burris J, Stewart NR, Mentewab A, Stewart CN (2007) Statistical tools for transgene copy number estimation based on real-time PCR. BMC Bioinform 8:S6 S6
10. Mohsenpour M, Tohidfar M, Jelodar NB, Jouzani GS (2015). Designing a new marker-free and tissue-specific platform for molecular farming applications. J Plant Biochem Biotechnol. 2015;24:433–440. https://doi.org/10.1007/s13562-014-0294-2
11. Pfaffl MW (2001) A new mathematical model for relative quantification in real-time RT–PCR. Nucleic Acids Res 29:e45
12. Schmittgen TD, Livak KJ (2008) Analyzing real-time PCR data by the comparative CT method. Nat Protoc 3:1101–1108
13. Doleschall M, Darvasi O, Herold Z, Doleschall Z, Nyirő G, Somogyi A et al (2022)

Quantitative PCR from human genomic DNA:the determination of gene copy numbers for congenital adrenal hyperplasia and RCCX copy number variation. PLoS One 17:e0277299
14. Quach H, de Bernardeaux G, Nguyen D, Sahay S, Hoang K. A simple and efficient CTAB plate-based protocol for genomic DNA extraction from crop plants. bioRxiv. Cold Spring Harbor Laboratory; 2026; 2001–2026.
15. Bustin SA, Benes V, Garson JA, Hellemans J, Huggett J, Kubista M, et al (2009) The MIQE guidelines: minimum information for publication of quantitative real-time PCR experiments. Oxford University Press
16. Taylor SC, Nadeau K, Abbasi M, Lachance C, Nguyen M, Fenrich J (2019) The ultimate qPCR experiment: producing publication quality, reproducible data the first time. Trends Biotechnol 37:761–774

Chapter 8

Application of SSR Markers for the Detection of Metal Tolerance in Plants

Khusboo Sahu, Yuganti Behera and Patitapaban Dash

Abstracts

The dynamic flow of heavy metals in ecosystems is disturbing the sustainability of environmental quality and the flow of resources in ecosystems. Detection of genotypes with metal-tolerance ability is a prerequisite for mitigating this burning problem, through restoration of degraded ecosystems, progressive crop productivity, and improvement in efficiency of phytoremediation programs. The application of simple sequence repeat (SSR) markers for tracing metal tolerance in plants is gaining importance among researchers. SSRs, as molecular markers, are highly informative and preferable compared to other molecular markers. It is gaining importance because of its codominant inheritance, polymorphic forms, genome-wide distribution, reliability, and reproducibility. This chapter presents a comprehensive protocol-based approach to facilitate the application of SSR markers for the detection and validation of metal tolerance in plants. The chapter is demonstrating the conceptual background, principles of the study, experimental design, protocol for experimental study, data analysis strategies, practical considerations of data analysis, and its importance for researchers working on plants under stress, molecular breeding, and environmental biotechnology.

Keywords Heavy metals, Metal-tolerance, SSR, Tolerance

1 Introduction

The steady growth of industrialization, overutilization of agrochemicals, mining of metallic resources, and disposal of metal wastes are responsible for irrational accumulation of heavy metals, such as cadmium, lead, arsenic, and chromium in the soil component of ecosystems [1]. Non-biodegradability is making the presence of heavy metals persistent in ecosystems. It is threatening the metabolism, health, and diversity of living organisms in ecosystems [2]. Plants exposed to hyperconcentrations of metals in the habitat sense the metal toxicity through a multifaceted network of biochemical, physiological, and molecular mechanisms. Plants are able to express metal chelation, sequestration, antioxidant synthesis, and gene expression on polluted sites [3].

Nitish Kumar (ed.), *Molecular Markers in Plants*, Methods in Molecular Biology, vol. 3097,
https://doi.org/10.1007/978-1-0716-5663-1_8,

Metal-tolerant plants are limited in nature due to their unique pattern of gene expression and molecular regulations [4]. The traditional phenotypic screening procedures are a time-consuming method, and to a certain extent, mislead the interpretation, through interference from environmental stimuli. This chapter on SSRs, representing DNA-based markers, is a more reliable and alternative method for the detection of metal tolerance in species at early stages of growth.

SSRs (microsatellites), ranging from 1 to 6 bp (base pairs), are favored as molecular markers for their co-dominant inheritance, polymorphic forms, genome-wide distribution, and reproducibility, compared to other molecular markers [5]. SSR markers allow highly reliable discrimination between heterozygous and homozygous alleles. It is an ideal tool for the detection of variation in plants under stress conditions, during molecular breeding, and biodiversity studies. It is a reliable tool for detection, molecular characterization, and utilization of metal-tolerant resources [6].

Simple sequence repeats (SSRs), representing repeated motifs (1–6 bp), are distributed throughout the genome [7]. Variation in DNA sequence, due to DNA replication slippage, reflects variation in repeat numbers within the genotype. SSR primers pair with unique DNA sequences flanking the SSR regions in targeted genomic DNA [8]. PCR amplifies the targeted SSR locus in genomic DNA, using SSR primers, and helps in analyzing genetic variations among plant species. The variation in amplicons of the PCR product is dependent on SSRs present in genomic DNA samples. The gel electrophoresis allows the separation and proper distribution of amplicons from the cathode towards the anode, based on the properties, such as molecular size and charge. Staining shows a series of bands in the gel used for electrophoresis.

2 Materials

2.1 Plant Materials

1. 100 mg of fresh leaves to be taken per sample for DNA extraction.

2.2 Materials/Reagents for DNA Extraction and Purification

The chemical reagents of analytical grade have to be used for effective isolation of genomic DNA from targeted plant samples [9].

1. CTAB (cetyltrimethylammonium bromide) extraction buffer (*see* **Note 1**).
2. Isopropanol (C_3H_8O)/ethanol (C_2H_5OH), store at 4°C.
3. Tris–EDTA buffer (TE buffer): 1 M Tris–HCl (pH 8.0), and 0.5 M EDTA (*see* **Note 2**).
4. Phenol (C_6H_5OH): Chloroform ($CHCl_3$): Isoamyl alcohol ($C_5H_{12}O$) (25:24:1 v/v/v), store at 4°C (*see* **Note 3**).

5. Phenol (C_6H_5OH).
6. Liquid nitrogen.
7. Sodium acetate (CH_3COONa).
8. Sodium hydroxide (NaOH).
9. Hydrochloric acid (HCl).

2.3 Materials/Reagents for Amplification of DNA Samples Using SSR Primer

1. PCR master-mix (PCR buffer, SSR forward, and reverse primers, *Taq* (*Thermus aquaticus*) DNA polymerase, dNTPs) [10].
2. Standardized microsatellite (SSR) forward and reverse primers (18–24 bp), specific for the targeted species (fluorescently labeled) [11].
3. Agarose/polyacrylamide gel components (2%) (*see* **Note 4**).
4. DNA ladder (>100 bp).

2.4 Materials/Reagents for Gel Electrophoresis

1. 50× TAE Buffer [12] (*see* **Note 5**).
2. Bromophenol blue ($C_{19}H_{10}Br_4O_5S$).
3. Ethidium bromide ($C_{21}H_{20}BrN_3$).
4. Agarose

3 Methods

The workflow for analysis of SSR marker-based, detection of metal tolerance in plants, involves extraction of genomic DNA from plant samples, purification of extracted DNA by removal of unwanted products, purity assessment of extracted genomic DNA, design/selection/screening of SSR primers to amplify specific SSR locus in genomic DNA, using PCR, and gel electrophoresis of PCR products to interpret genetic variation, among plant species.

3.1 Extraction of Purified Genomic DNA

Genomic DNA is the source material for predicting the metal tolerance in plants using SSR markers as tools. One can get a single band for homozygosity and two bands for heterozygosity during electrophoresis of genomic materials. The extracted DNA needs purification before downstream applications. The commonly used methods for the extraction of purified genomic DNA are the CTAB method and the SDS/phenol–chloroform method.

3.1.1 CTAB Method

The CTAB method can be used for genomic DNA extraction from sampled fresh plant leaves, following the standard protocol [9], with minor modifications for better yield and purity of products.

1. Surface sterilization of plant samples, using 90% alcohol/0.1% $HgCl_2$ (*see* **Note 6**).

2. Crush, sterilized plant samples (100 mg), using liquid nitrogen, to obtain a fine powder of the samples.
3. Mix the sample in its powder form with pre-warmed CTAB buffer, and incubate the mixture at 65°C, in a water bath, with intermittent stirring, for 60–90 min (*see* **Note 7**).
4. Process the output, using an equal volume of chloroform: isoamyl alcohol (24:1 v/v) for 15 min.
5. Centrifuge precipitates, with 10,000 rpm (rotations per minute), at 4°C, for 20 min.
6. Add prechilled isopropanol/ethanol (three times the volume of collected supernatant) for precipitation of DNA.
7. Transfer crude genomic DNA into a 2 ml Eppendorf tube, containing 70% ethanol.
8. Wash DNA, with 70% isopropanol/ethanol, followed by air drying (*see* **Note 8**).
9. Store the DNA in ethanol under freezing conditions for further use.
10. After defreezing, wash the crude genomic DNA with 70% ethanol twice.
11. Air-drying of DNA is required.
12. Dissolve the air-dried DNA in an adequate volume of TE buffer.
13. After dissolution of DNA in buffer, add chilled RNase-A solution.
14. Incubate the mixture at 37°C for 1 h in a water bath with intermittent stirring.
15. After incubation, add an equal volume of phenol: chloroform: isoamyl alcohol (25:24:1 v/v/v), and mix thoroughly for 15 min.
16. Centrifuge the mixture at 10,000 rpm (4°C), for 20 min.
17. Collect the upper clear supernatant in a clean and dry Eppendorf tube.
18. Repeat the washing process twice with chloroform:isoamyl alcohol (24:1 v/v).
19. Precipitate DNA by adding prechilled absolute ethanol (2.5 times the volume of the last collected supernatant).
20. Gently, mix it for better precipitation of DNA.
21. Transfer DNA into a 2 ml Eppendorf tube, having 70% ethanol.
22. Wash DNA, with 70% isopropanol/ethanol, followed by air drying.

23. Dissolve the air-dried DNA in a minimum volume of TE buffer.
24. After dissolution, store DNA at 4°C, for immediate use, or at −20°C, for use after an interval.

3.1.2 SDS/Phenol–Chloroform Method

The SDS (sodium dodecyl sulfate)/phenol–chloroform method can be used for genomic DNA extraction from sampled fresh/frozen samples, following the standard protocol [13], with minor modifications for better yield and purity of products.

1. Surface sterilization of plant samples, using 90% alcohol/0.1% $HgCl_2$.
2. Crush, sterilized plant samples (0.5 mg), using liquid nitrogen, to obtain a fine powder of the samples.
3. Add digestion buffer (0.5 ml) to the Eppendorf tube (*see* **Note 9**).
4. Incubate the solution at 56°C for 60–90 min, followed by overnight incubation at 37°C.
5. Mix the solution for 2–3 min after adding 0.5 ml Tris saturated phenol (pH 8.0).
6. Centrifuge at 10,000 rpm, for 10 min, at 4°C.
7. After the collection of aqueous phases in another Eppendorf tube, add an equal volume of phenol: chloroform: isoamyl alcohol (25:24:1 v/v/v), and mix thoroughly.
8. Centrifuge at 10,000 rpm, for 10 min, at 4°C.
9. Collect aqueous phase in another Eppendorf tube, using a micropipette.
10. Transfer DNA into another 2 ml Eppendorf tube, having 70% ethanol.
11. Wash twice, with 70% ethanol, and make that dry.
12. Dissolve DNA in 1× TE buffer, at room temperature, for a few minutes.
13. Remove residual phenol, using 0.5 ml chloroform: isoamyl alcohol (24:1 v/v), twice.
14. Collect aqueous phase in another Eppendorf tube, using a micropipette, and add 50 μl 3 M sodium acetate (pH 5.2).
15. Precipitate DNA by adding twice the volume of chilled isopropanol/ethanol.
16. After dissolution, store DNA at 4°C, for immediate use, or at −20°C, for use after an interval.
17. Make up the volume to 500 μl by adding 1× TE buffer to the DNA.

18. After thorough mixing, add RNase A (2–3 μl) to digest RNA in that solution.
19. Incubate at 37°C for 30 min.
20. Add 2.3 μl proteinase-K (100 mg/ml), and incubate for 30 min.
21. Repeat the washing process with phenol: chloroform, and chloroform:isoamyl alcohol, once each.
22. Precipitate DNA by adding 50 μl 3 M sodium acetate (pH 5.2) and twice the volume of chilled absolute ethanol.
23. Wash DNA with 70% isopropanol/ethanol, followed by air drying.
24. Dissolve the air-dried DNA in a minimum volume of 1× TE buffer.
25. After dissolution, and a few minutes at room temperature, store the DNA at 4°C, for immediate use, or at −20°C, for use after an interval.

3.2 Qualitative and Quantitative Analysis of Isolated DNA Before Molecular Analysis

SSR marker-based analysis of metal tolerance in plants can mislead the interpretation if the source material, genomic DNA, is contaminated by primary/secondary metabolites in plants. Hence, assessment of the quality and quantity of purified genomic DNA is essential before further molecular analysis and application. The quality, integrity, and suitability of genomic DNA for further molecular analysis can be determined on the basis of the spectrophotometric method/gel electrophoresis method.

3.2.1 Spectrophotometric Method

1. Check the ratio of absorbance at 260–280 nm (*see* **Note 10**).

3.2.2 Gel Electrophoresis Method

1. For electrophoresis of targeted plant genomic DNA, use a 0.8% agarose gel (*see* **Note 11**).
2. Bromophenol blue can be used as a loading dye.
3. Use ethidium bromide (0.2 μg.ml^{-1}) as a stain for visualization of DNA fragments under UV light.
4. Use a UV transilluminator or gel documentation unit for visualization of bands on the gel (Fig. 1).

3.3 Design/Selection/Screening of SSR Primers

Select SSR primers, and systematically screen those primers on the basis of effectiveness, to detect genetic variation associated with stress tolerance in plants [11, 14].

1. Select SSR primers on the basis of targeted regions (Table 1).
2. Screening of primers considers representative tolerant and susceptible genotypes.

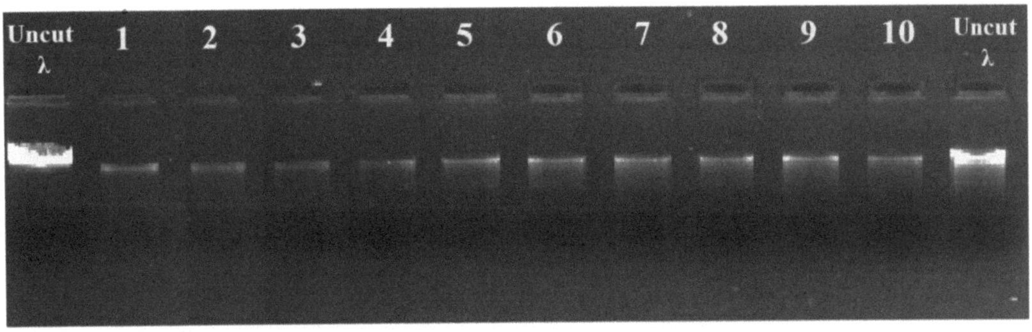

Fig. 1 Electrophoresis of genomic DNA of targeted plant species in a 0.8% agarose gel. Left uncut λ: DNA marker with a prominent band at 200 ng/μl, Lanes 1–10: purified genomic DNA of targeted plant species. Right uncut λ: DNA marker with a prominent band at 100 ng/μl. Distinct and clear bands with minimal smearing indicate good quality, integrity, and suitability of genomic DNA for molecular analyses

Table 1
Glances of SSR primer pairs

Sl. No.	Primer name	Forward sequence (5′ to 3′)	Reverse sequence (5′ to 3′)
1	MiIIHR01	GGATGCACAACAACAAGCAC	TCAGCAAGCAATCCCTTCTT
2	MiIIHR02	CCCCAACATTTCATAAACACA	CCTCCTTACATGCCTCCTTG
3	MiIIHR03	GTCGATGCCTGGAATGAAGT	AAGCATCGAACAGCTCCAAT
4	MiIIHR04	CGTTTTTGACCCTCTTGAGC	CCGCATACTTCCCTTCACAT
5	MiIIHR05	CTCTCCCTCACTTGCTCCAC	AGACCACCGACAACGAAAAC
6	MiIIHR06	CGCCGAGCCTATAACCTCTA	ATCATGCCCTAAACGACGAC
7	MiIIHR07	GCCACTCAGCTAAATAGCCTCT	TGCAGTCGGTAAAGTGATGG
8	MiIIHR08	TGCTCTCTACTGCCCCGTAT	GTCACACCAATCGGGAATCT
9	MiIIHR09	GTTGTGACCGAGGCCTTAAA	CTTTGACATCGCTGATCTGG
10	MiIIHR10	CGATTCAAGACGGAAAGGAA	TTCAAGCACAGACGACCAAC

3. Retain the primers producing a clear and replicable polymorphic band.
4. Favorable GC content ranges between 40% and 60%.
5. Primer pairs with identical Tm values are generally preferred.
6. Annealing temperature is to be significantly different from the estimated Tm value.
7. Tools, such as Primer3, can be used for designing primers.

3.4 Amplification of a Specific SSR Locus in DNA Using PCR

PCR (polymerase chain reaction) is an iterative process. It involves thermal denaturation of template DNA, annealing of primers to single-stranded target sequence(s), and extension of annealing products by a thermostable, *Taq* DNA polymerase. Use PCR for rapid and exponential amplification of the targeted DNA sequences [10].

1. Purified genomic DNA samples for PCR reactions: 20–25 µl.
2. Preparation of 25 µl master-mix per unit PCR tube (*see* **Note 12**).
3. Keep the PCR tubes for reactions inside the thermocycler.
4. Allow 36 cycles of PCR. In each cycle, maintain temperature for initial denaturation at 95°C (5 min), for denaturation at 92°C (1 min), for annealing at 52°C (1 min), for primer extension at 72°C (7 min) in a thermocycler.
5. Annealing temperature is specific for each SSR primer.
6. Keep the PCR product at 4°C.

3.5 Gel Electrophoresis and Visualization of PCR Products for the Interpretation of Genetic Variation

The gel electrophoresis allows the separation and proper distribution of fragments based on properties, such as molecular size and charge. It follows proper staining and visualization of bands in the gel. It helps in better assessment of amplification of DNA fragments [15].

1. Prepare agarose gel with 1× TAE buffer, and add ethidium bromide (0.2 µg.ml^{-1}), at 55°C–60°C.
2. Remove the comb from the agarose gel after 30 min.
3. Take 1× TAE buffer, and fill the electrophoresis tank.
4. Load the wells in the gel with DNA/PCR products along with the loading dye.
5. Run the PCR products along with the DNA ladder (>100 bp) on Agarose/polyacrylamide gel components (2%), at 60 V, for 2–3 h.
6. 2% agarose gel can be used for electrophoresis of genomic DNA amplified using an SSR primer. Polyacrylamide gel can be used for better resolution of shorter fragments of DNA.
7. Bromophenol blue can be used as a loading dye.
8. Use ethidium bromide (0.2 µg.ml^{-1}) as a stain for visualization of DNA fragments under UV light.
9. Use a UV transilluminator or gel documentation unit for visualization of bands on the gel (Fig. 2).

3.6 Data Analysis and Interpretation

The experimental data analysis and interpretation involve the scoring of SSR alleles, genetic diversity analysis, and association with metal tolerance in plants, following established analytical methods [16].

3.6.1 Scoring of SSR Alleles

1. Score alleles based on fragment size differences.
2. Analyze the presence or absence of bands for the detection of markers.
3. Calculate the heterozygosity and richness of the alleles.

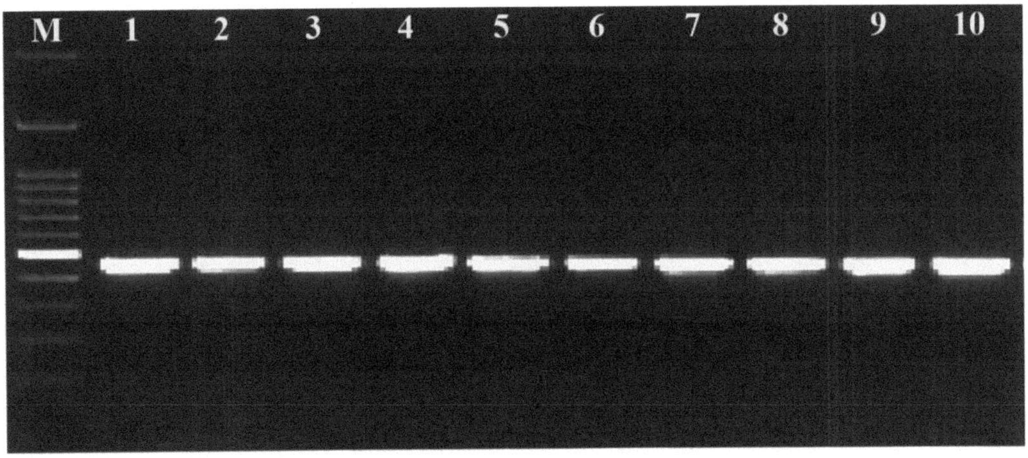

Fig. 2 Amplification of the genomic DNA of targeted plant species with the SSR primer was observed in 2% agarose gel. Lane M: DNA molecular marker (bp ladder), Lanes 1–10: Genomic DNA of plant species amplified with SSR primer. Distinct and clear bands indicate successful amplification

3.6.2 Genetic Diversity Analysis

1. Perform polymorphism information content (PIC) analysis for statistically quantifying genetic marker-based variation.
2. Determine the observed and expected heterozygosity for the estimation of genetic similarity and cluster analysis.

3.6.3 Association with Metal Tolerance in Plants

1. Identify the SSR loci linked with the metal tolerance traits by comparison of allele frequencies between the sensitive and tolerant species.
2. Validate it through the replicated trials.

4 Notes

1. CTAB, 2% (w/v), DNA extraction buffer (100 μl): 1.4 M NaCl (35 μl, 4 M NaCl solution), 0.1 M Tris–HCl, pH 8.0 (10 μl, 1 M Tris–HCl solution, pH 8.0), 0.02 M EDTA (Ethylenediaminetetraacetic acid), pH 8.0, (4 μl, 0.5 M EDTA, pH 8.0), 2% CTAB solution (20 μl, 10% CTAB solution), 0.04 M β-mercaptoethnol (2 μl, 2% β-mercaptoethnol). Adjust the volume using 29 μl, nuclease-free, double-distilled water (DDW).

   CTAB is a cationic surfactant and decreases surface tension at the time of DNA isolation. Its involvement in conformational changes of biomolecules, such as polysaccharides, proteins, and lipids, is useful for lysis of the cell wall and membrane. Na^+ ions from NaCl allow better precipitation of DNA, due to neutralization of –ve charges. EDTA, a chelating agent, prevents degradation of DNA by removing cofactors of DNase, such as Mg^{++} and Ca^{++}. SDS is useful for the degradation of the lipoprotein layer of the cell and bringing DNA out of the nucleus by denaturation of membrane proteins. β -mercaptoethnol is a reducing agent that can prevent oxidative DNA damage. Besides, it denatures proteins by breaking disulfide bonds, present between sulfur-containing amino acid residues.

2. Tris–EDTA (TE) buffer (100 ml): 10 mM Tris–HCl, pH 8.0 (1 ml, 1 M Tris–HCl solution, pH 8.0), 1 mM EDTA, pH 8.0 (0.2 ml, 0.5 M EDTA, pH 8.0). Adjust the volume using 98.8 ml, nuclease-free, double-distilled water (DDW). Keep photosensitive TE buffer in an amber bottle.
3. Addition of organic solvents, Phenol:chloroform:isoamyl alcohol (25:24:1 v/v/v), helps in the removal of cellular debris and membrane proteins during DNA extraction. Chloroform denatures proteins and helps in the separation of DNA by dissolving other organic constituents. Isoamyl alcohol helps in the phase separation of solvents during extraction.
4. Polyacrylamide gel can be used for electrophoresis of smaller DNA fragments with better resolution. Ethidium bromide (0.2 $\mu g.ml^{-1}$) can be used as a stain for visualization of DNA fragments under UV light. Agarose gel is preferable for cost-effective electrophoresis of DNA fragments showing variation between a few hundred base pairs (2%) to 18 kbp (0.8%).
5. 50× TAE buffer for gel electrophoresis: Tris-base (242 g), Glacial acetic acid (57.1 ml), 0.5 M EDTA, pH 8.0 (100 ml) [17].
6. Mercuric chloride ($HgCl_2$) does not dissolve in water easily. Therefore, prepare a 0.1% (w/v) solution by carefully dissolving the required volume in sterile distilled water before using it for surface sterilization.
7. Incubation at 65°C for 60–90 min, with stirring at intervals, can allow proper dissolution of CTAB.
8. Use prechilled isopropanol/ethanol as a dehydrating agent. It can allow precipitation of DNA.
9. Digestion buffer: 10–100 mM Tris-HCl (pH 8.0), 10-100 mM EDTA, 2% SDS, proteinase-K (100–200 μg/ml), 0.1–0.5 M NaCl, 0.04 M β-mercaptoethnol (2 μl, 2% β -mercaptoethnol). Tris–HCl (pH 8.0) maintains pH, EDTA protects DNA, SDS disrupts the membrane, NaCl solubilizes DNA, β-mercaptoethnol removes phenolic compounds, and Proteinase K actively degrades nucleases, along with other proteins, in the presence of SDS.

10. Optical density ratio (260:280) for pure DNA is 1.8.
11. Use 0.8% agarose gel for checking the purity of genomic DNA (18 kbp), 1% agarose gel for DNA fragments (500 bp–8 kbp), and 2% agarose gel for PCR products (˂500 bp). The concentration of gel is reciprocal to DNA size.
12. Master-mix for PCR reaction: 10× PCR buffer (2.60 μl), 25 mM dNTPs mix (0.21 μl), 5 pM forward primer (1.04 μ l), 5 pM reverse primer (1.04 μl), Taq DNA polymerase, concentration: 3 U/μl (0.18 μl). Adjust the volume using nuclease-free double-distilled water (19.93 μl).

Acknowledgments

The authors acknowledge the support provided by the Siksha 'O' Anusandhan (Deemed to be University), Bhubaneswar, India.

References

1. Sahu K, Dash P (2025) Chromium dynamics in soil and detoxification of chromite belts using rhizospheric soil–plant interface. Environ Monit Assess 197(6):654. https://doi.org/10.1007/s10661-025-14085-4
2. Alengebawy A, Abdelkhalek ST, Qureshi SR, Wang MQ (2021) Heavy metals and pesticides toxicity in agricultural soil and plants: ecological risks and human health implications. Toxics 9(3):42. https://doi.org/10.3390/toxics9030042
3. Ningombam L, Hazarika BN, Yumkhaibam T, Heisnam P, Singh YD (2024) Heavy metal priming plant stress tolerance deciphering through physiological, biochemical, molecular and omics mechanism. South Afr J Bot 168:16–25. https://doi.org/10.1016/j.sajb.2024.02.032
4. Kushwaha A, Rani R, Kumar S, Gautam A (2015) Heavy metal detoxification and tolerance mechanisms in plants: implications for phytoremediation. Environ Rev 24(1):39–51. https://doi.org/10.1139/er-2015-0010
5. Khan MM, Akram MT, Khan MA, Ali I, Qadri RWK, Al-Yahyai R (2025) Application of molecular markers in characterization and conservation of crop plants. In: Molecular markers and crop improvement: principle and application. Springer Nature Switzerland, Cham, pp 81–93. https://doi.org/10.1007/978-3-031-87401-7_7
6. Ahmad A, Wang JD, Pan YB, Sharif R, Gao SJ (2018) Development and use of simple sequence repeats (SSRs) markers for sugarcane breeding and genetic studies. Agronomy 8(11):260. https://doi.org/10.3390/agronomy8110260
7. Cuadrado A, Cardoso M, Jouve N (2008) Physical organisation of simple sequence repeats (SSRs) in Triticeae: structural, functional and evolutionary implications. Cytogenetic Genome Res 120(3-4):210–219. https://doi.org/10.1159/000121069
8. Raza Q, Riaz A, Saher H, Bibi A, Raza MA, Ali SS, Sabar M (2020) Grain Fe and Zn contents linked SSR markers based genetic diversity in rice. PLoS ONE 15(9):e0239739. https://doi.org/10.1371/journal.pone.0239739
9. Aboul-Maaty NAF, Oraby HAS (2019) Extraction of high-quality genomic DNA from different plant orders applying a modified CTAB-based method. Bull Natl Res Cent 43(1):1–10. https://doi.org/10.1186/s42269-019-0066-1
10. Schori M, Appel M, Kitko A, Showalter AM (2013) Engineered DNA polymerase improves PCR results for plastid DNA. Appl Plant Sci 1(2):1200519. https://doi.org/10.3732/apps.1200519
11. Zalapa JE, Cuevas H, Zhu H, Steffan S, Senalik D, Zeldin E, Simon P (2012) Using next-generation sequencing approaches to isolate simple sequence repeat (SSR) loci in the plant sciences. Am J Bot 99(2):193–208. https://doi.org/10.3732/ajb.1100394
12. Sambrook J, Russell DW (2006) Neutral polyacrylamide gel electrophoresis. Cold Spring Harb Protoc 2006(1):pdb-prot4028. https://doi.org/10.1101/pdb.prot100412

13. Karakas E, Ferrante P, Schafleitner R, Giuliano G, Fernie AR, Alseekh S (2023) Plant sample collection and shipment for multi-omic analyses and phytosanitary evaluation. Curr Protoc 3(12):e952. https://doi.org/10.1002/cpz1.952
14. Ravishankar KV, Mani BHR, Anand L, Dinesh MR (2011) Development of new microsatellite markers from mango (Mangifera indica) and cross-species amplification. Am J Bot 98(4):e96–e99. https://doi.org/10.3732/ajb.1000263
15. Wittmeier P, Hummel S (2022) Agarose gel electrophoresis to assess PCR product yield: comparison with spectrophotometry, fluorometry and qPCR. Biotechniques 72(4):155–158. https://doi.org/10.2144/btn-2021-0094
16. Verma H, Borah JL, Sarma RN (2019) Variability assessment for root and drought tolerance traits and genetic diversity analysis of rice germplasm using SSR markers. Sci Rep 9(1):16513. https://doi.org/10.1038/s41598-019-52884-1
17. Vinod KK (2004) Total genomic DNA extraction, quality check and quantitation. Tamil Nadu Agric UnivCoimbatore 14(3):109–121

Check for updates

Chapter 9

RAPD Marker Optimization and Genetic Diversity Assessment in *Averrhoa Genus*

Soumya S.L., Bindu R. Nair, Mukesh Sankar S., Satheesh George and Delse P. Sebastian

Abstract

The genus *Averrhoa* includes economically and nutritionally important fruit crops like *Averrhoa carambola* (star fruit) and *Averrhoa bilimbi* (bilimbi). However, it has not been studied enough genetically, even though it is important for horticulture. For germplasm conservation, cultivar identification, and crop improvement programs, it is important to be able to accurately measure genetic diversity. This chapter elucidates the optimization and utilization of random amplified polymorphic DNA (RAPD) markers for the analysis of genetic diversity in *Averrhoa* species. We used a modified CTAB method to get high-quality genomic DNA from young leaf tissues. Then, we screened primers and set standard PCR conditions to get clear and repeatable amplification profiles. Out of 30 random decamer primers from different RAPD series, we chose the ones that made unique banding patterns for further study. Agarose gel electrophoresis was used to separate the amplified products, and only bands that were consistent and well-defined were scored to make sure the results were reliable. The standardized RAPD protocol described in this chapter offers a swift, economical, and reproducible method for the molecular characterization of Averrhoa germplasm. The methodology can be effectively employed for evaluating genetic variability, enhancing breeding strategies, and bolstering conservation initiatives in underutilized tropical fruit crops.

Keywords RAPD markers, Genetic diversity, Molecular markers, Averrhoa , CTAB method, PCR optimization, Molecular characterization, Polymorphism analysis

1 Introduction

In many developing countries, tropical fruit trees are very important for food security, nutrition, and the livelihoods of people who live in rural areas. They provide important carbohydrates, vitamins, minerals, and bioactive compounds. But their genetic improvement has been slow compared with that of temperate fruit crops and annual field crops. This is because they take a long time to grow, have many different genes, don't get enough research funding, and don't have enough infrastructure. These limitations are especially apparent in underutilized fruit tree species, which frequently remain genetically

Nitish Kumar (ed.), *Molecular Markers in Plants*, Methods in Molecular Biology, vol. 3097,
https://doi.org/10.1007/978-1-0716-5663-1_9,

unexamined despite their considerable nutritional and economic potential. The increasing demand for better fruit quality, longer shelf life, and better fit with modern supply chains shows how important molecular-assisted breeding methods are [1, 2].

Averrhoa, a genus in the family Oxalidaceae, is an example of a tropical fruit tree that isn't used very much. People grow two main types of *Averrhoa*: *Averrhoa carambola* (star fruit) and *Averrhoa bilimbi* (bilimbi). These plants are useful for their nutritional, culinary, and medicinal properties, but scientists still don't know enough about their genetics. The development of better cultivars with desirable agronomic and postharvest traits is limited by a small genetic base and few breeding efforts [3, 4, 5, 6, 7].

Assessing genetic diversity is essential for the efficient use and preservation of underutilized fruit germplasm. Conventional morphological evaluation is frequently affected by environmental variables, rendering molecular marker systems more dependable for ascertaining authentic genetic relationships [2]. Among these, random amplified polymorphic DNA (RAPD) markers are especially useful for species that haven't been studied much because they don't need any genomic information ahead of time and can quickly find polymorphism [8, 9, 10, 11].

This chapter emphasizes RAPD marker optimization and its application in assessing genetic diversity within *Averrhoa*. By strengthening molecular characterization of underutilized fruit trees, breeding programs can broaden the genetic base, enhance cultivar development, and promote sustainable utilization of these valuable yet neglected tropical resources.

2 Materials

2.1 Materials for DNA Extraction

2.1.1 Plant Material

Young, healthy leaf samples of different species and accessions belonging to the genus *Averrhoa* (e.g., *Averrhoa bilimbi* L. and *Averrhoa carambola* L.) were collected from orchards and natural populations. Fresh leaves were collected from mature plants, immediately placed in sterile zip-lock bags containing silica gel for rapid desiccation, and transported to the laboratory for DNA extraction. Dried leaf samples were stored at room temperature in airtight containers until further processing. Alternatively, fresh leaf tissues were stored at −80°C prior to DNA isolation.

2.1.2 Chemicals and Reagents

1. CTAB Extraction Buffer: Cetyl trimethyl ammonium bromide (CTAB, 5%) (*see* **Note 1**), Tris–HCl (1 M, pH 8.0), sodium chloride (5 M NaCl), ethylenediaminetetraacetic acid (0.5 M

Table 1
Composition of the stock solution required for genomic DNA extraction

Sl. No.	Solutions	Amount (for 1 mL extraction buffer)
1	1 M Tris (pH 8)	100 μL
2	5 M NaCl	280 μL
3	5% C-TAB	400 μL
4	0.5 M EDTA (pH 8.0)	40 μL
5	PVP	10 mg
6	Distilled Water	175 μL
7	B-mercapto Ethanol	5 μL

EDTA, pH 8.0), polyvinylpyrrolidone (PVP) (*see* **Note 2**), β-mercapto ethanol (Table 1).

2. Sodium acetate (3 M).
3. Isopropanol is stored at 4°C.
4. Ethanol (80%) stored at 4°C.
5. Chloroform:isoamyl alcohol (24:1) stored at 4°C.
6. Saturated phenol:chloroform:isoamyl alcohol in a ratio of 25:24:1 and stored at 4°C.
7. DNase-free RNase.
8. Sterile distilled water.
9. Liquid nitrogen.

2.1.3 Extraction Buffer Composition (Per 1 mL)

See Table 1.

2.1.4 Consumables and Equipments

The procedure required 1.5 mL microcentrifuge tubes, micropipettes with sterile tips, prechilled mortar and pestle, agarose, ethidium bromide (1 μg/mL), and a 1 kb DNA ladder. Major equipment included a refrigerated centrifuge, laminar airflow chamber, bio spectrophotometer, UV transilluminator, gel documentation system, deep freezer (−70°C), and refrigerator (−20°C).

These materials supported DNA extraction, purification, quantification, and integrity assessment for downstream molecular applications.

2.2 Materials for Primer Selection and RAPD Analysis

The following materials were used for primer screening and RAPD-based amplification studies.

2.2.1 Primers

A total of 30 random decamer primers belonging to the OPA, OPAE, OPAN, OPBA, OPC, OPD, OPN, OPP, OPT, OPX, P, and UBC series were used in the study; these primers were procured from Integrated DNA Technologies, and additional oligonucleotides were synthesized by the Oligonucleotide Synthesis Laboratory at the University of British Columbia.

2.2.2 Reagents for PCR

Taq DNA Polymerase (1 U per reaction), dNTP mixture (200 μM each nucleotide), Thermostable PCR buffer (10 mM Tris–HCl, 1.5 mM $MgCl_2$, 50 mM KCl; pH 8.3), isolated genomic DNA (approximately 50 ng per reaction), sterile double distilled water, all PCR reagents (except primers) procured from Genei Consumables, PCR tubes (0.5 mL), micropipettes and sterile tips, ice box for reaction setup, and tube racks.

2.2.3 Equipment

Thermal cycler (PCR machine), microcentrifuge, vortex mixer, refrigerated storage unit (−20°C), experimental controls, negative control reaction (PCR mixture without template DNA) to monitor contamination (*see* **Note 9**).

These materials were employed to standardize primer concentrations, optimize PCR reaction components, and ensure reliable amplification with distinct and reproducible banding patterns (*see* **Note 7**).

2.3 Materials for Agarose Gel Electrophoresis

The following materials and equipment were used for the separation and visualization of amplified PCR products.

2.3.1 Reagents and Chemicals

Agarose, 1× Tris–Acetate–EDTA (TAE) buffer, ethidium bromide (DNA staining dye), 100 bp DNA ladder (procured from Genei).

2.3.2 Equipment

Horizontal gel electrophoresis unit, UV transilluminator, gel documentation system (Alpha Innotech Corporation), power supply unit, gel casting trays, and combs.

These materials were utilized for electrophoretic separation of amplified DNA fragments, estimation of fragment size using a 100 bp molecular weight marker, and photographic documentation for further analysis.

3 Method

3.1 DNA Extraction

Total genomic DNA was extracted from young leaf tissues using a modified CTAB protocol described by Doyle and Doyle (1987), with suitable refinements. The procedure was carried out as follows.

1. Approximately 100 mg of fresh, young leaf tissue was thoroughly washed, blotted dry, and ground into a fine powder using liquid nitrogen in a prechilled mortar and pestle.
2. The powdered tissue was mixed with 1 mL of prechilled CTAB extraction buffer containing Tris–HCl (pH 8.0),

NaCl, CTAB, EDTA (pH 8.0), PVP, and distilled water. β -mercapto ethanol was added freshly prior to use. The mixture was homogenized to form a uniform slurry.

3. The homogenate was transferred into 1.5 mL microcentrifuge tubes and incubated at 70°C for 1 h to ensure complete lysis of cells and release of nucleic acids.
4. After cooling to room temperature, the samples were centrifuged at 12,000 rpm for 10 min at 4°C using a refrigerated centrifuge (Eppendorf Centrifuge 5804 R). The supernatant was carefully collected.
5. An equal volume of chloroform:isoamyl alcohol (24:1) was added to the supernatant, mixed gently, and centrifuged to separate phases. This step was repeated three times to remove proteins and other contaminants.
6. The aqueous phase was transferred to a fresh tube, and one-tenth volume of 3 M sodium acetate, along with an equal volume of chilled isopropanol (*see* **Note 4**), was added. The mixture was incubated overnight at −20°C to precipitate DNA.
7. DNA was pelleted by centrifugation at 13,000 rpm for 15 min at 4°C. The pellet was washed with 80% ethanol, centrifuged again, air-dried in a laminar airflow chamber, and dissolved in sterile distilled water.
8. DNase-free RNase (*see* **Note 5**) was added to the crude DNA solution and incubated at room temperature for 30 min to remove RNA contamination.
9. The sample was extracted with saturated phenol:chloroform:isoamyl alcohol, followed by centrifugation (*see* **Note 3**). The aqueous layer was further purified with chloroform:isoamyl alcohol (24:1) and centrifuged.
10. DNA was reprecipitated by adding one-tenth volume of 3 M sodium acetate and an equal volume of isopropanol, followed by overnight incubation at −20°C.
11. The DNA pellet was collected by centrifugation, washed with 80% ethanol, air-dried, and finally dissolved in 100 μL of sterile distilled water. The samples were labeled as stock genomic DNA and stored at −70°C until further use.
12. Quantification and quality assessment: DNA concentration and purity (A260/A280 ratio) were determined using a biophotometer (*see* **Note 6**).
13. DNA quality was verified by electrophoresis on 0.8% (w/v) agarose gel prepared in 1× TAE buffer. The gel was stained with ethidium bromide and visualized under UV light using

a gel documentation system (Alpha Innotech Corporation). A 1 kb DNA ladder was used to assess DNA integrity.

14. Working concentrations of 25, 50, and 75 ng/μL were prepared and evaluated to obtain clear and reproducible amplification patterns for downstream applications.

3.2 RAPD Primer Screening and PCR

An initial set of thirty arbitrary decamer primers representing the OPA, OPAE, OPAN, OPBA, OPC, OPD, OPN, OPP, OPT, OPX, P, and UBC series was evaluated for suitability in the study (Table 2). Preliminary screening was carried out using representative DNA samples from each species to identify primers capable of producing clear and reproducible amplification patterns. To optimize amplification efficiency and band clarity, three primer concentrations, such as 0.1, 0.3, and 0.5 μM were tested before selecting the most effective concentration for further analysis.

For RAPD amplification, PCR was performed in a total reaction volume of 25 μL. The reaction mixture contained template DNA, appropriate PCR master mix components, and a selected decamer primer. Amplification was carried out in a thermal cycler (Eppendorf Mastercycler Gradient). The thermal cycling program began with an initial denaturation step at 95°C for 4 min. This was followed by 35 cycles consisting of denaturation at 94°C for 30 s, annealing at 37°C for 1 min (*see* **Note 8**), and extension at 72°C for 2 min. A final extension step was performed at 72°C for 5 min to ensure complete synthesis of amplified fragments.

Each amplification reaction was conducted at least twice to confirm reproducibility. Only clear, intense, and repeatable bands were considered for scoring and further interpretation. After amplification, PCR products were properly labeled and preserved at −20°C until electrophoretic separation.

The PCR amplicons were separated on a 1.3% agarose gel (*see* **Note 10**) prepared by dissolving 1.3 g of agarose in 100 mL of electrophoresis running buffer. Ethidium bromide (1 μL of a 10 mg/mL stock solution) was incorporated into the gel for DNA staining. Prior to loading, amplified samples were mixed with a suitable gel loading buffer containing tracking dyes to facilitate monitoring of migration during electrophoresis (*see* **Note 11**). Following electrophoretic separation, the DNA bands were examined and documented using a gel documentation system (Fig. 1).

Table 2
Primer information of selected primers

Sl. No.	Primer designation	Primer Sequence (5′→3′)	Presence/absence of bands
1	OPA-10	GTGATCGCAG	+
2	OPA-18	AGGTGACCGT	+
3	OPA-07	GAAACGGGTG	−
4	OPAE-03	CATAGAGCGG	+
5	OPAE-15	TGCCTGGACC	+
6	OPAN-01	ACTCCACGTC	+
7	OPB-10	CTGCTGGGAC	−
8	OPBA-14	TCGGGAGTGG	+
9	OPC-11	AAAGCTGCGG	+
10	OPD-07	GGCGGACTGT	+
11	OPD-14	CTTCCCCAAG	+
12	OPN-06	GAGACGCACA	+
13	OPN-09	TGCCGGCTTG	−
14	OPP-01	GTAGCACTCC	+
15	OPT-01	GGGCCACTCA	+
16	OPW-03	GTCCGGAGTG	−
17	OPX-17	GACACGGACC	+
18	OPX-18	GACTAGGTGG	−
19	P-1	AAGAGTTTGATCCTGGCT	+
20	P-4	GTAATACGACTCACTATAG	+
21	P-7	CGTCCTTCATCGGCTCTT	+
22	SF-04	GGTGATCAGG	+
23	SK-03	CCAGCTTAGG	+
24	UBC-210	CCGGGGTTTT	+
25	UBC-292	TGCCGAGCTG	+
26	UBC-465	AGCTGAAGAG	+
27	UBC-509	CTTCCCCAAG	+
28	UBC-514	AAGGCCCCTG	−
29	UBC-54	CCTGATTGCC	+
30	UBC-620	CCTTTCTCCC	−

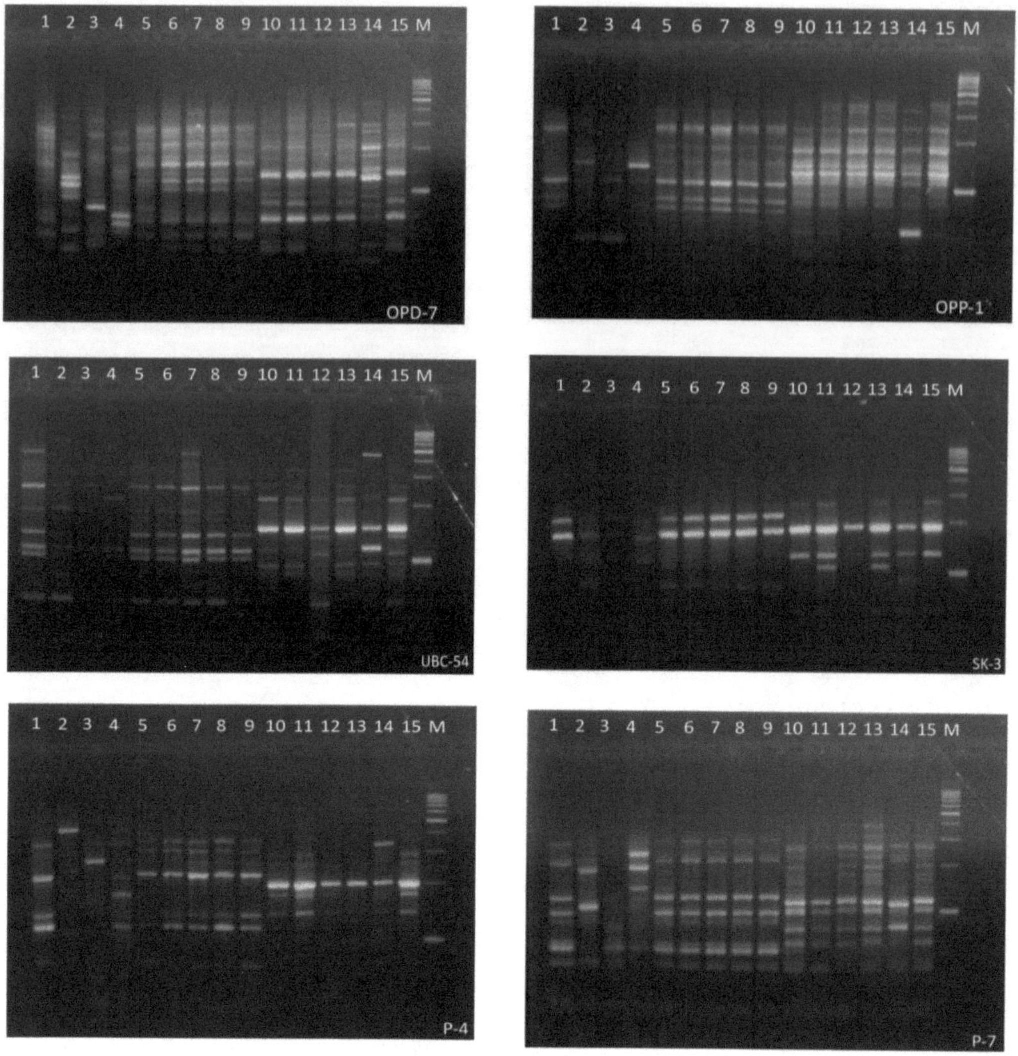

Fig. 1 RAPD amplification profiles of fifteen collections (lanes 1–15) of *Averrhoa bilimbi* and *Averrhoa carambola* were generated using primers OPD-7, OPP-1, UBC-54, SK-3, P-4, and P-7. Lane M represents the 1.5 kb DNA ladder used as a molecular size marker

4 Notes

1. CTAB Extraction Buffer should be dissolved at approximately 60°C–65°C to ensure complete solubilization. β-mercapto ethanol must be added freshly to the extraction buffer just before use to prevent oxidation of phenolic compounds.
2. Species of the genus *Averrhoa* may contain polyphenols and polysaccharides. The addition of PVP binds phenolic

compounds, while β-mercaptoethanol reduces oxidative damage, thereby improving DNA purity.

3. All centrifugation steps should be performed at 4°C to prevent enzymatic degradation of DNA. Care must be taken while transferring the aqueous phase to avoid disturbing the interphase.
4. Isopropanol used for precipitation should be prechilled. Overnight incubation at −20°C improves DNA yield, especially when working with plant tissues rich in secondary metabolites.
5. RNase digestion is essential to remove RNA contamination, which can otherwise interfere with accurate DNA quantification and PCR amplification.
6. A spectrophotometric A260/A280 ratio between 1.8 and 2.0 indicates acceptable DNA purity. Ratios below 1.8 may indicate protein contamination, while higher values may suggest RNA contamination.
7. RAPD reactions are highly sensitive to changes in template concentration, primer concentration, $MgCl_2$ levels, and thermal cycling conditions. Therefore, all reaction components should be standardized and prepared on ice to minimize nonspecific amplification.
8. Low annealing temperatures (37°C) are characteristic of RAPD assays and promote binding of arbitrary primers. However, strict consistency in thermal cycling parameters is necessary to ensure reproducibility.
9. A PCR reaction without template DNA must be included in each run to monitor contamination or nonspecific amplification.
10. A 1.3% agarose gel provides good resolution for RAPD fragments typically ranging between 200 and 3000 bp. Gel concentration may be adjusted depending on fragment size distribution.
11. Amplified products should be stored at −20°C until electrophoresis. Prolonged storage may reduce band intensity due to degradation.

Acknowledgments

The authors gratefully acknowledge St. Joseph's College (Autonomous), Devagiri, Calicut, and the University of Calicut, Malappuram, for institutional support and encouragement. The authors thank the Head, Department of Botany, University of Kerala, Kariavattom,

Thiruvananthapuram, for the research facilities provided. The authors extend heartfelt thanks to Kerala State Council for Science Technology Environment (KSCSTE), Government of Kerala, India, for financial support.

References

1. Rafalski JA, Tingey SV (1993) Genetic diagnostics in plant breeding: RAPDs, microsatellites and machines. Trends Genet 9(8):275–280. https://doi.org/10.1016/0168-9525(93)90013-8
2. Govindaraj MMVG, Vetriventhan M, Srinivasan M (2015) Importance of genetic diversity assessment in crop plants and its recent advances: an overview of its analytical perspectives. Genet Res Int 2015(1):431487. https://doi.org/10.1155/2015/431487
3. Soumya SL, Nair BR (2018) Genetic Fingerprinting of Two Species of Averrhoa Using RAPD and SRAP Markers. Agric Res 7:402–415. https://doi.org/10.1007/s40003-018-0339-1
4. Williams JG, Kubelik AR, Livak KJ, Rafalski JA, Tingey SV (1990) DNA polymorphisms amplified by arbitrary primers are useful as genetic markers. Nucleic Acids Res 18(22):6531–6535. https://doi.org/10.1093/nar/18.22.6531
5. Welsh J, McClelland M (1990) Fingerprinting genomes using PCR with arbitrary primers. Nucleic Acids Res 18(24):7213–7218. https://doi.org/10.1093/nar/18.24.7213
6. Doyle JJ, Doyle JL (1987) A rapid DNA isolation procedure for small quantities of fresh leaf tissue. Phytochem Bull 19:11–15.
7. Soumya SL, Nair BR (2017) Internal transcribed spacer (ITS) sequence analysis of nuclear ribosomal DNA (nrDNA) in *Averrhoa* L. Int J Curr Res 9(2):45353–45359.
8. Soumya SL, Nair BR (2017) Internal transcribed spacer (ITS) sequence analysis of nuclear ribosomal DNA (nrDNA) in Averrhoa L. Int J Curr Res 9(2):45353–45359.
9. Soumya SL, Nair BR (2016) Assessment of heavy metals in *Averrhoa bilimbi* and *A. carambola* fruit samples at two developmental stages. Environ monit assess 188(5):291. https://doi.org/10.1007/s10661-016-5298-z
10. Soumya SL, Nair BR (2013) Floral morphological features and variability in two species of *Averrhoa* L.(Oxalidaceae).
11. Soumya SL, Nair BR (2014) Changes in the biochemical profile of fruits of two species of *Averrhoa* during development. Int J Pharm Pharm Sci 6(4):572–577.

Chapter 10

Development and Validation of SNP Markers for Plant Genomic Studies

Kannan Shrinithi and Tennyson Jebasingh

Abstract

One of the most useful techniques for studying genetic variations in plants is single-nucleotide polymorphisms (SNPs), which are useful in developing molecular markers for genetic research. In this chapter, the development and validation of SNP markers using targeted resequencing and in silico analysis of genomic resources for plant genomic research and breeding are included. Key steps include genomic DNA isolation, whole-genome resequencing, raw read processing, reference-based alignment, SNP detection, and stringent filtering to identify high-confidence variants. The importance of cross-platform validation, experimental genotyping, and array evaluation is also discussed. Overall, standardized SNP development pipelines provide a robust framework for genetic diversity analysis, linkage mapping, association studies, and marker-assisted selection, thereby accelerating crop improvement programs.

Keywords Single-nucleotide polymorphisms, Genomic DNA isolation, SNP development pipelines

1 Introduction

The main goal of genetic crop improvement is to enhance economic traits like yield and stress resistance through the selection of superior plants. Conventional phenotypic selection relies on morphological traits, which can be misleading due to genetic complexity and environmental factors. This method is subjective and time-consuming, requiring extensive field testing. In contrast, biochemical markers such as proteins and isozymes, which reflect genetic variation more reliably, offer a more stable alternative for early identification and selection of genotypes. However, these markers also face limitations like restricted coverage and low variability, which affect their effectiveness in plant breeding [1].

DNA markers, which provide higher information content and reliability than biochemical markers, have revolutionized molecular biology over the last three decades. By analyzing DNA variations across nuclear and organellar genomes (including chloroplast DNA),

Nitish Kumar (ed.), *Molecular Markers in Plants*, Methods in Molecular Biology, vol. 3097,
https://doi.org/10.1007/978-1-0716-5663-1_10,

they provide valuable insights into the genetic makeup and diversity of organisms. They fall into one of three classes: (i) Hybridization-based markers (e.g., restriction fragment length polymorphism (RFLPs)) that reveal polymorphism but are labor-intensive and require a large amount of DNA; (ii) PCR-based markers (e.g., random amplified polymorphic DNA (RAPDs)) that are inexpensive and do not require prior sequence information, though they are less reproducible; and (iii) markers that combine RFLP and PCR principles (e.g., simple sequence repeats (SSRs), amplified fragment length polymorphisms (AFLPs)), such as simple sequence repeats (SSRs) and amplified fragment length polymorphisms (AFLPs), which offer high resolution and polymorphism but are expensive and time-consuming because of the complex assay requirements. DNA markers are essential to marker assisted selection (MAS) in plant breeding, despite their limitations [1].

Because of their high informational content and polymorphic nature, single-nucleotide polymorphisms (SNPs) have emerged as important molecular markers in genomics. Even though SNP analysis has upfront costs, its automation potential and lower time requirements make it more cost-effective. SNPs are defined by a minimum allele frequency of 1% in genomic DNA, where alleles vary within a population. Compared to more complex markers like microsatellites, SNPs have a lower information density at each site because they are primarily biallelic and offer numerous analysis loci. Their low rates of mutation guarantee evolutionary stability, which is advantageous for studies into being predominantly biallelic [2]. SNPs provide many analysis loci, offsetting the lower information density at each site compared to more complex markers like microsatellites. Their low mutation rates ensure evolutionary stability, beneficial for research into complex traits and population genetics.

In maize, there is one SNP (single-nucleotide polymorphism) per 70 base pairs, whereas in wheat, one SNP per 20 base pairs in certain regions. The density of SNPs in maize enhances its utility for genetic mapping and trait-association studies, and barley offers utility with about one SNP per 189 base pairs for accurate cultivar identification. Although soybeans have lower polymorphism, specific genomic regions remain valuable for SNP discovery. Other crops, like grapevine and sunflower, benefit significantly from SNP analyses, whereas tomato faces challenges due to its lower polymorphism levels. Overall, SNP markers are becoming vital in genomic studies and crop improvement [3].

Various genotyping assays for SNPs, pivotal in crop genetics, include direct sequencing, EST database mining, cleavage assays, and molecular beacons. The selection of the assay is guided by study objectives and resource availability. These assays employ detection methods from traditional gel-based systems to advanced high-throughput technologies like microarrays, mass spectrometry, and flow cytometry.

SNPs have significant applications in crop genetics, including association studies, economic gene tagging, genotyping, and diversity analysis among plant species [4].

2 Materials

2.1 Plant Material and DNA Isolation

1. Extraction buffer (3×): 3% CTAB (w/v), 1.4 M NaCl, 0.8 M Tris–HCl, pH 8.0, 0.5 M EDTA, pH 8.0 (*see* **Note 1**).
2. 0.3% 2-β-Mercaptoethanol.
3. Chloroform:isoamyl alcohol (24:1 v/v).
4. 6 M NaCl.
5. 3 M potassium acetate.
6. Ice cold 100% isopropyl alcohol.
7. 70% ethanol.
8. 1× TE buffer (10 mM Tris–HCl, pH 8.0; 1 mM EDTA, pH 8.0) (*see* **Note 1**).
9. Agarose.

2.2 SNP Pipelines Software and Tools

1. FastQC: It is used to assess the quality of raw sequencing reads by detecting issues such as low base quality, GC bias, adapter contamination, and overrepresented sequences.
2. MultiQC: It is used to compile and summarize results from multiple quality control tools into a single combined report for easy comparison across samples.
3. Trimmomatic: It is used to trim adapter sequences and remove low-quality bases or reads from raw NGS data to improve sequencing accuracy.
4. Fastp: It is used for fast quality filtering, adapter trimming, and preprocessing of NGS reads while also generating quality control reports.
5. BWA: Burrows–Wheeler Aligner is a widely used bioinformatics tool for aligning short sequencing reads to a reference genome.
6. SAMtools: It is used to process SAM/BAM alignment files by converting, sorting, indexing, and filtering mapped sequencing reads for downstream analysis.
7. GATK: Genome Analysis Toolkit is used for accurate variant discovery and genotyping, including SNP and indel calling, from high-throughput sequencing data.
8. VCFtools: It is used to filter, analyze, and manipulate variant data stored in VCF files for population genetics and SNP analysis.

9. NCBI BLAST: It is used to compare nucleotide or protein sequences against databases to identify sequence similarity, gene identity, and functional annotation.

3 Methods

Different varieties of plant samples, such as founder lines, intermediate ancestors, and important breeding parents, were collected. These types of varieties will cover maximum genetic diversity and help to discover useful breeding and mapping of SNPs. Steps involved in SNP marker development and validation are presented in Fig. 1.

3.1 DNA Extraction

1. Preheat the extraction buffer in a water bath at 65°C.
2. Add 0.3% 2-β-mercaptoethanol to the buffer just before the extraction.
3. Use liquid nitrogen and a prechilled motor and pestle to grind 50 mg of the plant sample, and add 800 μl of the preheated extraction buffer to the sample and swirl gently.
4. Collect the sample in a 2-ml microcentrifuge tube, incubate in a water bath at 65°C for 1 h (*see* **Note 2**).
5. Add an equal volume of chloroform:isoamyl alcohol (24:1 v/v) and mix.
6. Centrifuge at 12,000 rpm for 15 min at room temperature (RT).
7. Transfer the aqueous phase to a new 1.5-ml Eppendorf tube.
8. Repeat the extraction Steps (5–7) when necessary, until the upper aqueous phase is clear.
9. Then add half the volume of 6 M NaCl and mix well. Add 1/10 the volume of 3 M potassium acetate and simultaneously mix with 500 μl ice cold 100% isopropyl alcohol (2/3 volume). Invert gently to precipitate DNA threads.
10. Incubate at −20°C for 30 min.
11. Centrifuge at 12,000 rpm for 5 min, discard supernatant (*see* **Note 3**).
12. Wash the DNA pellet with 70% ethanol and centrifuge at 13,000 rpm for 5 min.
13. Discard 70% alcohol (*see* **Note 3**) and air dry at room temperature for 15 min.
14. Resuspend the DNA pellet in 25 μl 1× TE buffer and store at −20°C till further use.

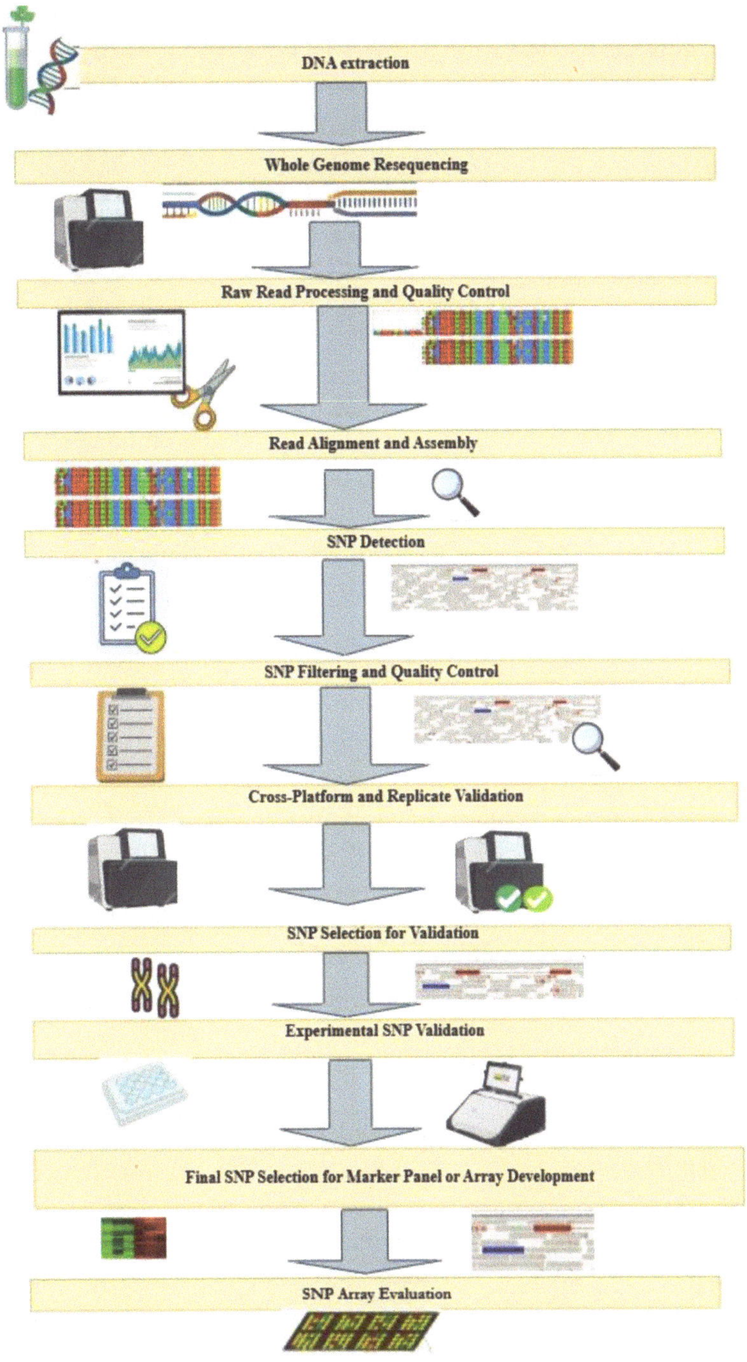

Fig. 1 Steps involved in SNP marker development and validation

15. DNA quality is assessed by 0.8% agarose gel electrophoresis. The concentration and purity can be measured using a NanoDrop spectrophotometer [5] (*see* **Note** 4).

3.2 Identification of SNP by Whole-Genome Resequencing

1. The individual biological samples have to be chosen based on available pedigree information and prior estimates of genetic relatedness to maximize genetic diversity within the panel.
2. To enable cost-effective SNP discovery, the individual biological samples are organized into multiple pools for low-depth whole-genome resequencing.
3. For each accession, high-quality genomic DNA is used for sequencing library construction. Briefly, 1–3 μg of genomic DNA is fragmented by sonication to generate DNA fragments of suitable size for high-throughput sequencing.
4. Fragmented DNA is subjected to end-repair and 3′ adenylation, followed by ligation of platform-specific sequencing adapters containing sample-specific index sequences.
5. Following adapter ligation, libraries have to be amplified by limited-cycle PCR (8–12 cycles, depending on DNA input) to enrich for fragments containing adapters at both ends.
6. Library concentration and fragment size distribution are assessed using spectrophotometric and microfluidic electrophoresis-based quality control methods.
7. Libraries are normalized and pooled in equimolar proportions to generate pooled libraries for sequencing.
8. Cluster generation and sequencing are performed according to the manufacturer's recommendations on a high-throughput sequencing platform.
9. Sequencing is carried out using paired-end or single-end reads with read lengths appropriate for the platform and experimental design (*see* **Note 5**).
10. For a subset of samples, alternative sequencing platforms and library preparation strategies may be employed to increase read length or sequencing depth where it is required [6].

3.3 Processing of Raw Reads and Quality Control

1. Filtered reads are aligned to the species-specific reference genome assembly using suitable short-read alignment tools.
2. Raw sequencing data has to be processed using standard base-calling and demultiplexing software to generate high-quality sequence reads for each accession.
3. Sequence reads are subjected to quality filtering and adapter removal using FastQC-guided trimming with Trimmomatic, followed by reporting of summary with MultiQC (*see* **Notes 6–9**).
4. Reads failing to meet minimum quality score and length thresholds need to be discarded prior to downstream analysis [7].

3.4 Read Alignment and Assembly

1. Quality-filtered reads aligned to the appropriate species-specific reference genome assembly using BWA–MEM.
2. Alignment parameters are optimized to maximize mapping accuracy while minimizing mismatches and ambiguous alignments, and it will be aligned to the indexed reference genome using the BWA–MEM algorithm with default or optimized parameters depending on genome characteristics (*see* **Note 10**).
3. Alignment output files in SAM format are converted to BAM format using SAMtools, and the alignment quality metrics, including mapping rate, properly paired reads, insert size distribution, and mapping quality scores, can be assessed (*see* **Notes 11 and 12**).
4. PCR duplicates generated during library preparations are removed using tools such as Picard Mark Duplicates to reduce biases in variant calling.
5. In cases where a complete reference genome assembly is not available, de novo assembly or reference-guided assembly strategies are employed prior to downstream variant analysis.
6. Final processed, sorted, duplicate-marked, and indexed BAM files are generated and used as input for subsequent SNP detection and variant calling [7].

3.5 Detection of SNP

1. SNP detection is performed using Genome Analysis Toolkit (GATK) HaplotypeCaller, a variant calling software to generate raw variant call format (VCF) files for each accession.
2. Variant calling is conducted using default parameters, with modifications to the minimum base quality score and read depth thresholds as required (*see* **Note 13**).
3. Candidate SNPs are required to meet minimum criteria for base quality, read depth, and allele balance.
4. Putative SNPs located in repetitive or low-complexity genomic regions are identified and removed using repeat annotation and custom filtering scripts.
5. Multiallelic variant sites and indel polymorphisms are separated from biallelic SNPs for downstream SNP-specific analyses.
6. Filtered, high-confidence SNPs are retained in standardized VCF format for subsequent SNP filtering, validation, and downstream genomic analyses [8].

3.6 Filtering of SNP and Quality Control

1. SNPs are filtered to retain only those supported by a minimum number of sequencing reads per allele.
2. SNPs with excessively high read depth, indicative of repetitive regions or copy number variation, are excluded.

3. SNPs are retained if the minor allele is supported by a minimum number of independent reads.
4. SNPs located in genomic regions with high copy number or ambiguous mapping are excluded from further analysis.
5. Filtered SNPs passing all quality criteria are designated as high-confidence candidate SNPs for downstream validation [8].

3.7 Cross-Platform and Replicate Validation

1. For accessions, sequences from different platforms or from different laboratories, SNP calls are compared to confirm reproducibility and reduce technical bias (*see* **Note 14**).
2. SNPs showing consistent genotype calls across independent datasets are prioritized as reliable markers [9].

3.8 SNP Selection for Validation

1. A representative subset of candidate SNPs is selected for experimental validation based on genomic distribution, predicted functional location (genic or intergenic), and minor allele frequency.
2. SNPs are selected to ensure uniform physical spacing across chromosomes and to include regions associated with traits of interest.
3. Additional SNPs are selected within known quantitative trait loci or candidate gene regions where is applicable [9].

3.9 Experimental SNP Validation

1. A diverse validation panel of plant accessions, including cultivars, breeding lines, and related species, is assembled for SNP validation.
2. Genomic DNA is extracted from validation panel samples and quantified prior to genotyping.
3. Selected SNPs are genotyped using a suitable high-throughput SNP genotyping assay platform following the manufacturer's protocols.
4. Genotype calling and cluster analysis are performed using dedicated genotyping analysis software.
5. SNPs are evaluated for call rate, clustering quality, and consistency across samples [10].

3.10 Final SNP Selection for Marker Panel or Array Development

1. Candidate SNPs are converted to platform-specific assay design formats and evaluated for technical suitability.
2. SNPs with poor assay design scores or incompatible nucleotide combinations are excluded.
3. Redundant SNPs and duplicated probe sequences are removed.
4. SNPs are filtered to retain those located on well-assembled chromosomes and within annotated genomic regions.

5. Pre-validated SNPs from external sources or prior studies are incorporated after applying comparable quality filters.
6. Final SNPs are selected to achieve uniform genome-wide coverage and appropriate marker density [10].

3.11 SNP Array Evaluation

1. The final SNP marker set is evaluated using independent validation panels representing diverse genetic backgrounds.
2. Genomic DNA extraction and quantification for evaluation panels are performed using standardized protocols.
3. Genotyping is conducted using the selected SNP array or high-throughput genotyping platform following manufacturer-recommended procedures.
4. Genotype clustering, quality scoring, and call rate assessment are performed using dedicated analysis software.
5. SNPs failing to meet predefined quality thresholds for clustering, call rate, or reproducibility are excluded.
6. The final validated SNP dataset is used for downstream genetic analysis, including diversity assessment, linkage mapping, and association studies [10].

4 Notes

1. Extraction and TE buffers for DNA extraction should be autoclaved and stored at room temperature.
2. Every 20 min, mix the sample by inverting the tube and then cool it down to room temperature.
3. Completely drain off the supernatant.
4. Working concentration of samples should be 20–50 ng/μL for downstream processes.
5. Moderate to high depth (10×–30×) for accurate SNP discovery.
6. To assess base quality distribution, GC content, adapter contamination, and sequence duplication levels, perform FastQC before and after trimming.
7. To remove low-quality bases (e.g., sliding window trimming with a minimum average Phred quality score of Q20) and to discard reads shorter than a minimum length (e.g., 50 bp) after trimming, and use trimmomatic parameters to optimize.
8. Adapter sequences specific to the sequencing platform are provided to Trimmomatic to ensure accurate adapter removal.

9. MultiQC is used to aggregate FastQC and trimming reports across all samples for efficient project-level quality assessment. The software tools should all be consistent versions for reproducibility of quality control results.
10. Alignment parameters such as mismatch penalties, seed length, and clipping penalties are adjusted to accommodate species-specific genome characteristics and sequencing error profiles.
11. Mapping quality (MAPQ) scores are assessed to filter out low-confidence alignments before SNP detection.
12. Genome coverage and alignment statistics are evaluated to ensure sufficient representation of the genome for reliable SNP discovery.
13. Preliminary filtering is to remove low-quality variants based on variant quality score, mapping quality, and strand bias metrics for raw variant calling.
14. SNPs with inconsistent genotype calls across platforms or replicates are flagged and removed from further consideration.

References

1. O'Neill R, Snowdon R, Köhler W (2003) Population genetics: aspects of biodiversity. Progress in botany: genetics physiology systematics ecology. Springer, Berlin, Heidelberg, pp 115–137
2. Sachidanandam R et al International SNP Map Working Group Cold Spring Harbor Laboratories (2001) A map of human genome sequence variation containing 1.42 million single nucleotide polymorphisms. Nature 409(6822):928–933
3. Bhattramakki D, Ching A, Morgante M, Dolan M, Register J, Smith H et al (2000) Conserved single nucleotide polymorphism (SNP) haplotypes in maize. In: Plant and animal genome VIII conference, pp 9–12
4. Tabassum Jehan TJ, Suman Lakhanpaul SL (2006) Single nucleotide polymorphism (SNP)—methods and applications in plant genetics: a review. Indian J Biotechnol 5(4): 435.
5. Aboul-Maaty NAF, Oraby HAS (2019) Extraction of high-quality genomic DNA from different plant orders applying a modified CTAB-based method. Bull Natl Res Cent 43(1):1–10
6. Verde I, Bassil N, Scalabrin S, Gilmore B, Lawley CT, Gasic K et al (2012) Development and evaluation of a 9K SNP array for peach by internationally coordinated SNP detection and validation in breeding germplasm. PLoS ONE 7(4):e35668
7. Mehravi S, Ranjbar GA, Mirzaghaderi G, Severn-Ellis AA, Scheben A, Edwards D, Batley J (2021) De Novo SNP discovery and genotyping of Iranian Pimpinella species using ddRAD sequencing. Agronomy 11(7):1342
8. McKenna A, Hanna M, Banks E, Sivachenko A, Cibulskis K, Kernytsky A et al (2010) The Genome Analysis Toolkit: a MapReduce framework for analyzing next-generation DNA sequencing data. Genome Res 20(9):1297–1303
9. Shen Y, Wan Z, Coarfa C, Drabek R, Chen L, Ostrowski EA et al (2010) A SNP discovery method to assess variant allele probability from next-generation resequencing data. Genome Res 20(2):273–280
10. Marth GT, Korf I, Yandell MD, Yeh RT, Gu Z, Zakeri H, Stitziel NO, Hillier LD, Kwok P-Y, Gish WR (1999) A general approach to single-nucleotide polymorphism discovery. Nat Genet 23(4):452–456

Chapter 11

Exploring the Role of Single Nucleotide Polymorphisms in Modern Plant Genomic Research

Ravimurugan Subashini and Tennyson Jebasingh

Abstract

Single-nucleotide polymorphisms (SNPs) represent the most frequent type of genetic variation in plant genomes. Recently, SNPs have become essential markers for various studies conducted using plant genomes. With recent advances in high-throughput genomic sequencing techniques as well as advanced genotyping platforms, efficient strategies have emerged for discovering, validating, and using SNPs for genomic analysis and improvement of numerous plant species. This chapter aims to offer an overview of methods used for conducting studies using SNPs for plant species, particularly providing procedures for exploring SNPs from plant genomes via genomics sequencing techniques. Additionally, strategies for SNP genotyping, quality assessment, and analysis approaches, including population genomics, genome-wide association studies, and genomic prediction, are also discussed. This chapter will enable plant geneticists, molecular breeders, and genomics researchers to have a complete guide for using SNPs for various applications for different plant species.

Keywords Single nucleotide polymorphisms, Plant genomics, Next-generation sequencing, Marker-assisted selection, Crop improvement

1 Introduction

Molecular markers have brought a revolution in the genetic analysis of plants and have enabled researchers to use molecular breeding on a large scale. In the past three decades, there has been a remarkable development in marker systems and platforms for their identification. Molecular markers have made it easier to detect allelic variation in the genome of a species [1, 2]. A genetic marker can be defined as a variation that might occur due to a mutation or change in the observable genetic locus. A brief DNA sequence, like the region around a single base pair change, is called a single nucleotide polymorphism (SNP), and it serves as an effective genetic marker for genomic studies [3]. DNA markers have many applications in plant molecular genetic studies [4]. SNPs exist as single-nucleotide variations that can be observed in the genomic sequences of different individuals within a

Nitish Kumar (ed.), *Molecular Markers in Plants*, Methods in Molecular Biology, vol. 3097,
https://doi.org/10.1007/978-1-0716-5663-1_11,

specific population. The most common molecular markers in biology are SNPs, which demonstrate different distribution patterns across various species. SNPs represent the basic form of genetic variation, which are used to identify genetic markers that serve essential functions in genetic mapping and marker-assisted breeding, map-based cloning [5]. SNPs are essential markers for plant breeding and crop improvement since they are the most prevalent type of genetic variation found in crop genomes [6]. Methods such as next-generation sequencing (NGS), genotyping-by-sequencing (GBS), chip-based NGS, and allele-specific PCR are some of the most recently developed high-throughput genotyping technologies that have been used to identify very informative SNPs by employing a genome-wide coverage. In this context, SNP discovery can be defined as the process by which next-generation sequencing data is mapped to genome reference sequences and then used to identify base variations. Currently, many SNPs techniques are being developed and aimed to achieve an instant, large-scale analysis. Most of such SNP technologies, like the conventional ones, still require a marker-specific amplification reaction, marker-specific oligonucleotide primers, or the use of probes such as Taqman or Molecular Beacon and sequencing, but SNP markers are biallelic and show genetic stability over many generations. Additionally, SNP markers are automatable, high-throughput, and therefore a very popular and convenient technique for molecular plant breeding [7]. SNPs are widely applied in plants for the analysis of genetic resources, genetic diversity, genetic diagnostics, phylogenetic studies, association mapping using disequilibrium, and identification of varieties/cultivars, among other applications of genetic markers. Recent developments in high-throughput genotyping methods, such as allele-specific PCR, genotyping-by-sequencing (GBS), chip-based next-generation sequencing (NGS), and NGS, have made SNPs the marker of choice for genotyping [8]. Some of the fundamental uses of molecular markers, particularly SNPs in plants, include genetic mapping, comparative mapping, framework mapping, region-specific marker saturation, map-based gene cloning, bulk segregant analysis, marker-assisted selection, varietal/line identification, genetic diversity studies, and novel allele detection [9]. In the case of SNPs, plant biologists could benefit from the investment in technology development that has already been made by the broader community. Much progress has been made in the areas of SNP discovery, SNP assay development, and the use of haplotype diversity for association mapping. On the other hand, plant genomes have some unique advantages for researchers using SNP-based technologies [10]. Consequently, the discovery of SNPs has become increasingly dependent on computer-based methods, which have been driven by the rapid accumulation of publicly available genomic information. Nevertheless, the complexity of plant genomes creates difficulties in the discovery of informative SNPs, requiring different approaches for their discovery in specific crops. The cost of genotyping remains an important consideration in the adoption of markers, especially in breeding programs that require the genotyping of large segregating populations. The advent of high-throughput genotyping platforms and technologies has made the construction of dense genetic linkage maps easier, thus improving the use of SNPs in marker-assisted breeding. NGS has also improved the speed, efficiency, and economy of SNP discovery and genotyping. NGS-based methods have been extremely useful for the identification, validation, and use of genome-wide SNP markers, as well as the development of array-based genotyping platforms in a variety of plant species [11]. In this chapter, the conceptual framework, computational infrastructure, and analytical approaches used in genome-wide SNP discovery, filtering, annotation, and interpretation in plant systems are discussed. Special attention is given to the integration of SNP information with population genomics, genetic diversity analysis, genome-wide association analysis, and genomic selection approaches. Overview of the workflow for SNP-based plant genomic analysis, highlighting preparation of genomic data, SNP detection, SNP annotation, and applications, including population genomics, genome-wide association analysis, and predictions of breeding, is described in Fig. 1.

2 Materials

2.1 Plant Genomic Datasets

High-quality paired-end or single-end FASTQ files generated either from publicly available repositories (e.g., NCBI SRA, and ENA) or generated from genomic DNA extracted from experimental plant

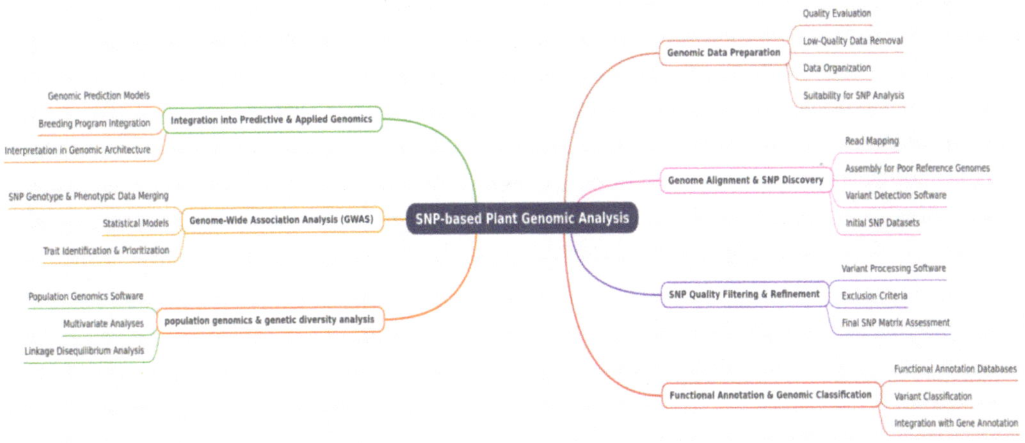

Fig. 1 Overview of the workflow for SNP-based plant genomic analysis, highlighting preparation of genomic data, SNP detection, SNP annotation, and applications, including population genomics, genome-wide association analysis, and predictions of breeding

accessions through whole-genome sequencing, reduced-representation sequencing like GBS, or transcriptome sequencing data with minimum recommended coverage of ≥10× for diploid species are needed to facilitate accurate SNP identification and subsequent genomic analysis.

2.2 Reference Genome Resources

Publicly available, curated genome assemblies retrieved from genome repositories such as NCBI Genome, Ensembl Plants, Phytozome, Gramene, or species-specific genome databases, in the form of FASTA-formatted genome sequence files together with gene annotation files in GFF3 or GTF format for precise SNP detection and functional annotation.

2.3 Computational Infrastructure

High-performance computing system or workstation running a Linux-based operating system, with ≥16 GB RAM, multicore processor, ≥1 TB storage capacity, and installed bioinformatics software to process large-scale genomic data.

2.4 Sequence Quality Control Tools

Software packages like FastQC (v0.11.9 or later) and Trimmomatic (v0.39 or later) are used for evaluating and filtering raw sequencing reads according to base quality scores, read length, and adapter contamination before downstream analysis.

2.5 Read Alignment and Variant Detection Software

Bioinformatics software like BWA–MEM (v0.7.17 or later), Bowtie2 (v2.5 or later), or HISAT2 (v2.2 or later) for aligning sequence reads to reference genomes or assemblies and for detecting single-nucleotide polymorphisms across the genome, software like SAMtools (v1.17 or later), GATK (v4 or later), or FreeBayes (v1.3 or later) are needed.

2.6 Variant Filtering and Data Processing Tools

Computational tools such as VCFtools (v0.1.16 or later), PLINK (v1.9 or later), for filtering SNP datasets according to read depth, genotype quality, minor allele frequency, and missing data values to produce high-confidence variant matrices.

2.7 Functional Annotation Resources

Gene annotation databases and variant effect prediction software like SnpEff (v5.0 or later), ANNOVAR, or Ensembl Variant Effect Predictor (VEP) for assigning SNPs based on their genomic position and putative functional significance.

2.8 Population Genomics and Statistical Analysis Software

Software such as STRUCTURE (v2.3.4), ADMIXTURE (v1.3), and TASSEL (v5.0 or later) for the analysis of genetic diversity, population structure, linkage disequilibrium, and relatedness in plant accessions.

2.9 Phenotypic Datasets

Trait data for genotyped plant populations, formatted for integration with SNP data in association and prediction analyses. Structured trait data file in CSV or tab-delimited format containing sample identifiers corresponding to the genotype matrix.

2.10 Genome-Wide Association and Genomic Prediction Frameworks

Software like GEMMA, MLMM, rrBLUP, BLUPF90, or BGLR for genome-wide association studies and genomic selection analyses using SNP data.

3 Methods

3.1 Preparation of Genomic Data for SNP Analysis

1. Plant genomic sequencing data are initially evaluated for quality using sequence quality control tools (e.g., FastQC and Trimmomatic).
2. Low-quality reads identified by FastQC as having poor per-base sequence quality (Phred score $<$ 20), adapter contamination, or overrepresented sequences should be removed, and reads should be trimmed using Trimmomatic with a sliding window approach of four bases with an average Phred score cutoff of 20, removing leading and trailing bases with quality scores $<$ 20, and removing reads $<$ 50 bp after trimming.
3. Sequencing data are organized according to sample identity and population structure for downstream alignment and variant analysis.
4. The prepared dataset with mean per-base Phred quality score $\geq$30 and minimum sequencing depth $\geq$10× should be suitable for carrying out high-resolution and reproducible SNP analysis (*see* **Note 1**).

3.2 Genome Alignment and SNP Discovery

1. Quality-filtered sequence reads are mapped onto a good reference genome using read alignment and mapping tools like BWA, Bowtie2, and HISAT2, depending on the genome size, ploidy level, and availability of the reference genome used for sequencing strategy discovery (*see* **Note 2**).
2. For species with poor-quality reference genomes, sequencing data are assembled or aligned by using suitable reference-guided or de novo assembly approaches.
3. Identify potential SNPs across the genome with variant calling tools such as GATK HaplotypeCaller, SAMtools/BCFtools mpileup, or FreeBayes, based on sequence mismatches between the aligned reads and the reference genome.
4. Initial SNP datasets are derived using basic confidence criteria related to read depth and alignment quality.

3.3 SNP Quality Filtering and Dataset Refinement

1. Potential SNPs are filtered by VCFtools or PLINK to exclude low-confidence SNPs based on the following criteria: minimum read depth (DP) $\geq$ 10, genotype quality (GQ) $\geq$ 20, mapping quality (MQ) $\geq$ 30, minor allele frequency (MAF) $\geq$ 0.05, and missing data per SNP $\leq$ 10% (*see* **Note 3**).

2. SNPs with high rates of missing data, poor genotype quality, and very low frequencies of alleles may be excluded from the data, and other filtering criteria, such as exclusions of variant calls from misalignment in genomic duplicated regions, may be used (*see* **Note 4**).
3. The final SNP matrix is assessed for genome-wide coverage and marker density for suitability in population-level analyses.

3.4 Functional Annotation and Genomic Classification of SNPs

1. High-confidence SNPs are functionally annotated using functional annotation databases (e.g., snpEff, ANNOVAR, or Ensembl Variant Effect Predictor) to identify their genomic location with respect to genes and regulatory regions and to prioritize biologically relevant variants (*see* **Note 5**).
2. Variants are classified into coding and noncoding variants and further categorized based on predicted functional impact.
3. The annotated SNP dataset is then combined with gene annotation data to enable the biological interpretation of the observed sequence variations. This will facilitate the integration of genotype data with other functional and phenotypic data (*see* **Note 6**).

3.5 Population Genomics and Genetic Diversity analysis

1. Refined SNP datasets are analyzed using population genomics software (e.g., STRUCTURE, ADMIXTURE, or TASSEL) to estimate genetic diversity parameters and population structure (*see* **Note 7**).
2. Multivariate analyses performed using principal component analysis (PCA) and multidimensional scaling (MDS) methods are carried out to plot genetic relationships among plant accessions and to identify population stratification.
3. Linkage disequilibrium data generated by calculating pairwise r^2 values between SNP loci are analyzed to evaluate the genomic distribution of SNPs and their suitability for association analysis.

3.6 Genome-Wide Association Analysis

1. SNP genotype matrices are merged with corresponding phenotypic datasets for association analysis.
2. Statistical models based on the mixed linear model (MLM) framework incorporating population structure and relatedness are employed using associated tools such as GEMMA, TASSEL, or MLMM to identify significant associations between SNPs and phenotypic traits (*see* **Note 8**).
3. Regions of the genome that are associated with the traits of interest are identified and ranked according to a genome-wide significance threshold of $p \leq 0.05$ after Bonferroni correction or an FDR-adjusted q-value ≤ 0.05, in addition to biological significance.

3.7 Integration of SNP Data into Predictive and Applied Genomics

1. The genome-wide SNP data are integrated into the genomic prediction models using software such as rrBLUP, BLUPF90, or BGLR to predict the breeding values and the performance of traits (*see* **Notes 9** and **10**).
2. The SNPs that are linked to agronomically valuable traits are integrated into the breeding programs to accelerate the improvement of crops.
3. The data obtained from the analysis using SNPs is interpreted in the context of the genomic architecture of plants.

4 Notes

1. Single-nucleotide polymorphisms (SNPs) are especially useful for plant genomic research because of their genome-wide distribution, codominant model of inheritance, and high degree of reproducibility, making them the marker of choice for high-resolution genetic analysis in various plant species.
2. The discovery approach of choice for SNPs should depend on the size of the genome, ploidy level, and presence of reference sequences. Whole-genome resequencing provides a comprehensive survey of SNPs, while reduced-representation methods such as GBS are cost-effective alternatives for large-scale populations [12].
3. High-quality input sequence data are essential for reliable SNP calling. Suboptimal read quality or depth of sequencing can lead to false positives or missing data, which can be problematic for downstream analyses such as association mapping or genomic prediction.
4. The accuracy of alignment is especially important for plant genomes, which harbor repetitive sequences, segmental duplications, or homoeologous sequences. Strategic choices of read alignment parameters can help reduce misalignment and false SNP calls [13].
5. Functional annotation of SNPs helps to distinguish among coding, noncoding, synonymous, and non-synonymous SNPs, thus making it easier to prioritize SNPs of possible biological or agronomic interest.
6. The integration of SNP information with gene models and functional databases improves the understanding of genotype–phenotype associations, especially in the context of candidate gene studies and trait analysis.
7. Population structure and relatedness are important factors in SNP analyses. Neglecting these aspects can result in the

detection of false associations in genetic diversity analyses and genome-wide association studies [14].

8. SNP information is particularly useful in GWAS because of its high density of genomic distribution, which increases the resolution of loci associated with traits, compared to previous marker technologies [15].
9. The effectiveness of SNPs in genomic selection is influenced by marker density and genome coverage. High-density SNP arrays enhance the accuracy of predictions, especially for complex traits governed by multiple loci.
10. Genomic prediction models based on SNPs can take advantage of large and well-annotated reference populations, as the accuracy of predictions increases with enhanced allelic diversity and recombination information.

References

1. Wang DG, Fan JB, Siao CJ, Berno A, Young P, Sapolsky R et al. (1998) Large-scale identification, mapping, and genotyping of single-nucleotide polymorphisms in the human genome. Science 280(5366):1077–1082
2. Ferguson ME, Hearne SJ, Close TJ, Wanamaker S, Moskal WA, Town CD et al. (2012) Identification, validation and high-throughput genotyping of transcribed gene SNPs in cassava. Theor Appl Genet 124(4):685–695
3. Al-Samarai FR, Al-Kazaz AA (2015) Molecular markers: an introduction and applications. Eur J Mol Biotechnol 9(3):118–130
4. Collard BC, Mackill DJ (2009) Start codon targeted (SCoT) polymorphism: a simple, novel DNA marker technique for generating gene-targeted markers in plants. Plant Mol Biol Reptr 27(1):86–93
5. Kumar R, Das SP, Choudhury BU, Kumar A, Prakash NR, Verma R,..., Mishra VK (2024) Advances in genomic tools for plant breeding: harnessing DNA molecular markers, genomic selection, and genome editing. Biol Res 57(1):80
6. Mammadov J, Aggarwal R, Buyyarapu R, Kumpatla S (2012) SNP markers and their impact on plant breeding. Int J Plant Genomics 2012(1):728398
7. Amiteye S (2021) Basic concepts and methodologies of DNA marker systems in plant molecular breeding. Heliyon 7(10):e08093
8. Thakur R, Dogra S, Pandey RK, Laishram N, Singh A, Kumar R et al. (2025) Molecular breeding advancements and role of molecular markers in ornamental plants. HortScience 60(12):2321–2330
9. Batley J, Edwards D (2007) SNP applications in plants. Association mapping in plants. Springer New York, New York, NY, 95–102
10. Rafalski A (2002) Applications of single nucleotide polymorphisms in crop genetics. Curr Opin Plant Biol 5(2):94–100
11. Kumar S, Banks TW, Cloutier S (2012) SNP discovery through next-generation sequencing and its applications. Int J Plant Genomics 2012(1):831460
12. Elshire RJ, Glaubitz JC, Sun Q, Poland JA, Kawamoto K, Buckler ES, Mitchell SE (2011) A robust, simple genotyping-by-sequencing (GBS) approach for high diversity species. PLoS ONE 6(5):e19379
13. Li H, Durbin R (2009) Fast and accurate short read alignment with Burrows–Wheeler transform. Bioinformatics 25(14):1754–1760
14. Yu J, Pressoir G, Briggs WH, Vroh Bi I, Yamasaki M, Doebley JF et al. (2006) A unified mixed-model method for association mapping that accounts for multiple levels of relatedness. Nat Genet 38(2):203–208
15. Huang X, Han B (2014) Natural variations and genome-wide association studies in crop plants. Annu Rev Plant Biol 65(1):531–551

Check for updates

Chapter 12

Development and Application of Random Amplified Polymorphic DNA (RAPD) for the Improvement of Cotton Crops

Sowparnika Binu, Janani Sree Sivakumar, Vasanth Krishnan, Kanakachari Mogilicherla and Dhandapani Gurusamy

Abstract

Random amplified polymorphic DNA (RAPD) markers have played a significant role in accelerating cotton (Gossypium spp.) improvement by enabling rapid genome-wide assessment of genetic variation without the requirement for prior sequence information. RAPD has been widely applied in cotton germplasm characterization, cultivar identification, hybrid confirmation, and the early development of molecular linkage maps. Its simplicity, low cost, and the ability to amplify multiple polymorphic loci in a single reaction made RAPD particularly valuable during the early phases of cotton molecular breeding. The successful implementation of RAPD depends strongly on high-quality genomic DNA, and optimized extraction protocols, such as those designed specifically for cotton tissues, ensure reliable amplification. Although more advanced marker systems such as SSRS and SNPS have since become dominant, RAPD continues to be a useful tool for preliminary screening and for breeding programs operating in resource-limited environments. This chapter highlights the workflow, development, and applications of RAPD in cotton improvement, supported by the evidence from key studies that established its relevance in molecular plant breeding.

Keywords Cotton, RAPD, Molecular markers, DNA extraction, PCR, Genetic diversity, Germplasm, Breeding, Polymorphism

1 Introduction

Cotton (*Gossypium* spp.) is one of the most economically significant fiber crops worldwide and forms the backbone of numerous agricultural economies due to its immense contribution to the textile, oilseed, and livestock feed industries. The global importance of cotton is rooted not only in its extensive cultivation but also in the breadth of its industrial uses, which span textile manufacturing, seed oil extraction, biofuel production, and animal feed production. Despite its commercial prominence, cotton improvement faces multiple

Nitish Kumar (ed.), *Molecular Markers in Plants*, Methods in Molecular Biology, vol. 3097,
https://doi.org/10.1007/978-1-0716-5663-1_12,

constraints, including a narrow genetic base in cultivated species, vulnerability to pests and diseases, susceptibility to climatic variability, and the ongoing demand for superior fiber traits. Much of the genetic stagnation observed in cultivated cotton crops, particularly *Gossypium hirsutum*, is attributable to domestication bottlenecks and the continuous selection pressure exerted during breeding, resulting in the loss of numerous alleles of potential agronomic importance [1]. As cotton breeders increasingly aim to integrate desirable traits such as drought tolerance, pest resistance, fiber fineness, and early maturity, there is a compelling need for the advanced molecular tools that can dissect genetic diversity at a finer resolution than conventional phenotypic methods. The transition from morphological traits to molecular markers revolutionized plant breeding by shifting the focus from phenotype-based selection towards DNA-level characterization. Early marker systems, such as isozyme markers and restriction fragment length polymorphism (RFLP), played an essential role in the initial phases of molecular breeding but were hindered by labor-intensive workflows, high DNA requirements, and limited applicability in genetically complex polyploid crops like cotton [2]. These limitations generated strong interest in PCR-based marker systems, which provided faster, cheaper, and more efficient access to genetic information. With the introduction of random amplified polymorphic DNA (RAPD) markers in the early 1990s, cotton researchers gained access to a tool capable of producing genome-wide polymorphic profiles rapidly and without prior sequence information [3]. The development of RAPD technology marked a major milestone in molecular marker research because it introduced a simplified, widely applicable method for detecting DNA polymorphism using short arbitrary primers under low-stringency PCR conditions. Unlike probe-based technologies such as RFLP, which require cloning, sequencing, and hybridization steps, RAPD relies on the ability of arbitrary ten primers to anneal at multiple genomic locations, generating reproducible amplification patterns that reveal genetic variation through the presence or absence of DNA fragments [4]. RAPD markers also facilitated phylogenetic analyses, species differentiation, intra- and interspecific diversity studies, and the identification of unique genotypes within breeding programs. Furthermore, because cotton tissues contain high levels of polysaccharides, phenolic compounds, and oxidizing agents that can inhibit PCR reactions, the success of RAPD was strongly dependent on efficient DNA extraction. A breakthrough contribution in this regard was the cotton-specific DNA extraction method developed by Paterson, Brubaker and Wendel [5], which incorporated antioxidants, PVP and the CTAB method to effectively neutralize inhibitory compounds and ensure high-quality DNA suitable for PCR amplification [6].

2 Materials

Cotton samples (leaf/seedlings/boll stages/fibers etc.), genomic DNA extraction buffer (Table 1), liquid nitrogen, mortar and pestle, microfuge tubes, micropipettes, microtips, ice bucket, water bath, thermo cycler, DEPC-treated nuclease free water, 70% ethanol, centrifuge, PCR buffer, $MgCl_2$, $dNTP_S$, Taq DNA polymerase, arbitrary ten RAPD primers, Template DNA, PCR reaction mix, 0.2 mL PCR tubes, lysis buffer, β-mercaptoethanol, chloroform:isoamyl alcohol (24:1), isopropanol, wash buffers, TE buffer, TAE buffer, agarose powder, DNA gel loading dye, DNA ladder (100 bp), ethidium bromide/SYBR safe, gel documentation system, MOPS buffer (*see* **Note 5**) [7].

1 M Tris: 100 mL of 1 M Tris was prepared by dissolving 12 g of Tris base in 60 mL of distilled water. After dissolving, the pH was adjusted to 8.0 using concentrated HCl. Then, the volume was made up to 100 mL with distilled water and autoclaved at 15 psi pressure for 20 min.

0.5 M EDTA: 100 mL of 0.5 M EDTA was prepared by dissolving 18.6 g of EDTA in 75 mL of distilled water. NaOH pellets can be added to dissolve the EDTA, as it takes time to dissolve completely. The pH was adjusted to 8.0 using 1 N NaOH. Finally, the volume was made up to 100 mL with distilled water and autoclaved at 15 psi pressure for 20 min.

5 M NaCl: 5.84 g of NaCl was added to 80 mL of distilled water and distilled completely, finally made up to 100ml and autoclaved at 15 psi pressure for 20 min.

5% CTAB: 100 mL of 5% CTAB was prepared by dissolving 5 g of CTAB in 80 mL distilled water, and thereby 2 g of NaCl was added to it. Then the volume was made up to 100 mL and autoclaved at 15 psi pressure for 20 min.

Chloroform: Isoamyl Alcohol (24:1): 100 mL of 24:1 Chloroform:Isoamyl alcohol was prepared by adding 4 mL of isoamyl alcohol in 96 mL of chloroform and mixing. This mixture has to be made ice cold before use.

Table 1
Genomic DNA extraction buffer (for 100 mL) (*see* Note 16)

Chemicals required	Volume/weight
Milli-Q water/Nuclease-free water	38 mL
2% CTAB	20 mL (5% stock)
1.4 M NaCl	28 mL (5 M stock)
100 mM Tris. HCl	10 mL (1 M stock)
20 mM EDTA	4 ML (0.5 M stock)

0.1× TE: 20 mL of 0.1× TE was prepared by adding 0.02 mL of 1 M Tris (pH-8.0) and 0.004 mL of 0.5 M EDTA (pH-8.0) to 18 mL of distilled water. After dissolving properly, the volume was made up to 20 mL and autoclaved at 15 psi pressure for 20 min.

Preparation of 50× TAE buffer: Dissolve 242.0 g of Tris base in 700 mL Milli-Q water with gentle heating. Add 57.1 mL glacial acetic acid, followed by 100 mL of 0.5 M EDTA (pH 8.0). Mix thoroughly and make the final volume to 1000 mL with Milli-Q water. Adjust pH to 7.6–7.8. Autoclave the buffer and store in a room temperature.

Ethidium bromide (EtBr) preparation: Ethidium bromide (EtBr) is a potent mutagen used to visualize nucleic acids, typically prepared as a 10 mg/mL stock solution in water (stored in the dark) and used at a 0.5–1.0 μg/mL working concentration for agarose gels. It is added directly to molten agarose or used for post-electrophoresis staining to detect DNA/RNA via fluorescence under UV light.

- **Stock solution preparation (10 mg/mL)**
  - Dissolve 1.0 g of EtBr powder in 100 mL of distilled water. Store in a dark, light-proof container or wrapped in aluminum foil at 4°C and handle with gloves or use a fume hood to avoid inhaling dust
- **Working solution preparation (0.5–1.0 μg/mL):**
  - **In-gel staining**: Add 5–10 μL of the 10 mg/mL stock to 100 mL of molten agarose (<60°C) for a final concentration of 0.5–1.0 μg/mL.
  - **For post-staining**: Dilute the stock solution in electrophoresis buffer (1× TAE or TBE) to a final concentration of 0.5–1.0 μg/mL and immerse the gel for 20–30 min.

Preparation of 10 mL $MgCl_2$ stock solution: Weigh 0.0508 g of $MgCl_2.6H_2O$ using an analytical balance. Transfer the salt into a sterile volumetric tube, and dissolve in 8 mL of nuclease-free water. Mix thoroughly until completely dissolved and adjust the final volume to 10 mL to obtain a 25 mM $MgCl_2$ stock solution. Filter and sterilize the solution through a 0.22 μm membrane, and store at 20°C for RAPD PCR reactions in cotton improvement studies.

3 Methods

3.1 Sample Collection and Preparation

1. **Selection of plant material:** Young, fully expanded cotton leaves (15–30 days after emergence) are selected because they contain high concentrations of intact nuclei and minimal secondary metabolites that interfere with DNA extraction (*see* **Note 1**) [1].

2. **Avoiding environmental stress:** Leaves are collected from plants grown under uniform, nonstress conditions to avoid DNA degradation caused by drought, heat, or pest stress, which can negatively affect RAPD amplification [2].
3. **Use of sterile tools:** Stainless steel scissors or scalpels are sterilized using 70% ethanol before collecting samples to prevent contamination with foreign DNA or microbes that may inhibit PCR reactions [3].
4. **Immediate storage in ice:** Fresh leaf samples are placed immediately into ice-cold containers to slow enzymatic activities such as DNase degradation, preserving genomic DNA integrity for downstream analysis (*see* **Note 2**) [1].
5. **Labeling and sample tracking:** Each leaf sample is labeled with genotype name, collection date, and field location to ensure accurate mapping of RAPD profiles to specific cotton accessions [2].
6. **Transport under cold conditions:** Samples are transferred to the laboratory in insulated ice boxes to maintain cold chain integrity, essential for high-quality DNA extraction required RAPD [1].
7. **Surface cleaning before processing:** Leaves are washed with sterile distilled water and briefly dried to remove dust, microbes, and chemical residues that could interfere with PCR amplification [3].
8. **Preparation of tissue for grinding:** Approximately 100–200 mg of clean leaf tissue is cut into small pieces and flash-frozen in liquid nitrogen to facilitate easy grinding into a fine powder for efficient DNA extraction.
9. **Storage prior to extraction:** If immediate DNA extraction is not possible, samples are stored at −80°C to prevent oxidative damage and to ensure high-molecular-weight DNA for RAPD analysis [1].
10. **Use of controls and standards:** Additional control samples from known cotton varieties are collected to compare genetic profiles and validate the reproducibility of RAPD markers within the experiment [3].

3.2 Extraction of Genomic DNA

3.2.1 Extraction of Genomic DNA Through Modified CTAB Method (for Plants Containing High Polysaccharide and Polyphenol Components) [8–13]

1. 1.0 g of plant leaf tissue should be ground with liquid nitrogen using a chilled mortar and pestle, and the powder should be transferred to a polypropylene tube (30 mL) using a chilled spatula (50 mg of PVP for 0.5 g tissue should be added when grinding the tissue) (*see* **Notes 3 and 4**).

2. 10 mL of DNA extraction buffer (preheated at 60°C) should be added, mix the tubes by inversion, keep at 60°C for 45 min, and mix once every 15 min. Take the tubes and bring to room temperature (RT) (*see* **Note 6**).
3. Add 10 mL of chloroform:Isoamyl alcohol (24:1) and mix by inversion until it forms an emulsion.
4. Centrifuge the tubes for 10 min at 6000 rpm (25°C), then repeat the steps once more to remove cloudiness from the samples.
5. Carefully transfer the supernatant to fresh tubes, add and mix ½ volume at 5 M NaCl, add 2.5 volumes of 95% ethanol (pre-chilled at 4°C), mix the tubes, and then keep at −20°C for an hour.
6. Spin at 6000 rpm for 10 min (25°C), discard the aqueous solution, wash the pellet with 70% ethanol, and spin for 5 min. Air-dry the pellet for 1 h, add 300 μL of 0.1× TE to dissolve the pellet, and keep for 4°C overnight (*see* **Note 19**).
7. Dissolved DNA should be transferred to 1.5 mL Eppendorf tube, and 5 μL of RNase A (10 mg/mL) should be added and incubated for 1 h at 37°C water bath.
8. Extract an equal volume of neutral phenol twice with the ether solution to remove the phenol solution completely.
9. Add 1/10 volume of 3 M sodium acetate, pH 5.2, and 2.5 volumes of 90% ethanol to the solution for precipitation of genomic DNA. Keep the tubes overnight at −70°C.
10. Next day, spin the tubes for 12,000 rpm, 10 min at 4°C, discard the supernatant, and wash the pellet twice with 70% ethanol (pre-chilled). Finally, dry the pellet and dissolve it with 100 μL of 0.1× TE buffer
11. The quality and quantity of the extracted genomic DNA should be analyzed by 0.8% agarose gel electrophoresis and a Nanodrop spectrophotometer, respectively, prior to the RAPD reaction setup.

3.2.2 Extraction of Genomic DNA by the CTAB Method

1. Weigh 1 g of the sample, and freeze it in liquid nitrogen.
2. Grind the frozen tissue into a fine powder using a chilled mortar and pestle in the presence of liquid nitrogen.
3. Approximately 350 g of the powder should be distributed in each of three micro fuse tubes (micro fuse tubes must be chilled in liquid nitrogen before and after weighing. The spatula should be pre-chilled and used to transfer the ground powder. An equal volume (w/v) of pre-hot (65°C) 2× CTAB buffer should be added.

4. Mix well by inversion in a vigorous manner, and an equal volume of 700 µL v/v of ice-cold chloroform:Isoamyl alcohol (24:1) should be added.
5. Mix thoroughly by inversion to form an emulsion, and centrifuge for 5 min at 14,000 rpm.
6. Then, using the cut tips, the upper aqueous content should be transferred to new microfuge tubes.
7. Measure the volume. 1/5 volume of 5% CTAB solution should be added, and mixed well by gentle inversion.
8. Equal volume of chloroform:Isoamyl alcohol (24:1) should be added. Then mix thoroughly to form an emulsion and centrifuge for 5 min at 14,000 rpm.
9. Using cut tips, the upper aqueous content should be transferred to new centrifuge tubes, then add an equal volume of CTAB precipitation buffer, and mix gently by inversion.
10. Centrifuge for 1 min at 14,000rpm, discard the supernatant, and keep the tube by inverting it in tissue paper and drying it for 30 min.
11. Dissolve the pellet in 200 µL of high-salt TE buffer, and gently tap the tubes. Keep it overnight at 4°C fridge.
12. The next day, tap the tube gently and make sure the pellet dissolved completely, and then 2.5 volumes of ice-cold 95% ethanol should be added, and mix gently by inversion.
13. Centrifuge for 10 min at 14,000 rpm at 4°C and discard the supernatant.
14. Add 500 µL of 70% ethanol to the pellet, and centrifuge for 5 min at 14,000 rpm at 4°C.
15. Air-dry or vacuum-dry the pellet and dissolve in 50 µL of 0.1× TE.
16. Leave it for some time at room temperature or 4°C, then analyze the isolated genomic DNA by 0.8% agarose gel electrophoresis.

3.3 Quantification of Genomic DNA by Using Nanodrop Spectrophotometer

1. **Cleaning the pedestals:** Open the sampling arm and clean the upper and lower pedestals with a dry, wipe Kimwipe, preferably.
2. **Blanking:** Pipette 2 µL of the blanking buffer (the same buffer used to resuspend your DNA, such as TE or water) onto the lower pedestal. Close the arm and click blank in the equipment.
3. **Sample loading:** Wipe the pedestals clean. Pipette 2 µL of your DNA sample onto the pedestal, lower the arm, and enter the Sample ID.

4. **Measurement:** Click measure. The software will display the concentration (in ng/μL) and purity ratios (260/280 and 260/230).
5. **Cleanup:** Wipe the pedestal with a clean wipe between samples and perform a final clean with water after all measurements are finished.
6. **Interpretation:** Optimum range of A260/A280 and A260/A230 are 1.8–2.0 and 2.0–2.2, respectively (see Table 2) (Fig. 1A).
7. **Contamination:** Range below 1.8 and 2.0 in A260/A280 and A260/A230 indicates protein, phenol contamination and salt contamination present in the sample, respectively.

3.4 Agarose Gel Electrophoresis [10]

1. **Preparation of agarose solution:** Agarose (0.5 g) is weighed and melted thoroughly in an appropriate volume of 1× TAE buffer until the solution becomes fully transparent and uniform.
2. **Addition of DNA stain:** Once the melted agarose cools slightly (still in liquid form), 5 μL of ethidium bromide is added to the solution. This stain intercalates with DNA fragments and enables direct visualization of RAPD bands post-electrophoresis.
3. **Casting the agarose gel:** The agarose TAE mixture is poured gently into a clean gel casting tray fitted with a comb. The gel

Table 2
Optimum DNA purity range

Absorbance ratio[a]	Optimal range	Interpretation
A260/A280	<1.8	Presence of residual protein, phenol, or other extraction reagents that absorb at or near 280 nm
A260/A280	1.8–2.0	Indicates high-purity DNA with minimal protein contamination
A260/A280	>2.0	Possible RNA contamination and other nucleic acid impurities in the DNA sample, as RNA typically exhibits an A260/A280 ratio of approximately 2.0
A260/A230	<2.0	Contamination by carbohydrates, phenol, EDTA, guanidine salts, or other organic compounds
A260/A230	2.0–2.2	Optimal high-purity DNA with minimal organic or salt contamination
A260/A230	>2.2	Generally, not considered problematic; however, it may result from the use of a blank or buffer with very low absorbance at 230 nm or from minor measurement variability

[a]The A260/A280 ratio is used to assess protein contamination in DNA samples because proteins exhibit strong absorbance at 280 nm. The A260/A230 ratio is used to evaluate organic and salt contamination, since many extraction reagents and organic compounds absorb strongly at 230 nm

is left undisturbed to solidify. Casting the gel in a horizontal tray is essential for proper band separation.

4. **Preparation of electrophoresis:** After solidification, adhesive tapes or side seals from the tray are removed. The gel is placed carefully into the electrophoresis chamber and fully submerged in 1× TAE running buffer, ensuring uniform electrical contact.
5. **Sample preparation with loading dye:** DNA samples are mixed with 6× gel loading dye, then briefly centrifuged to ensure proper mixing. The tracking dye helps visualize sample loading and migration during electrophoresis.
6. **Loading of samples into wells:** Using a micropipette, DNA samples along with a 100 bp DNA ladder were slowly dispensed into the wells to avoid well damage or sample leakage. Careful loading maintained accurate band patterns (*see* **Note 21**).
7. **Running the electrophoresis:** The system is connected to a power supply, and electrophoresis is carried out at a constant voltage, ensuring even migration of amplified DNA fragments. Proper voltage maintains a clear separation of DNA bands.
8. **Visualization of bands:** After completion of the run, the gel is transferred to a UV transilluminator, where bands stained with ethidium bromide are observed (Fig. 1B) [11].

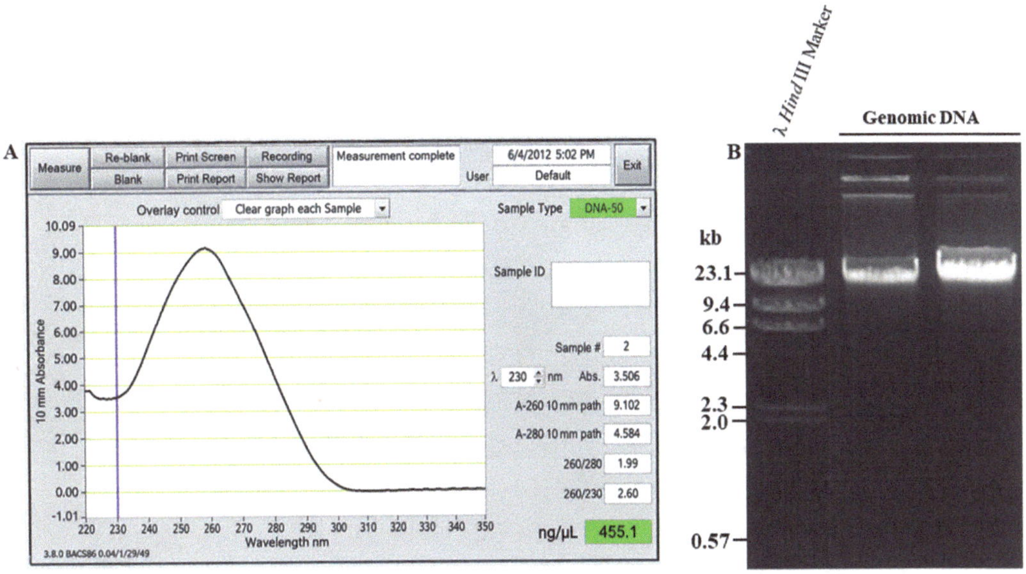

Fig. 1 Analysis of genomic DNA extracted from the cotton tissues by Nanodrop Spectrophotometer and Agarose gel electrophoresis. (**A**) Image shows the quantity and purity analysis of extracted cotton genomic DNA by Nanodrop1000 Spectrophotometer (Thermo Fisher Scientific, USA), (**B**) Agarose gel image shows the quantity and integrity of genomic DNA extracted from the cotton tissues [11]

Table 3
List of primers commonly employed for the RAPD

S. No.	Primer	Sequence (5′–3′)	Reference
1.	OPM-01	GTTGGTGGCT	[14,15]
2.	OPM-02	ACAACGCCTC	
3.	OPM-03	GGGGGATGAG	
4.	OPM-04	GGCGGTTGTC	
5.	OPM-05	GGGAACGTGT	
6.	OPM-06	CTGGGCAACT	
7.	OPM-07	CCGTGACTCA	
8.	OPM-08	TCTGTTCCCC	
9.	OPM-09	GTCTTGCGGA	
10.	OPM-10	TCTGGCGCAC	
11.	OPM-11	GTCCGGAGTG	
12.	OPM-12	GGGACGTTGG	
13.	OPM-13	GGTGGTCACT	
14.	OPM-14	AGGGTCGTTC	
15.	OPM-15	GACCTACCAC	
16.	OPM-16	GTAACCAGCC	
17.	OPM-17	TCAGTCCGGG	
18.	OPM-18	CACCATCCGT	
19.	OPM-19	CCTTCAGGCA	
20.	OPM-20	AGGTCTTGGG	
21.	OPA-8	GTGACGTAGG	[16]
22.	OPA-11	CAATCGCCGT	
23.	GL Decamer A-02	TGCCGAGCTG	
24.	GL Decamer A-07	GAAACGGGTG	
25.	GL Decamer K-20	GTGTCGCGA	
26.	GL Decamer K-18	CCTAGTCGAG	
27.	GL Decamer J-20	AAGCGGCCTC	
28.	GL Decamer J-07	CCTCTCGACA	
29.	GL Decamer J-03	TCTCCGCTTG	
30.	GL Decamer C-20	ACTTCGCCAC	
31.	GL Decamer C-19	GTTGCCAGCC	
32.	GL Decamer C-18	TGAGTGGGTG	
33.	GL Decamer C-15	GACGGATCAG	

(Continued)

Table 3 (Continued)

S. No.	Primer	Sequence (5′–3′)	Reference
34.	GL Decamer C-14	TGCGTGCTTG	[16]
35.	GL Decamer C-12	TGTCATCCCC	
36.	GL Decamer C-11	AAAGCTGCGG	
37.	GL Decamer C-10	AGTCTGGGTG	
38.	GL Decamer C-09	CTCACCGTCC	
39.	GL Decamer C-08	TGGACCGGTG	
40.	GL Decamer C-06	GAACGGACTC	
41.	GL Decamer C-03	GGGGGTCTTT	
42.	GL Decamer C-02	GTGAGGCGTC	
43.	GL Decamer C-01	TTCGAGCCAG	
44.	GL Decamer A-20	GTTGCGATCC	
45.	GL Decamer A-11	CAATCGCCGT	
46.	GL Decamer A-10	GTGATCGCAG	

Table 4
List of software used for RAPD primer designing

S. no	Name of the software	Link
1.	Primo Random	https://www.changbioscience.com/primo/primor.html
2.	Primer3	http://primer3.ut.ee/
3.	OligoAnalyzer	https://www.idtdna.com/pages/tools/oligoanalyzer
4.	GelAnalyzer	http://www.gelanalyzer.com/
5.	PyElph	https://sourceforge.net/projects/pyelph/

3.5 RAPD Primer Designing

Since RAPD analysis does not require prior sequence information, commercially available Random Decamer primers or primers generated using Random Primer design tools can be used for the amplification of cotton genomic DNA (see Table 3).

3.5.1 Designing RAPD Primers

1. A wide range of bioinformatics tools and software are available for generating and analyzing RAPD primers.
2. These tools are internet-based, free and simple to use. After amplification, gel analysis software is used for band scoring and for converting RAPD profiles into a binary data matrix.

3. Some of the tools that are been used for designing RAPD primers are given in Table 4.
4. Download and install the required software, complete the registration using a valid email ID if necessary, and execute the program according to the instructions provided in the user manual for the respective tool [14].

3.6 PCR Reaction Setup for RAPD (see Notes 17, 20, 22, and 23)

1. For the preparation of master mix on ice, we have to combine 10× PCR buffer (1× final), $MgCl_2$ (1.5–3.0mM), dNTPs (200 μM each), RAPD primer (10 pmol), and nuclease-free water. Mix gently to avoid bubble formation (see Table 5).
2. Add 1.0 μL of Taq DNA polymerase to the master mix (or individually to each tube) as the second-to-last step to prevent premature amplification.
3. Aliquot the master mix into PCR tubes to achieve a final volume of 25 μL, then add 20-50 ng of template genomic DNA to each tube. Include a nontemplate control.
4. Using the following conditions, run the RAPD PCR program at 94°C for about 3 min, 40 cycles of 94°C for 1 min, 36°C for 1 min, 72°C for 2 min, and final extension at 72°C for about 7 min (see Table 6).
5. Load the PCR products onto a 1.5% agarose gel, run at 80–100 V, stain with safe dye, and visualize bands using a gel documentation system.

3.7 Agarose Gel Electrophoresis Analysis of RAPD PCR Products

1. Run RAPD PCR to amplify cotton genomic DNA using RAPD primers under optimized cycling conditions with proper controls.
2. Load 2–5 μL of PCR product on a quick agarose gel to confirm successful amplification.
3. Load 5–10 μL of each PCR product on a 1.2%–1.5% agarose gel with a DNA ladder and run at 80–100 V (*see* **Note 15**).

Table 5
Prepare the PCR reaction mix (50 μL reaction) (*see* Notes 7–9, 11, 14, and 18)

Chemicals required	Volume/weight
10× PCR buffer (with $MgCl_2$)	5.0 μL (1×)
dNTP mix (10 mM)	1.0 μL
Arbitrary Primer (acts as both forward and reverse primer)	2.0 μL
Taq DNA polymerase	1.0 μL
Template DNA	1.0 μL (50–100 ng)
Nuclease-free water	40.0 μL

4. Stain the gel with Ethidium bromide solution or an alternative nontoxic and environmentally safe dye such as GelRed® Nucleic Acid Gel Stain (Biotium, Inc., USA) and image the gel for visualization using a gel documentation system.
5. After visualizing the gel image, clear and unambiguous bands are recorded for each cotton variety, ensuring accurate scoring of genetic polymorphisms (Fig. 2) [17].
6. Prepare a binary matrix, calculate genetic similarity, and construct dendrograms to identify polymorphisms useful for cotton improvement.

3.8 Band Scoring

3.8.1 Band Scoring for Diversity Analysis

The amplified RAPD bands are scored based on presence or absence, and fragment size is determined relative to a suitable molecular size

Table 6
PCR reaction condition (*see* Note 10)

Steps	Condition
Initial denaturation	94°C for 5 min
Denaturation	94°C for 1 min
Annealing	50–60°C for 1 min
Extension	72°C for 1 min (1 min/kb)
Final extension	72°C for 7–10 min

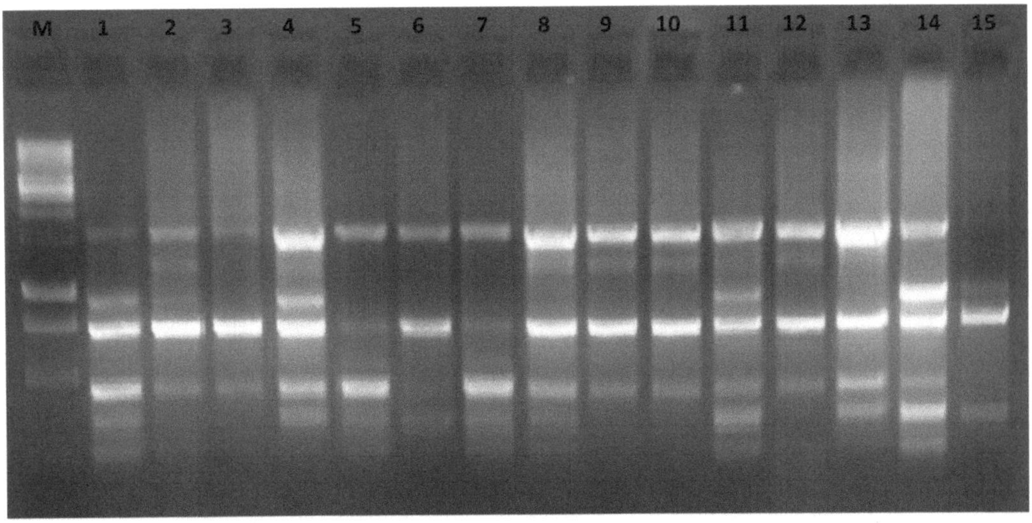

Fig. 2 Representative image of agarose gel electrophoresis showing the results of RAPD PCR with 15 cotton varieties using the OPM-08 primer. Lane M: kb marker, Lane 1–15: Different cotton varieties show multiple RAPD PCR amplification[Reprinted with permission from [18]]

marker (100 bp or 1 kb ladder). Only distinct bands are used for diversity analysis in cotton improvement studies (*see* **Notes 12 and 13**) [19].

3.8.2 Band Scoring and Analysis

- RAPD analysis was carried out to assess genetic diversity among cotton (*Gossypium hirsutum* L.) genotypes, while amplified DNA fragments produced distinct banding patterns after agarose gel electrophoresis. Only clear, reproducible and well-resolved RAPD bands were considered for scoring to ensure reliability of the results. Since RAPD markers are dominant in nature, each visible band was treated as a single allelic locus representing a phenotypic expression of the genome [20].
- The RAPD bands were scored based on their molecular size using gel electrophoresis image analysis software such as AlphaImager HP 3.4. The presence of a particular band in a genotype was scored as "1" and its absence as "0", resulting in a binary data matrix. Monomorphic bands, which appeared uniformly across all genotypes, were excluded from further analysis as they did not contribute to genetic differentiation. They pooled binary data from all primers to form the basis for estimating genetic diversity parameters.
- A total of 18 RAPD bands were generated by five primers, out of which 13 bands were polymorphic. The amplified fragments ranged in size from 180 bp to 800 bp, with an average of 3.6 bands per primer. The overall polymorphism percentage was 50.71%, indicating a moderate level of genetic variability among the cotton genotypes studied. Certain primers, such as OPB-17, showed 100% polymorphism, demonstrating high discriminatory power, while others showed comparatively lower polymorphism due to conserved genomic regions.
- The scored RAPD data were further analyzed to estimate gene frequency, gene diversity, and polymorphism information content (PIC). Gene diversity values ranged from 0.22 to 0.44 with an average of 0.34, suggesting moderate genetic variation among the cotton genotypes. PIC values averaged 0.29, indicating that the RAPD markers used were moderately informative for genetic diversity analysis.
- Genetic similarity among cotton genotypes was calculated using the Dice similarity coefficient, and cluster analysis was performed using the UPGMA method. The resulting dendrograms grouped the nine cotton genotypes into two major clusters, reflecting their genetic relationships. Genotypes clustered together showed close genetic similarity, whereas those in different clusters were genetically distinct. Principal component analysis (PCA) further supported the clustering pattern by separating the genotypes into distinct groups on RAPD variation.

- The combined RAPD band scoring and analysis revealed moderate genetic diversity among the studied cotton genotypes and enabled the identification of genetically divergent lines. Such genotypes are valuable for parental selection in hybridization and varietal improvement programs. Overall, RAPD band scoring and analysis proved to be an effective molecular tool for germplasm characterization, genetic diversity assessment and improvement of cotton crops [16].

4 Notes

1. Young, healthy cotton leaves should be used as starting material. Older or stressed tissues contain higher levels of polysaccharides and phenolics, which reduce DNA yield and PCR.
2. Samples should be processed immediately or stored at −80°C to prevent DNA degradation. Prolonged storage at room temperature or repeated freeze-thaw cycles result in fragmented DNA and poor RAPD profiles.
3. Efficient grinding of tissue in liquid nitrogen is critical. Incomplete grinding leads to poor cell lysis and low DNA recovery.
4. A modified CTAB method containing PVP and β-mercaptoethanol is strongly recommended for cotton tissues. These components neutralize phenolic compounds that otherwise bind and degrade DNA.
5. All reagents used for DNA extraction should be fresh and of molecular biology grade. Oxidized or contaminated reagents often result in brownish DNA pellets and low A260/A280 ratios.
6. Gentle mixing during the lysis and extraction steps should be maintained to avoid mechanical shearing of high-molecular-weight DNA.
7. Template DNA concentration must be optimized for RAPD PCR. Excess DNA leads to nonspecific amplification, while low DNA concentrations result in weak or absent bands.
8. RAPD primers should be aliquoted and stored at −20°C to prevent degradation from repeated freeze-thaw cycles.
9. $MgCl_2$ concentration is a critical factor influencing RAPD banding patterns. Optimization within the range of 1.5–3.0 mM is recommended for each primer.
10. Strict consistency in PCR cycling conditions must be maintained. Minor variations in annealing temperature or cycle number can alter RAPD profiles.

11. Negative controls must be included in every PCR run to detect contamination.
12. Only clear, sharp and reproducible bands should be scored. Faint or inconsistent bands should be excluded to avoid false polymorphism.
13. RAPD markers are dominant in nature; therefore, results should be interpreted carefully, particularly in genetic diversity and population structure analyses.
14. Replicate PCR reactions are recommended to confirm the reproducibility of RAPD markers.
15. Agarose concentration between 1.2% and 1.5% is optimal for resolving RAPD fragments in the 180–1000 bp range.
16. The purity of water used for PCR must be strictly maintained. Even trace nuclease contamination in nuclease-free water can degrade template DNA and reduce amplification efficiency.
17. Avoid repeated opening of PCR tubes after amplification. Exposure to amplified products increases the risk of contamination in subsequent reactions.
18. Use the same batch of Taq DNA polymerase throughout the experiment. Different enzyme formulations may produce variation in banding profiles.
19. Ensure complete removal of ethanol after DNA washing steps. Residual ethanol in DNA preparations inhibits Taq polymerase activity.
20. Avoid using degraded primers. Primer degradation results in weak amplification and inconsistent band patterns.
21. Do not overload wells during gel electrophoresis. Excess sample volume may cause band distortion.
22. Include an internal reference genotype across all samples before PCR setup. Variation in template concentration may produce artificial polymorphism.
23. Maintain detailed laboratory records, including reagent lot numbers, primer code, $MgCl_2$ Concentration and thermal cycling parameters, to ensure long-term reproducibility of RAPD data.

References

1. Iqbal MJ, Aziz N, Saeed NA, Zafar Y, Malik KA (1997) Genetic diversity evaluation of some elite cotton (*Gossypium hirsutum* L.) varieties by RAPD analysis. Theor App Genet 94(1):139–144
2. Bardakci F (2001) Random amplified polymorphic DNA (RAPD) markers. Turk J Biol 25:185–196
3. Williams JGK, Kubelik AR, Livak KJ, Rafalski JA, Tingey SV (1990) DNA polymorphisms

amplified by arbitrary primers are useful as genetic markers. Nucl Acid Res 18:6531–6535
4. Welsh J, McClelland M (1990) Fingerprinting genomes using PCR with arbitrary primers. Nucl Acid Res 18:7213–7218
5. Paterson AH, Brubaker CL, Wendel JF (1993) A rapid method for extraction of cotton (Gossypium spp.) genomic DNA suitable for RFLP or PCR analysis. Plant Mol Biol Rep 11:122–127. https://doi.org/10.1007/BF02670470
6. Hadrys H, Balick M, Schierwater B (1992) Applications of random amplified polymorphic DNA (RAPD) in molecular ecology. Mol Ecol 1:55–63
7. Doyle JJ, Doyle JL (1987) A rapid DNA isolation procedure for small quantities of fresh leaf tissue. Phytochem Bull 19:11–15
8. Porebski S, Bailey LG, Baum BR (1997) Modification of a CTAB DNA extraction protocol for plants containing high polysaccharide and polyphenol components. Plant Mol Biol Rep 15(1):8–15
9. Sivakumar JS, Gurusamy D, Chellamuthu S, Rathnasamy SA, Krishnan V (2015) Characterization of transgenic cotton plants by Southern blotting. Methods Mol Biol 2973:171–181
10. Rogers SO, Bendich AJ (1994) Extraction of total cellular DNA from plants, algae and fungi. In: Gelvin SB, Schilperoort RA (eds) Plant molecular biology manual. Springer, Dordrecht. pp 183–190
11. Gurusamy D (2014) Transcriptome analysis of cotton (*Gossypium hirsutum* L.) during boll development under drought conditions and functional characterization of drought tolerance genes. Ph.D. thesis: https://hdl.handle.net/10603/215568.
12. Sivakumar JS, Gurusamy D, Chellamuthu S, Rathnasamy SA, Krishnan V (2025) Characterization of transgenic cotton plants by PCR. Methods Mol Biol 2973:157–170
13. Dhandapani G, Kanakachari M, Padmalatha KV, Phanindra MLV, Singh VK, Raghavendra Rao S, Kumar PA (2015) A gene encoding cold-circadian rhythm-RNA binding-like protein (CCR-like) from upland cotton (*Gossypium hirsutum* L.) confers tolerance to abiotic stresses in transgenic tobacco. Plant Mol Biol Rep 33(1):22–42
14. Rana MK, Bhat KV (2002) Genetic diversity analysis in Indian diploid cotton (*Gossypium* spp.) using RAPD markers. Indian J Genet and Plant Breed 62(1):11–14
15. Asif M, Rahman MU, Mirza JI, Zafar Y (2009) Parentage confirmation of cotton hybrids using molecular markers. Pak J Bot 41(2):695–701
16. Tingey SV, Del Tufo JP (1993) Genetic analysis with random amplified polymorphic DNA markers. Plant Phys 101:349–352
17. Tidke SA, Kiran S, Harke SN (2014) Analysis of genetic diversity in 20 cotton germplasm lines using random amplified polymorphic DNA marker. Asian J Plant Sci 13:184–189
18. Chaudhary L, Sindhu A, Kumar M, Kumar R, Saini M (2010) Estimation of genetic divergence among some cotton varieties by RAPD analysis. J Plant Breed Crop Sci 2(3):039–043
19. Verma SK, Dhanda M, Salar RK (2015) Analysis of genetic diversity among Asiatic cotton (*Gossypium arboreum L.*) cultivars and breeding lines using RAPD and SSR markers. Int J Adv Res Biol Sci 2(3):114–122
20. Bukhari SA, Iqbal MA, Naz S (2021) Studies on genetic diversity of cotton using RAPD markers. Pure Appl Biol 3(3):95–100

Chapter 13

Development and Application of Restriction Fragment Length Polymorphism (RFLP) for the Improvement of Cotton Crops

Sowparnika Binu, Janani Sree Sivakumar, Vasanth Krishnan, Kanakachari Mogilicherla and Dhandapani Gurusamy

Abstract

Restriction fragment length polymorphism (RFLP) has been an important molecular tool in advancing cotton (Gossypium spp.) improvement by enabling researchers to detect stable genetic variations across the genome. Using restriction enzymes and labeled probes, RFLP generates highly reproducible, co-dominant markers that help identify differences among cultivars, wild species, and breeding lines. This technique has been widely used to construct early genetic linkage maps, locate quantitative trait loci (QTL_S) associated with fiber quality, yield, and disease resistance, and study genome structure in tetraploid cotton. RFLP has also contributed to detecting introgression from wild relatives and supporting marker-assisted selection in breeding programs. Although a more advanced marker system has emerged, RFLP remains valuable for its accuracy and continues to serve as a reference method for validating genetic relationships in cotton genomics.

Keywords RFLP, Molecular markers, QTL mapping, Genetic diversity, Marker-assisted selection, Cotton improvement

1 Introduction

Cotton (Gossypium spp.) plays a central role in global agriculture as the leading natural fiber crop, making its genetic improvement crucial for meeting demands for higher yield, superior fiber quality, and stress tolerance. Traditional breeding methods, though effective, often struggle to resolve the genetic complexity of cotton due to its polyploid genome. As a result, molecular markers have become vital tools for identifying desirable alleles and improving selection efficiency in cotton breeding programs [1]. Restriction fragment length polymorphism (RFLP) was one of the earliest and most reliable DNA marker systems developed for plant genomic research. RFLP detects variation in the length of DNA fragments produced by

Nitish Kumar (ed.), *Molecular Markers in Plants*, Methods in Molecular Biology, vol. 3097,
https://doi.org/10.1007/978-1-0716-5663-1_13,

restriction enzyme digestion, followed by hybridization with labeled probes. This method provides co-dominant, highly reproducible markers that enable accurate analysis of genetic diversity, gene mapping, and detection of allelic variation [2]. In cotton research, RFLP markers have played an essential role in genome characterization and the construction of early genetic linkage maps. These linkage maps helped identify chromosomal relationships, locate genes controlling key agronomic traits, and understand the genome structure of *G. hirsutum* and *G. barbadense*. RFLP markers enabled researchers to distinguish between A- and D-genome contributions in tetraploid cotton, providing crucial insights for breeding [3]. The use of RFLP has significantly contributed to identifying quantitative trait loci (QTL_S) associated with fiber length, strength, lint percentage, and yield components in cotton. By linking DNA markers to phenotypic traits, RFLP facilitated early marker-assisted selection (MAS) in cotton breeding. Additionally, RFLP probes were instrumental in detecting genes conferring resistance to diseases such as bacterial blight caused by *Xanthomonas citri* pv. Malvacearum (formerly *X. campestris* pv. Malvacearum) [4]. Beyond trait mapping and diversity studies, RFLP has been applied to detect introgression from wild or exotic cotton species, verify hybrid purity, and distinguish closely related cultivars. Although newer marker systems have largely replaced RFLP due to its labor-intensive nature, the method remains important for generating high-quality genomic data and for validating markers in cotton genomics. Its co-dominant and reliable nature continues to make RFLP a reference standard in marker development [5].

2 Materials

Cotton samples from different cultivars (leaf, seedlings, boll, fibers, and etc.), Genomic DNA from *Gossypium hirsutum* cultivars and breeding lines, liquid nitrogen, pre-chilled mortar and pestle for high quality DNA grinding, microfuge tubes (1.5 mL and 2 mL), sterile micropipettes and aerosol resistant microtips, refrigerated centrifuge capable of 10,000–14,000 rpm, electrophoresis tank and power supply, water bath, ice bucket, Nanodrop spectrophotometer for DNA quantification, DNA extraction reagents, including CTAB (see Table 1), 2% PVP for removing polyphenols, β-mercaptoethanol for preventing DNA oxidation, extraction buffer (Tris-HCL, EDTA, NaCl), Chloroform:isoamyl alcohol (24:1), 70% ethanol, absolute ethanol, RNase A, TE buffer for DNA resuspension, restriction endonucleases (e.g., *Eco*RI, *Hind*III, *Hae*III) selected for generating polymorphic restriction patterns in cotton, nuclease free water, agarose powder for gel preparation, TAE electrophoresis buffer (see Table 2), ethidium bromide/SYBR safe for gel staining, loading dye (6×), high range DNA molecular weight marker, hybridization oven or incubator, nylon membrane, vacuum blotting or capillary blotting apparatus, SSC buffer (20×) (see Table 3), denaturation and neutralization solutions (see Tables 4 and 5), restriction digestion buffer supplied with enzyme (see Table 6), UV crosslinker for fixing DNA to membrane, pre hybridization buffer, cotton-specific RFLP probes labeled using random priming or translation kits (see Table 7), hybridization bottles, radiolabel or nonradioactive detection systems (e.g., DIG

Table 1
Genomic DNA extraction buffer (For 100 mL)

Chemicals Required	Volume/weight
Milli-Q water	38 mL
2% CTAB	20 mL (5% stock)
1.4 M NaCl	28 mL (5 M stock)
100 mM Tris. HCl	10 mL (1 M stock)
20 mM EDTA	4 ML (0.5 M stock)

Table 2
TAE buffer (50× stock and 1× working)

Component	Final concentration (1×)	Stock concentration (50×)	Quantity for 1L (50× stock)
Tris base	40 mM	2.0 M	242.28 g
Acetic acid (glacial)	20 mM	1.0 M	~57.3 mL
EDTA (pH 8.0)	1 mM	50 mM	100 mL of 0.5 M EDTA Stock
Nuclease-free water	–	–	Up to 1 L

Table 3
SSC buffer composition and preparation details (20× stock and 1× working solution)

Component	Final concentration (1× SSC)	Stock concentration (20× SSC)	Quantity for 1 L (20× stock)
Sodium chloride	0.15 M	3.0 M	175.3 g
Trisodium citrate	0.015 M	0.3 M	88.2 g
Nuclease-free water	–	–	Up to 1 L
pH (adjusted)	7.0	–	Adjust with HCL/NaOH after dissolution
Working solution (1× SSC)	0.15 M NaCl, 0.015 sodium citrate	–	Prepare by diluting 20× SSC stock 1:20 with water

Table 4
Denaturation solution (500 mL)

NaCl	29.22 gm (1 M)
NaOH	10 gm (0.5 M)

Table 5
Neutralization solution (500 mL)

NaCl	43.83 gm (1.5 M)
Tris	30.28 gm (0.5 M)

Table 6
Restriction digestion analysis of plant genomic DNA

Nuclease free H_2O	Final volume 50 μL
Genomic DNA	20 μg (variable in μL)
Restriction buffer	5 μL (1×)
Restriction enzymes	5 μL (50 units)
Nuclease-free water	Make up to 50 μL

Table 7
List of probes employed in cotton

S. no	List of probes	Purpose	Reference
1.	BNL	Used for constructing the first comprehensive cotton genetic maps and identifying fiber quality QTLs.	[6]
2.	TAM	Developed for mapping the cotton genome and identifying markers linked to disease resistance (e.g., Fusarium wilt).	[7]
3.	OSU	Assessing genetic diversity and analyzing environmental stress responses in *Gossypium hirsutum*.	[8]
4.	MSU	Structural analysis of the cotton genome and distinguishing between subgenomes.	[9]
5.	pAR	Specifically used for Arkansas-developed markers targeting regional yield and quality traits.	[10]

labeling kit), washing buffers of varying stringency (2× SSC, 0.5× SSC, 0.1× SSC), X-ray film or chemiluminescent detection system, autoradiography cassette, gel documentation system, and data analysis software for scoring RFLP bands (*see* **Notes 20 and 22**).

1 M Tris: 100 mL of 1 M Tris was prepared by dissolving 12 g of Tris base in 60 mL of distilled water. After dissolving, the pH was adjusted to 8.0 using concentrated HCl. Then, the volume was made up to 100 mL with distilled water and autoclaved at 15 psi pressure for 20 min.

0.5 M EDTA: 100 mL of 0.5 M EDTA was prepared by dissolving 18.6 g of EDTA in 75 mL of distilled water. NaOH pellets can be added to dissolve the EDTA, as it takes time to dissolve completely. The pH was adjusted to 8.0 using 1 N NaOH. Finally, the volume was made up to 100 mL with distilled water and autoclaved at 15 psi pressure for 20 min.

5 M NaCl: 5.84 g of NaCl was added to 80 mL of distilled water and distilled completely, finally made up to 100 mL and autoclaved at 15 psi pressure for 20 min.

5% CTAB: 100 mL of 5% CTAB was prepared by dissolving 5 g of CTAB in 80 mL distilled water and thereby 2 g of NaCl was added to it. Then the volume was made up to 100 mL and autoclaved at 15 psi pressure for 20 min.

Chloroform:Isoamyl alcohol (24:1): 100 mL of 24:1 Chloroform:Isoamyl alcohol was prepared by adding 4 mL of isoamyl alcohol in 96 mL of chloroform and mixing. This mixture has to be made ice cold before use.

0.1× TE: 20 mL of 0.1× TE was prepared by adding 0.02 mL of 1 M Tris (pH-8.0) and 0.004 mL of 0.5 M EDTA (pH-8.0) to 18 mL of distilled water. After dissolving properly, the volume was made up to 20 mL and autoclaved at 15 psi pressure for 20 min.

Preparation of 50× TAE buffer: Dissolve 242.0 g of Tris base in 700 mL Milli-Q water with gentle heating. Add 57.1 mL glacial acetic acid, followed by 100 mL of 0.5 M EDTA (pH 8.0). Mix thoroughly and make the final volume to 1000 mL with Milli-Q water. Adjust pH to 7.6–7.8. Autoclave the buffer and store in a room temperature.

Ethidium bromide (EtBr) preparation: Ethidium bromide (EtBr) is a potent mutagen used to visualize nucleic acids, typically prepared as a 10 mg/mL stock solution in water (stored in the dark) and used at a 0.5–1.0 μg/mL working concentration for agarose gels. It is added directly to molten agarose or used for post-electrophoresis staining to detect DNA/RNA via fluorescence under UV light.

- **Stock solution preparation (10 mg/mL)**

  Dissolve 1.0 g of EtBr powder in 100 mL of distilled water. Store in a dark, light-proof container or wrapped in aluminum foil at 4°C and handle with gloves or use a fume hood to avoid inhaling dust

- **Working solution preparation (0.5–1.0 μg/mL):**
  - **In-gel staining:** Add 5–10 μL of the 10 mg/mL stock to 100 mL of molten agarose (<60°C) for a final concentration of 0.5–1.0 μg/mL.

– **For post-staining:** Dilute the stock solution in electrophoresis buffer (1× TAE or TBE) to a final concentration of 0.5–1.0 μg/mL and immerse the gel for 20–30 min.

0.25 M HCl: HCl from a 12 M Stock. Take 20.8 mL of HCl in a flask. Then, add 979.2 mL of HCl and mix by swirling the flask.

SDS (sodium dodecyl sulfate): Dissolve 10 g of electrophoresis-grade SDS in 90 mL of deionized water. Heat gently to 68°C to dissolve. Adjust volume to 100 mL.

3 Methods

3.1 Selection of Cotton Genotypes: Germplasm Collection and Documentation

Collect diverse cotton genotypes from authenticated sources. Assign each genotype a unique accession code and record its origin, morphological description, and known performance traits.

a. **Seed treatment and experimental layout:** Treat seeds with a suitable fungicide to ensure uniform germination. Arrange the field trial using randomized complete block design (RCBD) or alpha lattice design with two to three replications to minimize environmental variation.

b. **Field establishment and early growth assessment:** Sow seeds at recommended spacing and record germination percentage, seedling vigor, and plant height during the early vegetative stage and remove off types of uniformity.

c. **Growth and developmental trait measurement:** Monitor major growth traits, including number of sympodial branches, phenology, plant type, and canopy structure. These parameters help identify early, medium or late-maturing genotypes.

d. **Screening for biotic stress tolerance:** Evaluate genotypes under natural or artificial infection of cotton leaf curl virus, bacterial blight and fusarium wilt. Use standard disease rating scales to identify resistant or moderately resistant lines.

e. **Screening for abiotic stress tolerance:** Expose genotypes to controlled drought, salinity, or heat conditions. Record physiological parameters such as relative water content, chlorophyll stability index, and survival percentage.

f. **Molecular marker-based characterization:** Extract genomic DNA and genotype the selected lines using standardized marker RFLP. Use polymorphic markers to assess genetic diversity and confirm the identity of superior genotypes.

g. **Statistical analysis and final selection:** Perform ANOVA, estimate heritability and generate stability parameters. Select superior genotypes based on integrated performance across agronomic, fiber, and molecular traits.

3.2 Extraction of Genomic DNA [10–16]

3.2.1 Extraction of Genomic DNA Through Modified CTAB Method (for Plants Containing High Polysaccharide and Polyphenol Component)

1. 1.0 g of plant leaf tissue should be ground using liquid nitrogen with a chilled mortar and pestle, and transfer the powder to a polypropylene tube (30 mL) using a chilled spatula (50 mg of PVP for 0.5 g tissue should be added when grinding the tissue) (*see* **Notes 1–3**).
2. 10 mL of DNA extraction buffer (preheated at 60°C) should be added, mix the tubes by inversion, keep at 60°C for 45 min, and mix once every 15 min. Take the tubes and bring to room temperature (RT) (*see* **Note 4**).
3. Add 10 mL of chloroform:Isoamyl alcohol (24:1) and mix by inversion until it forms an emulsion.
4. Centrifuge the tubes for 10 min at 6000 rpm (25°C), and repeat the steps one more time to remove the cloudiness from the samples.
5. Carefully transfer the supernatant to fresh tubes, add and mix ½ volume at 5 M NaCl, add 2.5 volumes of 95% ethanol (pre-chilled at 4°C), mix the tubes, and then keep at −20°C for an hour.
6. Spin at 6000 rpm for 10 min (25°C), discard the aqueous solution, wash the pellet with 70% ethanol, and spin for 5 min. Air dry the pellet for 1 h, add 300 μL of 0.1× TE to dissolve the pellet, and keep for 4°C overnight.
7. Dissolved DNA should be transferred to 1.5 mL Eppendorf tube, and 5 μL of RNase A (10 mg/mL) should be added and incubated for 1 h at 37°C water bath.
8. Extract an equal volume of neutral phenol twice with the ether solution to remove the phenol solution completely.
9. Add 1/10 volume of 3 M sodium acetate pH 5.2 and 2.5 volumes of 90% ethanol to the solution to precipitate genomic DNA. Keep the tubes overnight at −70°C.
10. Next day, spin the tubes for 12,000 rpm, 10 min at 4°C, discard the supernatant, and wash the pellet twice with 70% ethanol (pre-chilled). Finally, dry the pellet and dissolve it with 100 μL of 0.1× TE buffer.
11. The quality and quantity of the extracted genomic DNA should be analyzed by 0.8% agarose gel electrophoresis and a Nanodrop spectrophotometer, respectively, prior to the restriction digestion.

3.2.2 DNA Extraction by the CTAB Method

1. Weigh 1 g of the sample, and freeze it in liquid nitrogen.
2. Grind the frozen tissue into a fine powder using a chilled mortar and pestle in the presence of liquid nitrogen.

3. Approximately 350 g of the powder should be distributed in each of three micro fuse tubes (micro fuse tubes must be chilled in liquid nitrogen before and after weighing). The spatula should be pre-chilled and used to transfer the ground powder. An equal volume (w/v) of pre-hot (65°C) 2× CTAB buffer should be added.
4. Mix well by inversion in a vigorous manner, and an equal volume of (700 μL v/v of ice-cold chloroform:Isoamyl alcohol (24:1) should be added.
5. Mix thoroughly by inversion to form an emulsion, and centrifuge for 5 min at 14,000 rpm.
6. Then, using the cut tips, the upper aqueous content should be transferred to new microfuge tubes.
7. Measure the volume. 1/5 volume of 5% CTAB solution should be added, and mixed well by gentle inversion.
8. An equal volume of chloroform:Isoamyl alcohol (24:1) should be added. Then mix thoroughly to form an emulsion and centrifuge for 5 min at 14,000 rpm.
9. Using cut tips, the upper aqueous content should be transferred to new centrifuge tubes, then add an equal volume of CTAB Precipitation buffer, and mix gently by inversion.
10. Centrifuge for 1 min at 14,000rpm, discard the supernatant, and keep the tube by inverting it in tissue paper and drying it for 30 min (*see* **Note 5**).
11. Dissolve the pellet in 200 μL of high-salt TE buffer, then gently tap the tubes. Keep it overnight at 4°C fridge.
12. The next day, tap the tube gently and make sure the pellet dissolved completely, and then 2.5 volumes of ice-cold 95% ethanol should be added, and mix gently by inversion.
13. Centrifuge for 10 min at 14,000 rpm at 4°C and discard the supernatant.
14. Add 500 μL of 70% ethanol to the pellet, and centrifuge for 5 min at 14,000 rpm at 4°C.
15. Air-dry or vacuum-dry the pellet and dissolve in 50 μL of 0.1× TE. Leave it for some time at room temperature or 4°C, then analyze the 2 μL of isolated genomic DNA by 0.8% agarose gel electrophoresis.
16. Followed by the purity and quantity of extracted and gel analyzed genomic DNA, which is checked by Nanodrop spectrophotometer prior to the restriction digestion process (*see* **Note 6**) (refer **Chap. 12**: "Development and Application of Random Amplified Polymorphic DNA (RAPD) for the Improvement of Cotton Crops").

3.3 Southern Blotting

3.3.1 Mock Restriction Digestion Analysis of Genomic DNA

1. Prior to processing the RFLP experiment, the minimum quantity (1 μg) of extracted genomic DNA should be restricted with ten units of enzymes and gel analyzed.
2. The resulting high-quality DNA, which is completely restricted, can then be used for RFLP applications in cotton improvement, including marker development, genetic diversity studies, mapping of fiber quality traits, and identification of varietal differences important for breeding.

3.3.2 Restriction Digestion for RFLP Experiment [17]

1. After confirmation of complete digestion of genomic DNA by mock restriction digestion analysis, the RFLP restriction digestion setup (Table 6) should be assembled, and samples are prepared for electrophoresis. Clean, fully digested DNA fragments are essential for resolving RFLP polymorphisms (*see* **Note 7**) (Fig. 1).
2. A suitable agarose concentration is chosen depending on the expected RFLP fragment sizes. Lower percentage gels are preferred for resolving larger fragments commonly produced in cotton RFLP analysis (*see* **Note 8**).
3. Agarose is dissolved in an appropriate electrophoresis buffer (TAE/TBE) that maintains pH and ionic strength. The gel is poured into a casting tray with a comb to create a uniform well for sample loading.
4. Digested DNA is gently mixed with a tracking solution that increases sample density and allows visual monitoring of fragment migration. Care is taken to avoid shearing the long fragments characteristic of RFLP profiles.
5. Each DNA digest is carefully pipetted into assigned wells along with a suitable DNA size ladder. The size ladder provides reference fragment lengths for interpreting RFLP polymorphisms.
6. An electric field is applied across the gel to drive the migration of DNA fragments toward the positive electrode. Large fragments migrate slowly while smaller fragments move faster, producing a size-based separation.
7. During electrophoresis, migration of the tracking dye (bromophenol blue) is observed to ensure adequate separation. Proper spacing between bands is essential for distinguishing fragment length polymorphisms among cotton genotypes.
8. Once sufficient separation is achieved, the power is turned off, and the gel is removed from the chamber. At this stage, the DNA remains embedded in the agarose matrix for downstream blotting.

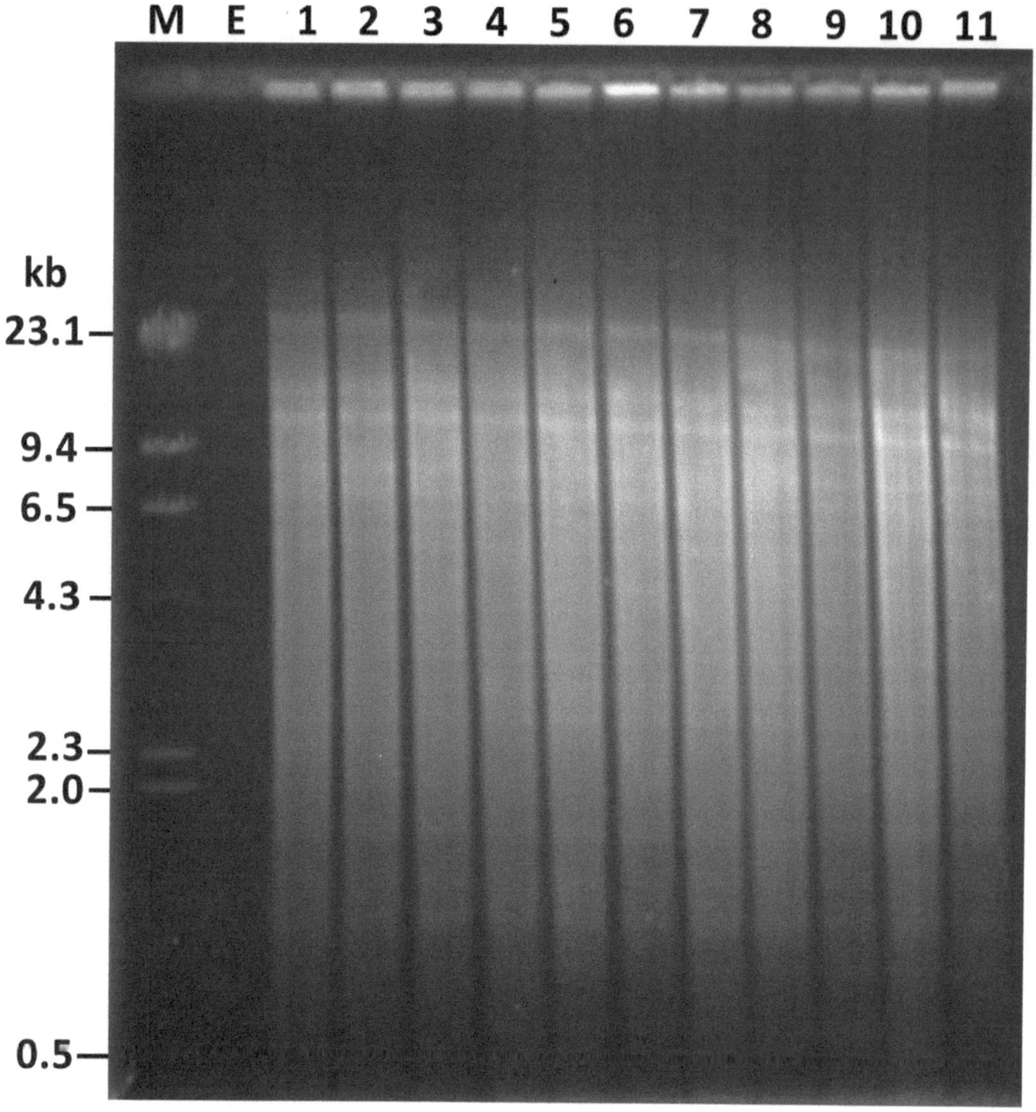

Fig. 1 Representative image shows separation of *Hind* III-digested cotton genomic DNA by 1× TAE agarose gel electrophoresis. Lane M: λ *Hind* III marker (ThermoFisher Scientific, USA), Lane E: Empty well, Lane 1–11: Different cotton germplasm restricted with *Hind* III enzyme (New England Biolabs, USA)

9. Before proceeding to Southern blotting, an optional visualization step is performed to confirm the presence of DNA and uniform digestion. Only intact, well-resolved digestion patterns are selected for the transfer.
10. The resolved gel is equilibrated to allow efficient capillary transfer of DNA onto a membrane. High-quality separation at this stage is critical for producing clear, interpretable RFLP banding patterns used in cotton genetic improvement.

3.3.3 Gel Treatment Before the Transfer of DNA [18]

After restriction enzyme digestion and agarose gel electrophoresis of cotton genomic DNA (see **Chap. 1**: "RAPD in Cotton"), the gel must be chemically treated before transferring DNA fragments onto a membrane. These treatments ensure efficient depurination, denaturation, neutralization, and transfer of DNA to the nylon or nitrocellulose membranes for hybridization with RFLP probes (*see* **Notes 8 and 9**).

3.3.4 Depurination

1. Cotton genomic DNA often produces large restriction fragments.
2. The gel is soaked in 0.25 M HCL for 10–15 min with gentle shaking (*see* **Note 10**).
3. This step partially removes purine bases, breaking large DNA fragments into smaller pieces.
4. Depurination improves the transfer efficiency of high-molecular-weight DNA fragments onto the membrane. The gel is rinsed briefly with distilled water after this step.

3.3.5 Denaturation of DNA in the Gel

1. The gel is incubated in denaturation solution (see Table 4), usually with 0.5 NaOH and 1.5 M NaCl for about 20–30 min. This treatment converts double-stranded DNA into single-stranded DNA, which is essential for probe hybridization.
2. Denaturation also disrupts hydrogen bonds between DNA strands, making fragments accessible for binding to the membrane (*see* **Note 11**).

3.3.6 Neutralization of the Gel

1. After denaturation, the gel is transferred to a neutralization buffer (commonly 1 M Tris-HCl, pH 7.4–7.5, with 1.5 M NaCl) for about 20–30 min.
2. Neutralization restores a stable pH and prevents damage to nylon or nitrocellulose membranes. This step prepares DNA for efficient and stable binding during transfer.

3.3.7 Preparation of the Gel for Transfer to the Membrane

The treated gel is equilibrated briefly in transfer buffer (such as 20× SSC). Air bubbles are removed from the gel surface to ensure uniform DNA transfer. The gel is now ready for capillary, vacuum, or electroblotting transfer onto a membrane (*see* **Notes 12–15**).

3.3.8 Importance of Gel Transfer

Proper gel treatment ensures clear and intact RFLP patterns. Accurate transfer enhances detection of polymorphisms linked to traits such as fiber quality, yield, and disease resistance. Reliable RFLP data support linkage mapping, QTL analysis, and marker-assisted selection in cotton breeding programs.

3.3.9 Probe Preparation

1. Probe preparation is a crucial step in RFLP analysis as the probe specifically detects complementary DNA sequences on the membrane. In cotton RFLP studies, probes are usually single-copy genomic clones or cDNA fragments associated with

agronomically important traits. The selected DNA fragment is first purified and denatured to obtain single-stranded DNA.

2. Labeling of the probe is carried out either by radioactive methods or by nonradioactive methods using digoxigenin or biotin. The labeled probe is then denatured by heating and immediately chilled on ice before hybridization to prevent re-annealing. Properly prepared probes ensure high specificity and sensitivity when detecting polymorphic restriction fragments in cotton genotypes.

3.3.10 Hybridization

1. Hybridization involves the binding of the labeled probe to its complementary DNA sequence immobilized on the membrane. Prior to the hybridization, the membrane containing transferred cotton genomic DNA is pre-hybridized in a buffer containing blocking agents such as Denhardt's solution and SDS (see Tables 8 and 9). This step minimizes nonspecific probe binding (*see* **Note 16**).

2. Hybridization is then carried out by adding the denatured probe to fresh hybridization buffer and incubating 60–65°C, depending on probe length and GC content. During this incubation, the probe anneals specifically to complementary restriction fragments. Efficient hybridization is essential for reliable detection of polymorphisms among cotton varieties or mapping populations (see Table 10) (*see* **Notes 17 and 18**).

Table 8

Denhardt's solution (50× Stock) (ThermoFisher Scientific, USA)

Composition	50× Stock
Ficoll	1%
Polyvinylpyrrolidone	1%
Bovine serum albumin	1%

Table 9

Pre-hybridization buffer composition

Composition	Concentration	Final volume 100 mL
SSC (saline-sodium citrate)	5×	25–30 mL of 20× SSC stock
Denhardt's solution	5×	10 mL of 50× Denhardt's stock
SDS (sodium dodecyl sulfate)	1%	5–10 mL of 10% SDS stock
EDTA (pH 8.0)	1 mM	0.2 mL of 0.5 M EDTA stock
Denatured salmon sperm DNA	100 μg/mL	1 mL of 10 mg/mL stock

Table 10
Post-hybridization buffer composition

Wash	Buffer composition
I	2× SSC+0.1% SDS
II	0.5× SSC+0.1% SDS
III	0.1× SSC+0.1% SDS

3.3.11 Washing

Post-hybridization washing removes unbound and nonspecifically bound probes, thereby improving signal clarity. Washing is performed stepwise using buffers of increasing stringency, typically SSC (saline sodium citrate) containing SDS. Initial washes at low stringency remove loosely bound probes, while subsequent high-stringency washes at elevated temperatures ensure that only perfectly matched probe-DNA hybrids remain on the membrane. Proper washing conditions are critical to reduce background noise and enhance the resolution of RFLP banding patterns in cotton DNA samples (*see* **Notes 19 and 21**).

3.3.12 Detection

Detection of hybridized probes depends on the labeling system used. In radioactive detection, the membrane is wrapped and exposed to X-ray film or a Phosphorimager screen to visualize autoradiographic bands. In nonradioactive systems such as DIG-labeled (manufacturer details) probes, detection involves incubation with enzyme-linked antibodies followed by a chromogenic or chemiluminescent reaction. The resulting bands represent restriction fragments containing sequences complementary to the probe. Clear and reproducible banding patterns enable the identification of polymorphisms among cotton genotypes, which is essential for genetic analysis and breeding applications (Fig. 2) (*see* **Notes 20 and 22**).

3.4 Data Analysis of RFLP [18]

1. RFLP data analysis involves scoring the detected banding patterns to identify polymorphisms among cotton samples. Each distinct band is treated as a marker locus, and the presence or absence of bands is scored across genotypes. Fragment sizes are estimated by comparing bands with molecular weight markers (*see* **Notes 23 and 25**).
2. Polymorphic bands are used to calculate genetic similarity or diversity using appropriate statistical methods such as T-REX, T-Align, PowerMark, PAST, etc. In cotton improvement programs, RFLP data are further used to construct genetic linkage maps and to identify quantitative trait loci related to fiber quality, yield, and stress tolerance.

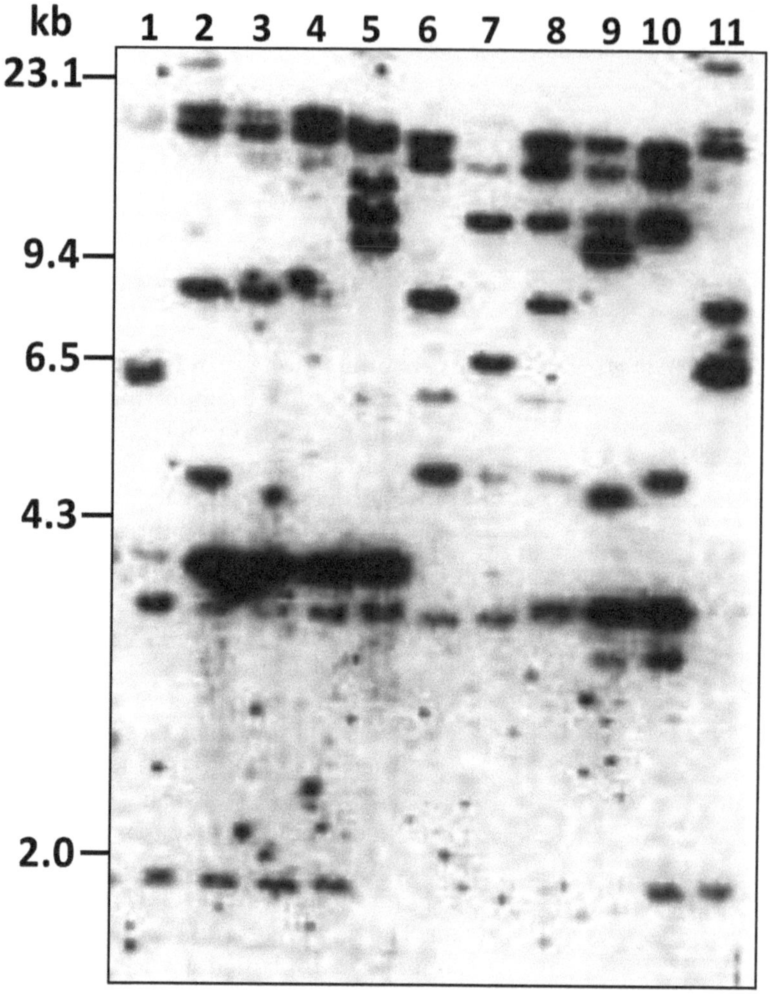

Fig. 2 Representative image of autoradiography shows the polymorphism of different genotypes. Genomic DNA were restriction digested with a restriction digestion enzyme, such as *Hind*III/*Eco*RI/*Eco*RV/*Bam*HI, and separated by agarose gel electrophoresis, processed with the Southern protocol, and then the gel was transferred onto a nylon membrane, followed by hybridization with a α-^{32}P dCTP labeled specific probe. Lane 1–11 represents various genotypes [19]

3. The stability and codominant nature of RFLP markers make them particularly valuable for long-term breeding and molecular characterization of cotton germplasm.

3.5 Scoring RFLP Patterns

Scoring of restriction fragment length polymorphism patterns refers to the systematic process of identifying, recording, and interpreting the banding patterns obtained after Southern blot hybridization. These patterns represent allelic differences among cotton genotypes and are crucial for genetic improvement programs such as linkage mapping, QTL identification, parental selection, and marker-assisted breeding (*see* **Note 27**).

3.6 Identification of Clear Bands

After hybridization and detection, each cotton genotype displays a unique set of DNA bands. Only sharp, well-defined bands are considered for scoring to avoid incorrect marker interpretation. Bands correspond to allele variants created by restriction site changes, insertions or deletions within cotton genomes.

3.7 Scoring Presence or Absence of Bands

Each band is treated as a genetic marker where bands are scored in a binary format, where 1 = band present and 0 = band absent. The presence or absence of scoring helps to differentiate polymorphic loci, essential for distinguishing elite cultivars from wild cotton accessions (*see* **Note 28**).

3.8 Determining Allelic Variants

RFLP bands represent different alleles. If multiple fragments hybridize with the same probe, they are scored as multi-allelic loci. These allelic variations help breeders identify useful genomic regions associated with Fiber fineness, Boll size, Lint percentage, Pest and disease resistance, Drought, and salinity tolerance.

3.9 Scoring Co-Dominant Nature of RFLP

Unlike dominant markers (RAPD/ISSR), RFLPs are codominant, allowing detection of (*see* **Note 24**).

1. Homozygotes, which will be with single band.
2. Heterozygotes, which will have different bands.

This gives precise data for cotton breeding programs where genome composition must be accurately tracked.

3.10 Comparing Band Sizes

Fragment sizes are compared by aligning cotton genotypes lanes with a DNA ladder. Accurate fragment size measurement enables the identification of new alleles, the estimation of genetic distance, and the mapping of important loci controlling fiber traits.

3.11 Generating a Marker Data Matrix

All scored bands (0/1) are transferred into a genotype-marker matrix. This matrix is used to construct genetic linkage maps and to perform QTL analysis of cotton yield and fiber quality traits. Assessing genetic diversity among breeding lines.

3.12 Scoring Polymorphic vs. Monomorphic Bands

1. Polymorphic bands show variation among cotton varieties, which is useful for the improvement programs.
2. Monomorphic bands show no variation, which is excluded from selection. This will ensure that only informative loci will contribute to the mapping and selection (*see* **Note 26**).

3.13 Statistical Analysis of Scored Data

Scored patterns are analyzed using the Jaccard similarity index, Nei's genetic distance, and cluster analysis (UPGMA). These analyses reveal genetic relationships among cotton genotypes and guide breeders in selecting diverse parents by hybridization.

3.14 Integration into Breeding Programs

Scored RFLP markers linked to desirable traits will allow for early selection in segregating populations, tracking of trait introgression from wild species, and identification of superior recombinants. This accelerates the development of improved cotton varieties with enhanced fiber traits, stress tolerance, and disease resistance.

3.15 Interpretation for Cotton Crop Development

The final scored RFLP patterns will help breeders to:

1. Identify genomic regions influencing key agronomic traits.
2. Develop molecular linkage maps for cotton breeding programs.
3. Select parents with complementary alleles.
4. Improve the genetic base of cotton through marker-assisted selection.

4 Notes

1. Always use young, actively growing cotton leaves for genomic DNA extraction, as mature leaves contain higher levels of polysaccharides and phenolics that inhibit restriction digestion.
2. Grinding of tissue must be done rapidly in liquid nitrogen; incomplete grinding results in poor cell lysis and low DNA yield.
3. The addition of PVP and freshly prepared β-mercaptoethanol is essential when working with cotton tissues to prevent oxidation of polyphenols that can irreversibly damage DNA.
4. Avoid vigorous vortexing after adding chloroform:isoamyl alcohol, as this can shear high-molecular-weight DNA required for RFLP analysis.
5. DNA pellets should not be over-dried, as this makes resuspension difficult and may affect restriction enzyme activity.
6. Always verify DNA quality by restriction digestion before proceeding to large-scale RFLP analysis. Partial digestion indicated contamination with polysaccharides or phenolics.
7. Use an excess of restriction enzyme and extend the incubation for cotton DNA, as plant genomic DNA often shows resistance to complete digestion.

8. Agarose concentration should be optimized, typically (0.7%–0.8%), to allow efficient separation of large restriction fragments common in cotton RFLP studies.
9. Avoid overheating during electrophoresis, as excessive heat can cause band distortion and reduce the resolution of polymorphic fragments.
10. Depurination using dilute HCL is recommended when fragment sizes exceed 15 kb, as this significantly improves DNA transfer efficiency during Southern blotting.
11. Ensure complete denaturation of DNA in the gel; incomplete denaturation will reduce probe hybridization efficiency and lead to weak signals.
12. The membrane used for blotting should always be slightly larger than the gel to ensure complete transfer of DNA fragments.
13. Avoid touching the membrane with bare hands or forceps directly, as oils and pressure marks interfere with DNA binding and signal detection.
14. Formation of air bubbles between the gel and membrane must be avoided, as bubbles prevent DNA transfer and cause blank regions on autoradiographs.
15. Nylon membranes generally offer greater binding capacity and durability than nitrocellulose and are preferred for repeated hybridizations.
16. Pre-hybridization is a critical step to reduce background noise; insufficient blocking often leads to nonspecific probe binding.
17. Probe concentration should be optimized; excessive probe results in high background, while insufficient probe leads to weak hybridization signals.
18. Hybridization temperature must be adjusted based on probe length and GC content to ensure probe-target binding specificity.
19. Post-hybridization washes should be performed stepwise with increasing stringency to balance signal strength and background reduction.
20. In case of high background on X-ray film or chemiluminescent images, increase wash stringency or reduce probe concentration rather than repeating the entire experiment.
21. If no signal is observed, check probe labeling efficiency, DNA transfer quality, and exposure time before repeating hybridization.

22. Autoradiography cassettes and films must be handled carefully and kept free of dust, as artifacts can be misinterpreted as polymorphic bands.
23. Only sharp and reproducible bands should be scored for RFLP analysis; faint or diffuse bands should be excluded to avoid scoring errors.
24. RFLP markers are co-dominant; therefore, heterozygous and homozygous banding patterns must be clearly distinguished during scoring.
25. Fragment sizes should always be estimated using an appropriate molecular weight marker run alongside the samples to ensure accurate allele identification.
26. Monomorphic bands provide limited breeding value and should be excluded from diversity and linkage analysis.
27. RFLP data matrices should be carefully checked for scoring consistency before statistical analysis to avoid errors in linkage mapping or diversity estimation.
28. Proper documentation of experimental conditions, probe details, and scoring criteria is essential for reproducibility and long-term utility of RFLP data in cotton improvement programs.

References

1. Reinisch AJ, Dong JM, Brubaker CL, Stelly DM, Wendel JF, Paterson AH (1994) A detailed RFLP map of the cotton genome. Genet 138:829–847
2. Sambrook J, Fritsch EF, Maniatis T (1989) Molecular cloning: a laboratory manual, 2nd edn. Cold Spring Harbor Lab Press
3. Williams JGK, Kubelik AR, Livak KJ, Rafalski JA, Tingey SV (1990) DNA polymorphisms amplified by arbitrary primers are useful as genetic markers. Nucleic Acids Res 18(22):6531–6535
4. Iqbal MJ, Khan IA, Javed M (2010) Genetic diversity and selection of superior cotton genotypes using morphological and molecular markers. Afr J Biotech 9(48):8235–8243.
5. Ali Q, Ahsan M, Ali F, Kanwal N (2019) A modified protocol for rapid DNA isolation from cotton (Gossypium spp.). Genet Mol Res 18(2):18218
6. Shappley ZW, Jenkins JN, Watson CE, Kahler AL, Meredith WR (1998) Establishment of molecular markers and linkage groups in two F2 populations of Upland cotton. Theor Appl Genet 92(7):915–919
7. Reddy OUK, Pepper AE, Abdurakhmonov I, Saha S, Jenkins JN, Brooks T, Bolek Y, El-Zik KM (2001) New dinucleotide and trinucleotide microsatellite marker resources for cotton genome research. J Cotton Sci 5(2):103–113
8. Verhalen LM, El-Zik KM, Thaxton PM (2003) Genetics of fiber quality. In: Cotton origin, history, technology, and production. John Wiley & Sons, Inc
9. Qureshi SN, Saha S, Kantety RV, Jenkins JN (2004) EST-SSR: a new class of genetic markers in cotton. J Cotton Sci 8:112–123
10. Stewart JM (2001) Genetics and cytogenetics. In: Compendium of cotton diseases. American Phytopathological Society
11. Porebski S, Bailey LG, Baum BR (1997) Modification of a CTAB DNA extraction protocol for plants containing high polysaccharide and polyphenol components. Plant Mol Biol Rep 15(1):8–15
12. Sivakumar JS, Gurusamy D, Chellamuthu S, Rathnasamy SA, Krishnan V (2015) Characterization of transgenic cotton plants by Southern blotting. In: Plant transformation methods and laboratory protocols. Springer US, New York, pp 171–181
13. Rogers SO, Bendich AJ (1994) Extraction of total cellular DNA from plants, algae and fungi.

In: Gelvin SB, Schilperoort RA (eds) Plant molecular biology manual. Springer, Dordrecht
14. Gurusamy D (2014) Transcriptome analysis of cotton (*Gossypium hirsutum* L.) during boll development under drought condition and functional characterization of drought tolerance genes. Ph.D. thesis. https://hdl.handle.net/10603/215568
15. Sivakumar JS, Gurusamy D, Chellamuthu S, Rathnasamy SA, Krishnan V (2025) Characterization of transgenic cotton plants by PCR. Methods Mol Biol 2973:157–170
16. Dhandapani G, Kanakachari M, Padmalatha KV, Phanindra MLV, Singh VK, Raghavendra Rao S, Kumar PA (2015) A gene encoding cold-circadian rhythm-RNA binding-like protein (CCR-like) from upland cotton (*Gossypium hirsutum* L.) confers tolerance to abiotic stresses in transgenic tobacco. Plant Mol Biol Rep 33(1):22–42
17. Zhang J, Stewart JM (2000) Economical and rapid method for extracting cotton genomic DNA. J Cotton Sci 4:193201
18. Reinisch AJ, Dong JM, Brubaker CL, Stelly DM, Wendel JF, Paterson AH (1994) A detailed RFLP map of cotton, *Gossypium hirsutum* × *Gossypium barbadense*: chromosome organization and evolution in a disomic polyploid genome. Genetics 138(3):829–847
19. Nagaraju J, Reddy K, Nagaraja G (2001) Comparison of multilocus RFLPs and PCR-based marker systems for genetic analysis of the silkworm, *Bombyx mori*. Heredity 86:588–597

Chapter 14

Development and Application of Expressed Sequence Tags (ESTs) for the Improvement of Cotton Crops

Gokul Murugesan, Tanu Priya Natarajan, Janani Sree Sivakumar, Dhandapani Gurusamy, Kanakachari Mogilicherla and Vasanth Krishnan

Abstract

Expressed sequence tags (ESTs), are short cDNA sequences obtained from expressed genes in any plant tissue of *Gossypium* spp. and represent a suitable approach for studying the transcriptome in the absence of a reference genome. In this study, protocols for EST development were established to explore expressed genes in cotton under diverse biotic and abiotic stress conditions. ESTs were generated by using two approaches: suppression subtractive hybridization (SSH) and next-generation sequencing (NGS) techniques. The process begins with RNA isolation, followed by mRNA purification, cDNA synthesis, and cloning, leading to the generation of single-pass EST libraries. High-quality ESTs were assembled into unigenes using complementary assembly strategies. Functional annotation of the resulting unigenes was carried out to enhance gene representation and to elucidate stress-responsive pathways. This approach supports key objectives in cotton breeding, including improvement of fiber quality and productivity, as well as enhanced resistance to biotic and abiotic stresses.

Keywords Cotton, Total RNA, Complementary DNA, Expressed sequence tags, Suppression subtractive hybridization, Cloning, Macroarray, Next generation sequencing

1 Introduction

Cotton, native to tropical and subtropical regions [1], stands as the world's most pertinent natural fiber crop and is often referred to as "white gold," accounting for over 50% of global clothing fiber production. It also ranks as the second most important oilseed crop [2]. Four species are commercially cultivated: the diploids *Gossypium herbaceum* L. and *Gossypium arboreum* L., native to Asia and Africa, and the tetraploids *Gossypium hirsutum* L. and *Gossypium barbadense* L., native to the Americas. *G. hirsutum* dominates global production, accounting for over 95%, followed by *G. barbadense* (2%–3%) and *G. arboreum* (1%–2%). Compared with other species, *G. hirsutum* exhibits superior yield potential, adaptability, and average fiber quality. In contrast, *G. herbaceum* is cultivated to a lesser

Nitish Kumar (ed.), *Molecular Markers in Plants*, Methods in Molecular Biology, vol. 3097,
https://doi.org/10.1007/978-1-0716-5663-1_14,

extent due to its inferior fiber quality, higher disease susceptibility, and low yield potential [3]. Cotton fiber cells undergo multiple developmental processes, including initiation, elongation, primary and secondary wall formation, and desiccation [4, 5]. Cotton cultivation faces significant constraints such as extended growth periods, delayed maturation, reduced boll size, and increased susceptibility to biotic and abiotic stresses during prolonged reproductive phases. These constraints manifest as seedling mortality, heat-induced yield and fiber quality losses, impaired anther dehiscence, and boll shedding [1]. Expressed sequence tags (ESTs) are short, single-pass sequences ranging from 200 to 800 base pairs, generated from randomly selected cDNA clones. They represent the transcribed portion of the genome and provide an efficient and economical alternative to whole-genome sequencing [6]. Despite cotton's superior fiber quality, which positions it as a premium textile resource, expressed sequence tag (EST) screening offers an effective approach for gene discovery and the development of polymorphic DNA markers to address key production and stress-related limitations [3].

2 Materials

Cotton Samples (Leaf/Bolls/Fibers), Total RNA, Diethyl pyro carbonate DEPC-treated water (*see* **Note 1**), liquid nitrogen, mortar and pestle, spatula, microfuge tubes, vortex mixer, chloroform–isoamyl alcohol, isopropanol, 70% and 75% ethanol, ice box with ice, centrifuge, centrifuge tubes, sterile Whatman Nos. 1 and 3 filter paper, microtips, pipettes, incubator, molecular-grade water, TRIzol reagent, RNA isolation kit (Sigma-Aldrich, USA), nanodrop spectrophotometer, RNA purification kit, cDNA synthesis kit, Clontech® PCR-Select™ cDNA Subtraction Kit, ethidium bromide, Clontech® PCR-Select™ Differential Screening Kit, BigDye® Terminator v3.1 Cycle Sequencing Kit, microplate centrifuge, heating block, cloning vector, restriction enzymes, DNA ligase, plasmid and gel extraction kits, gel electrophoresis system, ABI genetic analyzer, UV-transilluminator, plasmid DNA, sequencing analysis software, competent cells (*Escherichia coli* host), thermal cycler, water bath, illumina cluster station, and phosphorimager.

3 Methods

3.1 Total RNA Isolation

3.1.1 Total RNA Isolation by TRIzol Method [7]

1. Homogenization of samples: Liquid nitrogen was used to homogenize 500 mg of cotton tissue using a prechilled, sterile mortar and pestle (Table 1) (*see* **Note 2**).

Table 1
Sample selection for RNA isolation

Fiber development studies	
Used as a tester	**Used as a driver**
(Sample of interest; the one in which we want to identify specific or upregulated genes)	(Reference or control sample used to remove common sequences)
Developing fibers (e.g., 5, 10, 15, and 20 DPA—days post anthesis)	Ovules without fibers
Elongating fibers	Non-fiber tissues (Leaf/Root)
Secondary cell wall stage fibers	Earlier developmental stages
Reproductive development	
Flower buds, Anthers, Ovules	Vegetative tissues

2. Transfer of samples to microfuge tubes: The 500 mg powdered tissue was transferred into fresh, sterile microfuge tubes using a prechilled sterile spatula.
3. Samples mix with TRIzol and chloroform: To that tube, 2.5 mL of TRIzol reagent was added, vortexed for 1 min, incubated at room temperature for 5 min, followed by the addition of 0.5 mL of chloroform (*see* **Note 3**).
4. Incubation: Tubes were tightly closed, shaken vigorously for 15 s, and incubated at room temperature for 10 min.
5. Centrifugation and upper aqueous phase collection: Samples were centrifuged at 14,000 × *g* for 15 min at 2°C–8°C. The mixture separated into three phases: an upper aqueous phase (RNA), an interphase (DNA), and a lower organic phase (proteins).
6. Mixing with isopropanol: The upper aqueous phase was carefully transferred to a fresh microfuge tube, and 0.5 mL isopropanol was added per mL of TRIzol used in Step 3.
7. Incubation: Samples were incubated at room temperature for 10 min and centrifuged at 12,000 × *g* for 10 min at 2°C–8°C.
8. Observation of RNA precipitation: RNA precipitated as a visible pellet at the bottom of the tube.
9. First washing: The supernatant was discarded, and the pellet was washed with 700 μL of 75% ethanol by gentle inversion, followed by centrifugation at 7500 × *g* for 2 min at 2°C–8°C.
10. Second washing: Step 9 was repeated once more.
11. Dry the RNA pellet: Ethanol was discarded, and the pellet was air-dried by inverting the tube on the sterile filter paper and incubating at 37°C for 5 min to remove residual ethanol (*see* **Note 6**).

12. Resuspension of RNA pellet: The RNA pellet was dissolved in 10–20 μL RNase-free water and incubated overnight at 4°C.
13. Analysis of isolated total RNA: RNA quality and quantity were assessed by formaldehyde denaturing 1× MOPS agarose gel electrophoresis (1.2% gel) and Nanodrop spectrophotometer readings at A260/A280 and A260/A230 nm.

3.1.2 Total RNA Isolation by Kit Method [7]

Total RNA was extracted using a Spectrum™ Plant total RNA extraction kit (Sigma-Aldrich, USA) following the manufacturer's instructions.

1. Homogenization of samples: Take 1 g of cotton tissue, freeze immediately in liquid nitrogen, and grind using a mortar and pestle.
2. Transfer of samples to microfuge tubes: The sample was transferred to fresh sterile DEPC-treated/RNase-free 1.5 mL microfuge tubes.
3. Add lysis solution: Then 1000 μL (1 mL) of the lysis solution with 2-mercaptoethanol (ME) (10 μL of 2-ME added to 1000 μL of lysis solution) was pipetted out and added to 500 mg of tissue sample, vortexed immediately for 30 s, incubated at 56°C for 5 min, and brought to room temperature.
4. Centrifugation: Samples were centrifuged at 10,000–12,000 rpm for 5 min to pellet down cellular debris.
5. Filtration of lysate using column: The lysate was transferred using cut tips to a filtration column (blue retainer ring) placed in a 2.0 mL collection tube.
6. Centrifugation of column with lysate: Column was centrifuged at 10,000–12,000 rpm for 1 min, which removes the residual debris, and the flow-through lysate was retained.
7. Binding solution to the lysate: Then 500 μL of binding solution was added to the lysate obtained in the previous step, vortexed gently for 1 min to ensure equal distribution of the binding solution.
8. Bind the samples to the binding column: Then 700 μL of the mixture is added to the binding column (red retainer ring) kept in 2.0 mL collection tube and centrifuged at 12,000 rpm for 1 min which ensures perfect binding to RNA.
9. Repeat the binding process: Discard the flow through. Collection tubes should be gently tapped on sterile absorbent paper, which removes and drains the excess residual liquid. Again, the column should be placed in the collection tube, and the remaining mixture should be added to the column, repeating Step 8 for centrifugation and decanting steps.

10. First wash (Wash Solution I): Then 500 μL of wash solution I was added into the column and centrifuged at 12,000 rpm for 1 min. Discard the flow through and tap the collection tube on sterile absorbent paper (upside down) to drain the residual liquid. Columns were returned to the collection tube.
11. Second wash (Wash Solution II): Then 500 μL of wash solution II was added into the column and centrifuged at maximum speed for 30 s. Discard the flow through and tap the collection on sterile absorbent paper (upside down) to drain the residual liquid. Columns were returned to the collection tube.
12. Third wash (Wash Solution II): Again, 500 μL of diluted wash solution II was added into the column and centrifuged at maximum speed for 30 s. Discard the flow through and tap the collection tube briefly on sterile absorbent paper (upside down) to drain the residual liquid. Columns were returned to the collection tube (*see* **Note 4**).
13. Dry spin: The collection tube containing the column should be centrifuged at 12,000 for 1 min to dry.
14. First elution: Transfer the column to a fresh, sterile 2.0 mL collection tube. Ten microliter of elution solution is pipette out and directly added onto the center of the binding matrix inside the column. Allow the tube to settle for 1 min and follow by centrifuging at around 12,000 rpm (Elution 1) for 1 min to elute out the bound RNA in the column.
15. Second elution: Now again transfer the column to a fresh, sterile 2.0 mL collection tube. Ten microliter of elution solution is pipette out and directly added onto the center of the binding matrix inside the column, allow the tube to settle for 1 min, and followed by centrifugation at around 12,000 rpm (Elution 2) for 1 min to elute out the bound RNA in the column (*see* **Note 5, 7, 10, and 11**).
16. **Analysis of RNA quality and quantity**: RNA quality and quantity were assessed by formaldehyde denaturing 1× MOPS agarose gel electrophoresis and NanoDrop spectrophotometer readings at A260/A280 and A260/A230 nm (*see* **Notes 8 and 13**).

3.2 Analysis of Total RNA [8]

3.2.1 Denaturing Agarose Gel Electrophoresis for Total RNA Analysis

1. Weigh 1.2 g of agarose and add it to a 250 mL conical flask containing 97 mL of 1× MOPS buffer (1.2% of gel).
2. Heat the mixture in a microwave oven until the agarose is completely dissolved.

3. Add 3 mL of 37% formaldehyde to the dissolved agarose solution.
4. Mix the solution thoroughly and pour it into the gel casting tray fitted with a comb.
5. Allow the gel to solidify (approximately 30–45 min). After solidification, carefully remove the comb without disturbing the gel and wells.
6. Place the gel along with the casting tray into the electrophoresis tank containing 1× MOPS buffer.
7. Prepare the samples by mixing the components listed in Table 2 in sterile 0.5 mL RNase-free Eppendorf tubes.
8. Incubate the samples in a prewarmed water bath or heating block at 70°C for 5 min, followed by immediate incubation on ice.
9. Briefly centrifuge the samples and load them into the respective wells of the gel.
10. Run the gel at 60 V for 40–60 min in the gel electrophoresis system containing 1× MOPS buffer, until the dye migrates to three-fourths of the gel length.
11. Visualize the separated RNA bands under a UV transilluminator using a gel documentation system and capture the images (Fig. 1).

3.3 Purification of mRNA

Poly (A)+ RNA from each pooled sample was purified using the mRNA isolation Midi Kit Columns (Qiagen, Valencia, CA, USA), according to the manufacturer's instructions. The protocol is briefly described below, and for more details, refer to the manufacturer's protocol.

1. Prewarm Oligotex suspension and elution buffer: Prior to the procedure, Oligotex suspension was pre-warmed at 37°C, and the elution buffer was preheated to 70°C.
2. Mix the RNA with buffer and Oligotex: For 250 μg of total RNA, 250 μL of Buffer A and 25 μL of Oligotex suspension were added and mixed thoroughly by pipetting up and down.

Table 2
Sample preparation for total RNA (*see* Note 9)

Sample	Volume (10 μL)
Total RNA (1.0 μg)	2.0
2× RNA gel loading dye (Thermo Fisher Scientific™)	5.0
Nuclease-free water	3.0

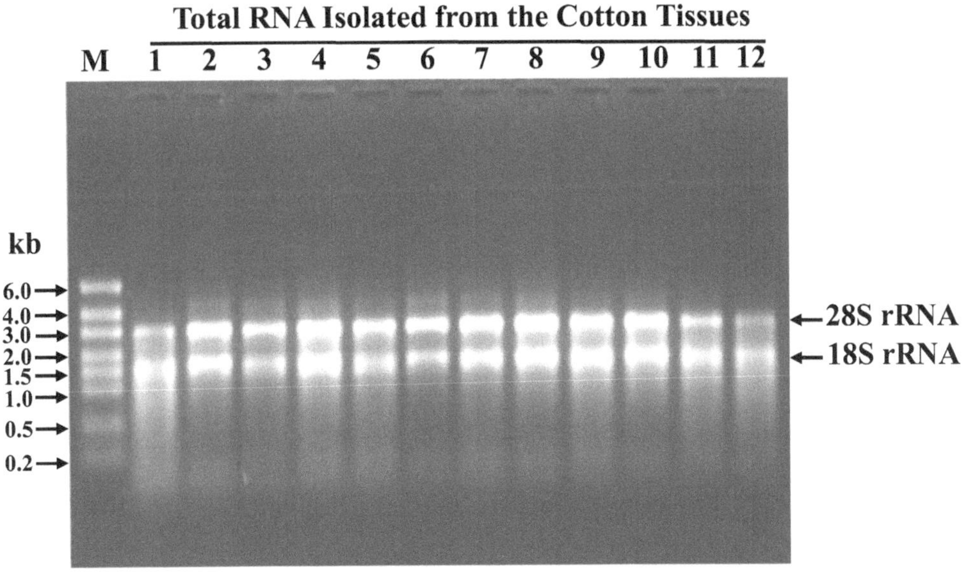

Fig. 1 1.2% MOPS (1×)—formaldehyde denaturing agarose gel electrophoresis analysis of total RNA isolated from the cotton tissues. Lane M: RiboRuler High Range RNA Ladder (Thermo Fisher Scientific Inc. USA), Lanes 1–12: Total RNA isolated from the cotton tissues [7,9]

3. Incubation: The samples were incubated in a water bath at 70°C for 3 min to disrupt the secondary structure of total RNA.
4. Incubation for hybridization: The samples were incubated at room temperature for 10 min to allow hybridization between the poly(A) tail of the mRNA and the oligo$(dT)_{30}$ on the Oligotex beads.
5. Oligotex–mRNA complex: The Oligotex–mRNA complexes were pelleted by centrifugation at 14,000–18,000 × *g* for 2 min, and the supernatant was discarded.
6. Washing Oligotex–mRNA complex: The pellet containing the Oligotex–mRNA complex was washed with wash buffer by gentle vortexing or pipetting.
7. Centrifugation of complex: The complexes were again pelleted by centrifugation at 14,000–18,000 × *g* for 2 min, and the supernatant was discarded.
8. Repeat washing: Steps 6 and 7 were repeated once to remove the contaminants.
9. Resuspension of Oligotex–mRNA complex with Elution buffer: Preheated elution buffer (20–100 μL at 70°C) was added to resuspend the Oligotex–mRNA complex, followed by centrifugation at 14,000–18,000 × *g* for 2 min.

10. The supernatant containing the eluted poly(A)+ mRNA was carefully transferred to a fresh RNase-free polypropylene tube.
11. Resuspension of residual Oligotex–mRNA complex with elution buffer: Step 9 was repeated, and both elutions were pooled to obtain a higher yield of poly(A)+ mRNA.

3.4 cDNA Synthesis [7]

cDNA synthesis for both tester and driver samples was carried out separately.

1. All components were centrifuged briefly before use for cDNA synthesis. The thermal cycler was preheated to 65°C.
2. The components listed in Table 3 were added to fresh 0.2 mL PCR tubes, and the tubes were kept on ice.
3. The reaction mixtures were incubated in the PCR tubes on a preheated thermal cycler at 65°C for 5 min and immediately transferred to ice for 1 min.
4. The components listed in Table 4 were then added to the respective tubes (*see* **Note 21**).
5. The samples were mixed thoroughly by a brief centrifugation and incubated in the thermal cycler at 25°C for 10 min, followed by incubation at 50°C for 50 min.

Table 3
Requirements for first-strand cDNA synthesis (primers and reagents from Thermo Fisher Scientific) (*see* Notes 12 and 23)

Requirements	Volume (13.0 μL)
mRNA RNA (2.0 μg)	2.0
10 mM dNTP mix (Thermo Fisher Scientific)	1.0
Random Primer (10.0 μM)	0.5
Oligo (dT) Primer (10.0 μM)	0.5
Nuclease-free water	9.0

Table 4
Requirements for second-strand cDNA synthesis (reagents and enzymes from Thermo Fisher Scientific)

Requirements	Volume (7.0 μL)
5× first-strand reaction mix (Thermo Fisher Scientific)	4.0
Dithiothreitol (DTT) (Thermo Fisher Scientific)	2.0
M-MLV reverse transcriptase (Thermo Fisher Scientific)	1.0

6. The reactions were terminated by incubation at 70°C for 15 min, followed by holding at 4°C.
7. The synthesized cDNA is analyzed by agarose gel electrophoresis and quantified using a Nanodrop spectrophotometer.

3.5 Agarose Gel Electrophoresis Analysis of cDNA (see Note 14)

1. Weigh 0.5 g of agarose and add it to a 150 mL conical flask containing 50 mL of 1× TAE buffer (refer Table 5).
2. Heat the mixture in a microwave oven until the agarose is completely dissolved.
3. Add 5 μL of ethidium bromide (10 mg/mL stock) to the dissolved agarose solution.
4. The above mixture is mixed well and poured into a gel casting tray, which is filled with a comb.
5. Allow the gel to solidify (approximately 30–45 min). After solidification, carefully remove the comb without disturbing the gel.
6. Place the gel along with the casting tray into the electrophoresis tank containing 1× TAE buffer.
7. Prepare the samples by mixing the components listed in Table 6 in sterile 0.5 mL Eppendorf tubes.
8. Mix the samples thoroughly by brief centrifugation or by pipetting up and down, and load them into the respective wells of the gel.

Table 5
TAE buffer preparation for gel electrophoresis (1000 mL) (pH 7.6–7.8)

Requirements	Volume (1000 mL)
Tris base (2 M)	242.2 g
Glacial acetic acid (1 M)	57.1 mL
EDTA (0.5 M)	100 mL

Table 6
Preparation of cDNA sample for gel electrophoresis analysis (*see* Note 20)

Sample	Volume (10 μL)
cDNA (1.0 μg)	2.0
6× DNA gel loading dye (ThermoFisher Scientific)	5.0
Nuclease-free water	3.0

9. Run the gel at 80 volts for 30–60 min in the gel electrophoresis system containing 1× TAE buffer until the dye migrates to three-fourths of the gel length.
10. Visualize the cDNA bands under a UV transilluminator using a gel documentation system, and capture the images (Fig. 2).

3.6 Construction of Subtractive Library [5]

3.6.1 *Rsa I Digestion*

Suppression subtractive library was constructed according to the user manual of the Clontech® PCR-Select™ cDNA Subtraction Kit PCR-Select™ cDNA Subtraction Kit (A Takara Bio Company, CA, USA).

Rsa I Digestion was performed separately for both tester and driver cDNAs to obtain shorter blunt-end fragments. The procedure was carried out using double-stranded driver and tester cDNA and control cDNA.

Follow these steps using the ds driver cDNA, ds tester cDNA, and the control cDNA (*see* **Note 15**).

1. Add the following reagents to each reaction mentioned in Table 7.
2. Vortex to mix, briefly centrifuge, and incubate at 37°C for 1.5 h.
3. To assess the efficiency of *Rsa* I digestion, set aside 5 μL of the digest mixture.

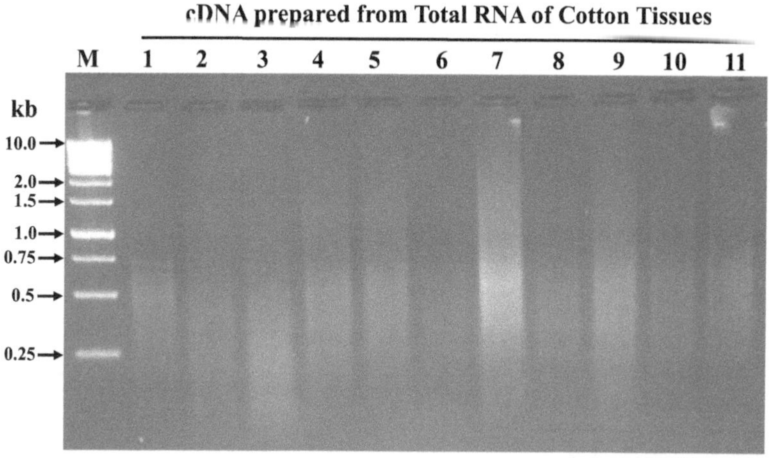

Fig. 2 Agarose (1.2%) gel electrophoresis analysis of cDNA synthesized from the total RNA of cotton tissues. Lane M: GeneRuler 1 kb DNA ladder (Thermo Fisher Scientific Inc., USA), Lanes 1–12: cDNA synthesized from the total RNA of cotton tissues [7,9]

Table 7
Sample preparation for *Rsa* I Digestion

Requirements	Volume (Per Rxn) total volume: 50 μL
ds cDNA	43.5
10× *Rsa* I Restriction Buffer	5.0
Rsa I (10 units/μL)	1.5

4. Terminate the reaction by adding 2.5 μL of the 20× EDTA/glycogen.
5. Add 50 μL of a phenol: chloroform:isoamyl alcohol mixture (25:24:1) (*see* **Note 14**).
6. Vortex vigorously and centrifuge at 14,000 rpm for 10 min at room temperature to separate phases.
7. Carefully transfer the aqueous layer to a fresh 0.5 mL tube.
8. Add 50 μL of chloroform:isoamyl alcohol (24:1) and repeat Steps 6 and 7.
9. Add 25 μL of 4 M NH_4OAc and 187.5 μL of 95% ethanol, then centrifuge as described in Step 6.
10. Discard the supernatant and gently add 200 μL of 80% ethanol over the pellets.
11. Centrifuge at 14,000 rpm for 5 min and carefully remove the supernatant.
12. Air-dry the pellets for 5–10 min, then dissolve them in 5.5 μL of H_2O, and store at −20°C.
13. Simultaneously, analyze the *Rsa* I-digested double-stranded cDNA by 1.2% agarose gel electrophoresis (Fig. 3).

3.6.2 Adaptor Ligation

In this step, two tester cDNA samples are separately ligated with two different adapters provided in the kit, whereas the driver cDNA is not ligated with any adapter.

1. The tester cDNA is aliquoted into two tubes. One tube is ligated with Adapter 1, and the other is ligated with Adapter 2R.
2. For sample preparation of tester cDNA (Samples 1 and 2), dilute 1 μL of each *Rsa* I-digested cDNA with 5 μL of sterile H_2O.
3. Preparation of Ligation master mix as shown in Table 8.
4. For tester cDNA Samples 1 and 2 prepare the ligation reactions setup as shown in Table 9.

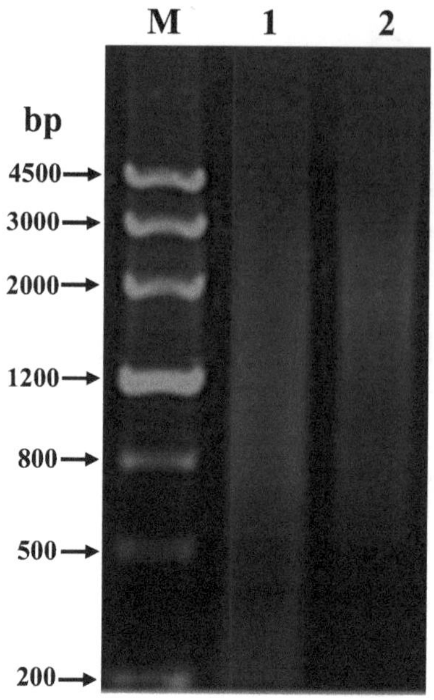

Fig. 3 Agarose (1.2%) gel electrophoresis analysis of double-stranded cDNA digested with *Rsa* I digestion. Lane M: 200–4500 bp seven bands DNA marker (EarthOx LifeSciences Pvt. Ltd., India), Lane 1: Double-stranded cDNA digested with *Rsa* I, Lane 2: Undigested double-stranded cDNA [10]

Table 8
Preparation of ligation master mix

Requirements	Volume (μL)
Sterile H_2O	3
5× Ligation buffer	2
T4 DNA ligase (400 units/μL)	1

Table 9
Preparation of ligation reaction setup

Requirements	Volume (Tester 1) (μL)	Volume (Tester 2) (μL)
Diluted tester cDNA	2	2
Adaptor 1 (10 μM)	2	–
Adaptor 2R (10 μM)	–	2
Master Mix	6	6
Total volume	10	10

5. Mix all components in 0.5 mL microcentrifuge tubes by gentle pipetting up and down.
6. Briefly provide a short spin to the samples and incubate at 16°C overnight.
7. After incubation, the reaction should be terminated by adding 1 μL of EDTA/glycogen mix.
8. Heat the samples at 72°C to inactivate the ligase.
9. Briefly centrifuge and store the unsubtracted and adapter-ligated tester cDNA samples at −80°C.
10. Dilute 1 μL of each unsubtracted tester cDNA with 1 μL of molecular-grade water. These diluted samples will be used as templates for PCR amplification.
11. The efficiency of adaptor-ligated tester cDNA is checked by PCR experiment using different primer combinations of Histone 3 3′ primer and PCR primer 1, and Histone 3 3′ and 5′ primer, and the PCR amplified products are analyzed by 1.0% agarose gel electrophoresis analysis (Fig. 4) (*see* **Note 16**).

3.6.3 First Hybridization (see Note 19)

1. Prior to the hybridization, allow the 4× hybridization buffer to warm to room temperature for at least 15–20 min.
2. For each tester, cDNA samples, set up the subtraction reaction by adding the components listed in Table 10.

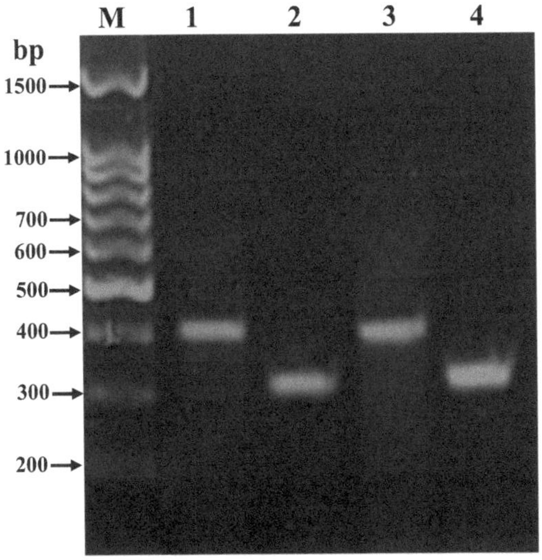

Fig. 4 Agarose gel electrophoresis analysis of PCR products amplified using adaptor-ligated tester cDNA templates. Lane M: GeneON 100 bp DNA Marker (GeneON BioScience, Germany), Lane 1: PCR products using Tester 1–1 (Adaptor 1-ligated) as the template and the Histone 3 3′ primer and PCR primer 1, Lane 2: PCR products using Tester 1–1 (Adaptor 1-ligated) as the template, and the Histone 3 3′ and 5′ primers, Lane 3: PCR products using Tester 1–2 (Adaptor 2R-ligated) as the template, and the Histone 3 3′ primer and PCR primer 1, Lane 4: PCR products using Tester 1–2 (Adaptor 2R-ligated) as the template, and the Histone 3 3′ and 5′ primers [10]

3. Briefly centrifuge the samples and incubate them in a thermal cycler at 98°C for 1.5 min.
4. Subsequently, incubate the samples at 68°C for 8 h.

3.6.4 Second Hybridization (see Note 21)

1. To perform the second hybridization, add the following components as mentioned in Table 11.
2. From the above mixture, transfer 1 μL into a fresh 0.5 mL microcentrifuge tube, which is used for performing the second hybridization.
3. Incubate the mixture at 98°C for 1.5 min in a thermal cycler to denature the driver cDNA.
4. Immediately add the samples from the first hybridization (Samples 1 and 2) to the tube containing the freshly denatured driver cDNA.
5. Briefly centrifuge the samples and incubate the reaction at 68°C overnight.
6. Add 200 μL of dilution buffer to the reaction and mix thoroughly.
7. Heat the samples at 68°C for 7 min in a thermal cycler and store at 20°C
8. The hybridized PCR-select cDNA subtraction samples are analyzed by agarose gel electrophoresis analysis (Fig. 5).

Table 10
Sample preparation for first hybridization

Requirements	Volume (Tester 1)(μL)	Volume (Tester 2)(μL)
Rsa I-digested Driver cDNA	1.5	1.5
Adaptor 1-ligated Tester 1	1.5	–
Adaptor 2R-ligated Tester 1	–	1.5
4× Hybridization Buffer	1.0	1.0
Final volume	4.0	4.0

Table 11
Sample preparation for second hybridization

Requirements	Volume (μL)
Driver cDNA	1
4× Hybridization buffer	1
Sterile H_2O	2

3.6.5 PCR Amplification (see Notes 24–27)

A) Primary PCR amplification

1. Aliquot 1 μL each of subtracted and unsubtracted tester cDNA into separate microcentrifuge tubes.
2. Prepare the primary PCR master mix using the requirements listed in Table 12.
3. Add the prepared primary PCR master mix to the tubes containing aliquoted subtracted and unsubtracted tester samples.
4. Briefly centrifuge the tubes to collect the contents.
5. Incubate the samples at 75°C for 5 min in a thermal cycler.
6. Perform PCR in a thermal cycler using the reaction conditions mentioned in Table 13.

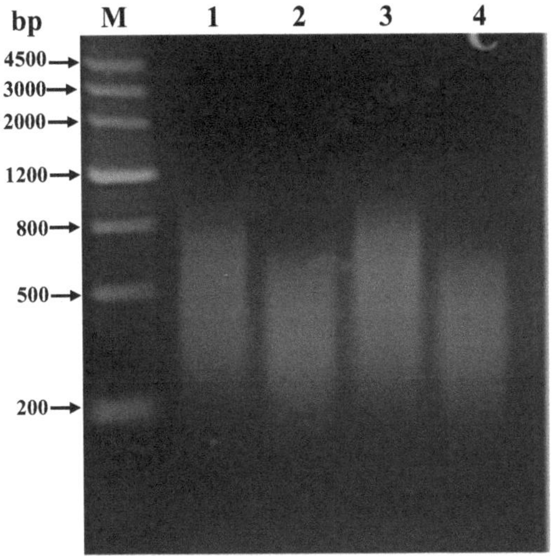

Fig. 5 Agarose gel electrophoresis analysis of PCR-select cDNA subtraction. Lane M: 200–4500 bp seven bands DNA marker (EarthOx LifeSciences Pvt. Ltd., India), Lane 2: Forward secondary PCR products of unsubtracted, Lane 3: Forward secondary PCR products of subtracted, Lane 4: Reverse secondary PCR products of unsubtracted, Lane 5: Reverse secondary PCR products of subtracted [10]

Table 12
Preparation of the primary PCR master mix (24.0 μL reaction)

Requirements	Volume
Sterile H_2O	19.5
10× PCR reaction buffer	2.5
dNTP Mix (10 mM)	0.5
PCR Primer 1 (10 μM)	1.0
50× Advantage cDNA Polymerase Mix	0.5
Total volume	24.0

7. Analyze 8 μL of each PCR product on a 2.0% agarose gel.
8. Dilute 3 μL of each primary PCR product with 27 μL of molecular-grade water.
9. Use 1 μL of the diluted primary PCR product as the template for secondary PCR amplification (*see* **Note 27**).

B) Secondary PCR amplification

1. The secondary PCR master mix should be prepared as shown in Table 14.
2. Mix the master mix thoroughly by brief centrifugation.
3. Add the prepared secondary PCR master mix to the aliquots of diluted primary PCR products in each tube.
4. Perform PCR in a thermal cycler using the conditions mentioned in Table 15.
5. Agarose (2%) gel electrophoresis of 8 μL of PCR products from each reaction, obtained at different cycle numbers, is performed to assess subtraction efficiency using the actin housekeeping gene (Fig. 6) (*see* **Notes 18 and 19**).
6. Store the PCR products at −20°C until further use.

Table 13
PCR reaction condition

Steps	Condition
Initial denaturation	94°C for 25 s
Denaturation	94°C for 10 s
Annealing	66°C for 30 s
Extension	72°C for 1.5 min
No. of cycles	27

Table 14
Prepare the secondary PCR Master Mix (24.0 μL reaction) (*see* Note 24)

Requirements	Volume (μL)
Sterile H_2O	18.5
10× PCR reaction buffer	2.5
Nested PCR primer 1 (10 mM)	1.0
Nested PCR primer 2 R (10 mM)	1.0
dNTP mix (10 mM)	0.5
50× Advantage cDNA Polymerase Mix	0.5
Total volume	24.0

Table 15
PCR reaction condition

Steps	Condition
Denaturation	94°C for 10 s
Annealing	68°C for 30 s
Extension	72°C for 1.5 min
No. of cycles	12

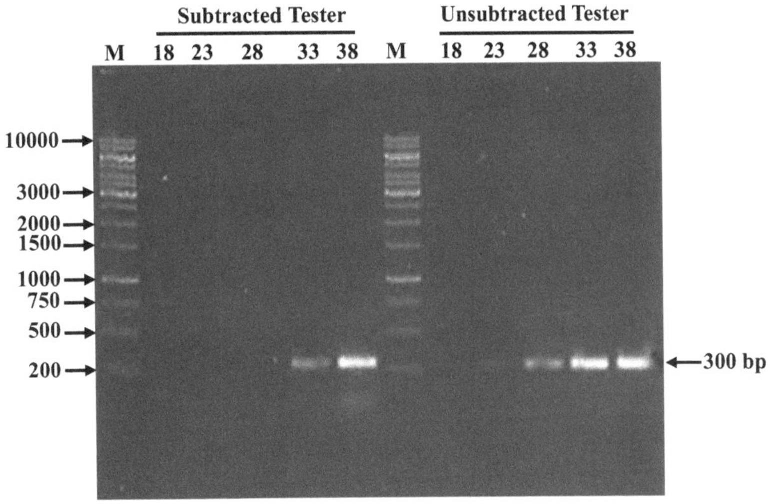

Fig. 6 Agarose gel electrophoresis analysis of PCR amplification shows the subtraction efficiency using the *Actin* housekeeping gene. The numbers indicate the number of PCR cycles at which the samples were collected. The amplification product (300 bp) was first seen at 33 cycles for the subtracted tester, whereas the same for the unsubtracted tester was seen even at 23 cycles, showing the efficiency of subtraction [5]

3.7 Differential Screening

3.7.1 Insertion into pGEM-T Easy Vector

1) Briefly centrifuge the pGEM-T Easy vector and control insert DNA.
2) In a fresh sterile 0.5 mL microfuge tube, prepare the reaction mixture as listed in Table 16.
3) Mix the reaction gently and incubate at room temperature for 1 h.

3.7.2 Transformations Using the pGEM®-T Easy Vector Ligation Reactions (Fig. 7)

1. After briefly centrifuging the ligation reactions, 2 μL from each ligated sample were aliquoted into fresh sterile 1.5 mL microcentrifuge tubes kept on ice.
2. In a separate tube kept on ice, 1 μL of uncut plasmid was added.

Table 16
Preparation the reaction mixture (*see* Note 17)

Requirements	Volume (standard reaction) (μL)	Volume (positive control) (μL)
2× Rapid ligation buffer, T4 DNA Ligase	5	5
pGEM®-T or pGEM®-T Easy Vector (50 ng)	1	1
PCR product	2	-
Control Insert DNA	–	2
T4 DNA ligase (3 Weiss units/μL)	1	1
Nuclease-free water to a final volume of 10 μL	1	1

3. Then 50 μL of competent cells were transferred into the tubes prepared in Steps 1 and 2.
4. The tubes were gently tapped to mix and incubated on ice for 20 min.
5. Following ice incubation, the samples were subjected to heat shock at exactly 42°C for 90 s in a water bath.
6. Immediately after heat shock, the tubes were returned to ice for 2 min.
7. Then 950 μL of Super Optimal broth with Catabolite repression (SOC) medium (Thermo Fisher Scientific) was carefully added to the ligation-transformed samples, and 900 μL was added to the uncut plasmid control (Table 17).
8. The samples were incubated at 37°C for 90 min with shaking at approximately 150 rpm.
9. Then 100 μL of each transformed culture was plated onto LB agar plates containing ampicillin, IPTG, and X-Gal, and incubated overnight at 37°C (Fig. 7).

3.7.3 Screening Transformants for Inserts

Blue (colonies which indicate an empty vector (no insertion), as the intact *lacZ* alpha produces a functional beta-galactosidase that converts the substrate X-Gal into a blue pigment), white (colonies which indicate successful insertion of the PCR product, which disrupts the *lacZ* alpha gene (insertional inactivation), preventing the functional enzyme from forming) colonies obtained after overnight incubation were identified by color screening on indicator plates and selected for further processing (Fig. 7).

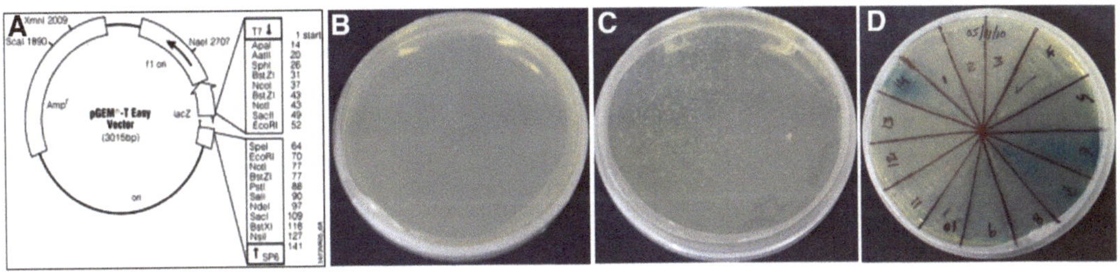

Fig. 7 Cloning of subtracted cDNAs into pGEM-T Easy vector and Screening of positive Clones. (**A**) pGEM-T Easy Vector Map. (**B**) Plate shows Luria Broth agar plate containing X-Gal (50 mg/L), IPTG (50 mg/L), and Ampicillin (100 mg/L) antibiotic. (**C**) Plate shows blue/white colonies on LB agar plate containing X-Gal, IPTG and Ampicillin of pGEM-T Easy vector ligated with our desired insert in the *E. coli* competent cells of DH5α strain. (**D**) Plate shows culture of colonies picked from the transformed plate (**C**) and streaked on fresh LB agar plate containing X-Gal, IPTG and Ampicillin for selection of clones [11]

Table 17
Composition of the SOC medium

Components	Quantity (per litre) (g)
Tryptone (2%)	20
Yeast extract (0.5%)	5
NaCl (10 mM)	0.584
KCl (2.5 mM)	0.186
$MgCl_2$ (10 mM)	2.03
$MgSO_4$ (10 mM)	2.46
Glucose (20 mM)	3.6

Adjust pH to 7.0, and make up the volume to 1 L with distilled water.

3.7.4 Amplification of cDNA Inserts (see Note 20)

1. The retrieved clones were cultured in a 96-well plate containing 100 μL of LB medium supplemented with ampicillin (100 mg/L) in each well and incubated at 37°C with continuous shaking for 2 h.
2. The cDNA inserts were amplified by PCR using 1 μL of each bacterial culture prepared in Step 1 as the template.
3. Aliquot 19 μL of the PCR master mix (preparation mentioned in Table 18) into each well of the plate, and add 1 μL of the corresponding bacterial culture to each well.
4. Briefly centrifuge the plate and perform thermal cycling under the following conditions mentioned in Table 19.
5. Following amplification, 5 μL of each PCR product was resolved on a 2.0% agarose gel containing ethidium bromide in 1× TAE buffer (Fig. 8).

Table 18
Prepare the PCR master mix (19.0 μL reaction)

Requirements	Volume (μL)
10× PCR reaction buffer	2.0
Nested Primer 1	0.6
Nested Primer 2 R	0.6
dNTP mix (10 mM)	0.4
H_2O	15.2
50× PCR enzyme mix	0.2
Total volume	19.0

Table 19
PCR reaction conditions

Steps	Conditions
Denaturation	94°C for 30 s
Annealing	95°C for 10 s
Extension	68°C for 3 min

3.7.5 Preparation of cDNA Dot Blots of the PCR Products (Fig. 9) (see **Notes 23, 28, 30**)

1. Combine 5 μL of PCR product with 5 μL of 0.6 N NaOH in a 96-well plate (*see* **Note 25**).
2. Mix thoroughly by briefly centrifuging the plate.
3. Transfer 2 μL of each mixture separately onto a nylon membrane.
4. Neutralize the blots with 0.5 M Tris–HCl (pH 7.5).
5. Wash the membrane thoroughly with sterile distilled water.
6. Incubate the membrane in an oven at 70°C for 2 h to allow DNA crosslinking to the membrane.

3.7.6 Random Primer Labeling of cDNA Probes

1. In separate 0.5 mL microcentrifuge tubes, add 3 μL of subtracted cDNA to one tube and 3 μL of unsubtracted cDNA to another tube.
2. Heat both tubes at 95°C for 8 min, then immediately place them on ice.
3. Add the following components to each reaction tube mentioned in Table 20.
4. After adding all components, incubate the tubes at 37°C for 30 min.

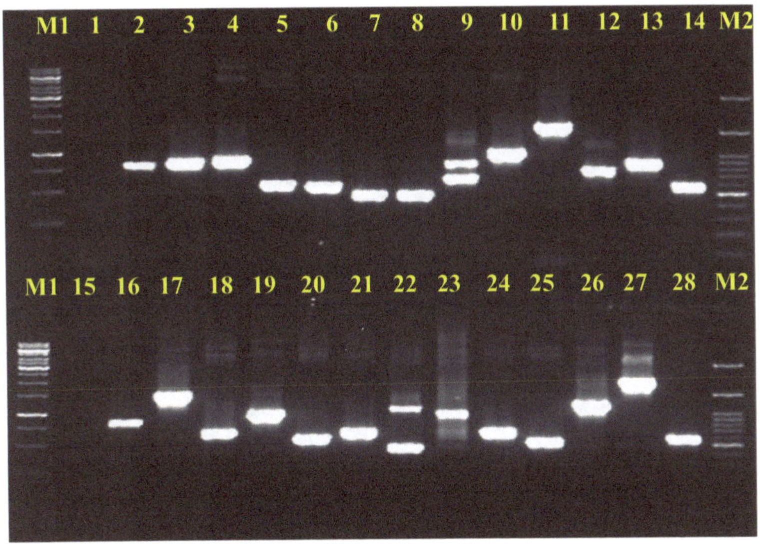

Fig. 8 Agarose gel electrophoresis shows screening of the SSH library. 2.0% agarose gel image shows the analysis of the SSH library by colony PCR with Nested Primer 1 and Nested Primer 2R. Lane M1: Thermo Scientific O'GeneRuler 1 kb DNA Ladder (250–10 kb), Lane M2: Thermo Scientific™ O'GeneRuler™ 100 bp Plus DNA Ladder (100–3000 bp), Lanes 1–28: Colony PCR confirmation of different *E. coli* clones

5. Terminate the reaction by adding 5 μL of the stop solution provided with the kit.
6. Unincorporated (unlabelled) cDNA was removed by TCA precipitation.

3.7.7 Hybridization with the Subtracted cDNA (see **Note 29**)

1. To 5 mL of hybridization solution (Table 21), add 50 μL of 20× SSC (Table 22) and 50 μL of blocking solution (Table 23), mix well, boil for 5 min, and immediately chill on ice (*see* **Note 26**).
2. Prehybridization was performed by placing the membrane in a hybridization container containing hybridization solution for 40–60 min with continuous agitation at 72°C.
3. The hybridization probe was prepared by adding 50 μL of 20× SSC and 50 μL of blocking solution to 100 ng of subtracted cDNA, followed by boiling for 5 min and chilling on ice.
4. The hybridization probe was not added directly to the hybridization container; instead, the membrane was immersed in the probe solution.
5. Hybridization was carried out at 72°C with continuous agitation for overnight (2–16 h).
6. Post-hybridization/washing should be carried out with low-stringency (2× SSC/0.5% SDS) and high-stringency (0.2× SSC/0.5% SDS) wash solutions, which were prewarmed to 68°C.

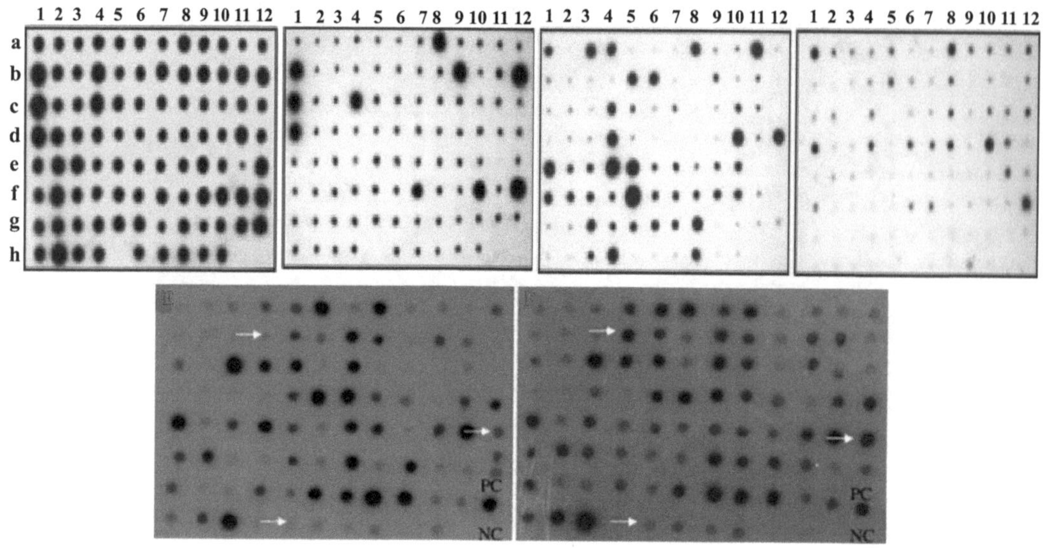

Fig. 9 Autoradiography shows differential screening of the cDNA library from the cotton SSH libraries using the dot blot (macroarray) technique. (**A–D**) Quadruplicate colony dot blots were prepared, and the membranes were hybridized with labeled probes. (**A**) Forward subtracted cDNA probes. (**B**) Reverse subtracted cDNA probes. (**C**) Unsubtracted tester cDNA probes. (**D**) Unsubtracted driver cDNA probes. (**E**) Unsubtracted leaf cDNA as probe. (**F**) Subtracted tester cDNA as probe for blot. The arrows indicate the clones that are differentially expressed. PC and NC indicate positive and negative controls, respectively [5,12]

Table 20
Random primer labeling reaction setup (*see* Note 28)

Requirements	Volume (per reaction) (μL)
Reaction buffer (dCTP)	3
Random Primer Mix	2
[α-^{32}P] dCTP	5
(50 μCi, 3000 Ci/mmol, aqueous solution) Klenow Enzyme (exo−)	1

7. The membranes were washed with low-stringency wash solution for 20 min, repeated four times at 68°C.
8. The membranes were subsequently washed with high-stringency wash solution for 20 min, repeated two times at 68°C.
9. Membranes were exposed to X-ray film overnight at −70°C using an intensifying 10×12 in. of cassette or alternatively analyzed using a phosphor imager.
10. The intensity of the detected signals corresponded to the relative abundance of differentially expressed transcripts in the subtracted cDNA library (Fig. 9) (*see* **Note 22 and 30**).

Table 21
Preparation of hybridization solution

Requirements	Volume (100 mL)
20× SSC	30 mL
10% SDS	5 mL
50× Denhardt's solution	10 mL
salmon sperm DNA (10 mg/mL)	1 mL
Nuclease-free water	54 mL

The hybridization solution should be stored at 4°C.

Table 22
Preparation of 20× SSC solution

Requirements	Quantity (per liter) (g)
NaCl	175.3
$Na_3Citrate{\cdot}2H_2O$	88.2

Adjust pH to 7.0 with 1 M HCl. Store at room temperature.

Table 23
Preparation of blocking solution (*see* Note 21)

Requirements	Quantity (100 mL) (g)
Ficoll 400	1.0
Polyvinylpyrrolidone (PVP)	1.0
Bovine serum albumin (BSA)	1.0

Dissolve in distilled water and make up to 100 mL. Filter sterilize and store at −20°C.

11. Colonies exhibiting strong tester-specific hybridization signals were selected as positive clones and the plasmids were isolated from an overnight-grown *E. coli* (DH5-α) culture and subjected to Sanger sequencing for further analysis (Fig. 10).

3.8 Sanger Sequencing Reaction Setup (see Note 22)

3.8.1 Template Preparation

Template preparation and reaction components are mentioned in Tables 24 and 25, respectively. Mix thoroughly and briefly centrifuge.

1. Purified PCR products along with a single sequencing primer (either forward or reverse) were used for sequencing reactions.

3.8.2 Cycle Sequencing Conditions

Cycle sequencing was performed in a thermal cycler using the following program mentioned in Table 26.

3.8.3 Post-Sequencing Cleanup (Ethanol Precipitation)

Sequencing reactions were purified to remove unincorporated dye terminators as follows:

1. Add 5 μL of 125 mM EDTA to each reaction.
2. Add 60 μL of 100% ethanol.
3. Seal the plate with aluminum tape and invert it four times to mix thoroughly.

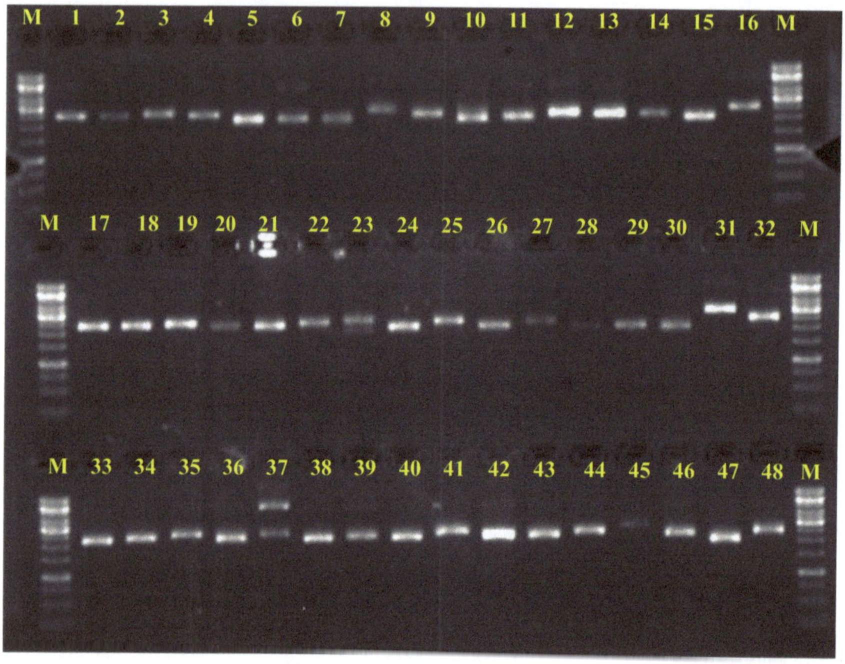

Fig. 10 Agarose gel electrophoresis analysis shows plasmid DNA isolated from the overnight-grown *E. coli* (DH5-α) culture for further sequencing study. Lane M: Thermo Scientific O'GeneRuler 1 kb DNA Ladder (250–10 kb), Lanes 1–48: Plasmid DNA isolated from the dot blot positive clones

Table 24

Template preparation for Sanger sequencing reaction setup.

Requirements	Volume (20 μL)
Terminator Ready Reaction Mix	8.0
Template	Variable
Primer	1.0
Deionized water	Variable
Nuclease-free water to a final volume of 20 μL	20.0

Table 25
Prepare the reaction components for the Sanger Sequencing Reaction Setup

Requirements	Volume (20 μL)
Ready Reaction Mix	4
5× Sequencing buffer	2
Sequencing primer (Adaptor primer/T7 Universal Primer)	1
Template	Variable
Nuclease-free water to a final volume of 20 μL	20

Table 26
PCR reaction condition

Step	Temperature (°C)	Time/cycles
Initial denaturation	96	1 min/1
Denaturation	96	10 s
Annealing	50	5 s/25
Extension	60	4 min
Hold	4	∞

After completion, the PCR products were briefly centrifuged

4. Incubate at room temperature for 15 min.
5. Centrifuge at 1400–2000 × *g* for 45 min.
6. Carefully discard the supernatant.
7. Wash the pellet with 60 μL of 70% ethanol.
8. Centrifuge at 1650 × *g* for 15 min at 4°C.
9. Resuspend the samples in injection buffer, cover with aluminum foil, and store at 4°C until further use.

3.8.4 Capillary Electrophoresis

Purified sequencing products were denatured at 95°C for 2 min, rapidly cooled on ice, and loaded onto an ABI Genetic Analyzer for capillary electrophoresis. Sequence data were collected and analyzed using sequencing analysis software.

3.8.5 Data Analysis

Raw chromatograms were examined for peak quality, base calling accuracy, and background noise. High-quality sequences were trimmed and aligned using standard bioinformatics methods. The processed sequences were further analyzed for gene identification and functional annotation.

3.8.6 Sequence Analysis and Functional Annotation

1. Nucleotide sequences obtained from SSH were screened for vector contamination using VecScreen (https://www.ncbi.nlm.nih.gov/tools/vecscreen/) or SeqTri (https://bio.tools/seqtrim).
2. All cleaned sequences are being made into a contig assembly by using the DNASTAR sequence analysis tool or SnapGene software.
3. Furthermore, similarity search was analyzed using BLASTX algorithm datasets with default parameters followed by computational annotation of the EST datasets was performed using the Blast2GO software (https://www.blast2go.com/) was used to obtain GO (Gene ontology) information from retrieved database matches. Annotation of all sequences was performed by using default parameters [13, 14].
4. eggnog -mapper software (https://bio.tools/eggnog) employed to identify the function of genes as to carry functional annotation.
5. Protein domain analysis was carried out for the retrieved sequences using InterProScan software (https://www.ebi.ac.uk/interpro/search/sequence/), and GO enrichment was assessed with the help of 'goslim_plant.obo'software (http://purl.obolibrary.org/obo/go#goslim_plant).

3.9 Next Generation Sequencing [15]

Here, NGS processes were carried out for both control and stressed-induced samples.

1. The high-quality total RNA is isolated by the above-mentioned protocol (Refer Sects. 3.1 and 3.2).
2. First- and second-strand cDNA is synthesized.
3. The synthesized cDNA was subjected to restriction digestion using *Nla*III.
4. The digested cDNA fragments were ligated with Illumina-specific adapters.
5. Initially, ligation was carried out with Adapter A, which provides the recognition site for the Type IIS tagging enzyme, *Mme*I.
6. This was followed by *Mme*I digestion of the samples.
7. Dephosphorylation and purification were performed to prevent unwanted ligation.
8. Next, Adapter B is ligated, which contains 2-bp degenerate 3′ overhangs.
9. This process resulted in complete Illumina cDNA library preparation.

10. The prepared cDNA library was enriched by PCR amplification using the following conditions mentioned in Table 27.
11. After PCR, the amplified products were resolved on a 12% PAGE gel.
12. The DNA band of approximately 85 bp was excised and purified from the gel, followed by ethanol precipitation.
13. The quality and concentration of the excised DNA were assessed using a NanoDrop spectrophotometer, and the DNA was diluted to 10 nM.
14. Clustering and sequencing were performed using an Illumina cluster station followed by analysis on an Illumina analyzer.
15. The prepared cDNA library was sequenced on an Illumina sequencing platform using paired-end chemistry [16,17].
16. Raw sequence reads were generated in FASTQ format.
17. Low-quality reads and adapter sequences were removed from the raw data.
18. High-quality reads were assembled de novo using Trinity software.
19. The assembled contigs representing expressed gene sequences were considered expressed sequence tags (ESTs).
20. The ESTs were annotated by sequence similarity searches using BLAST against the NCBI nonredundant protein database (*see* **Note 29**).
21. Gene Ontology (GO) terms and KEGG pathway annotations were assigned to predict the biological functions of the ESTs.

Table 27
PCR reaction condition

Temperature (15–17 cycles)	Time
Initial denaturation	98°C for 30 s
Denaturation	98°C for 10 s
Annealing	60°C for 30 s
Extension	72°C for 15 s
Final extension	72°C for 5 min
No. of. Cycles	15–17

4 Notes

1. All materials used for RNA isolation must be autoclaved and treated with DEPC-treated water, and ensure that the working area is sterile, free from RNase contamination, and cleaned with 70% ethanol.
2. Prior to isolation, the mortar, pestle, and spatula must be prechilled to avoid denaturation of sample transcripts and to prevent RNA degradation; a mask and gloves must be worn throughout the procedure.
3. Avoid prolonged vortexing to obtain good-quality RNA, as it may lead to genomic DNA contamination.
4. If a larger amount of sample is used, Wash solution II should be applied twice.
5. It is necessary to collect and quantify Elutions I and II in separate tubes.
6. Prolonged air-drying of the RNA pellet can lead to RNA degradation.
7. Isolated total RNA must be treated with DNase I to eliminate DNA contamination.
8. RNA quality should be assessed using the RNA Integrity Number (RIN).
9. Total RNA mixed with RNA loading dye should be incubated at 70°C for 5 min and immediately placed on ice before loading onto the gel for analysis.
10. The concentration of poly(A^+) mRNA highly depends on the volume of elution buffer added.
11. Do not discard the supernatant until satisfactory elution of poly(A^+) mRNA is confirmed.
12. For first-strand cDNA synthesis, the required components should be taken from the −20°C freezer and thawed on ice before use.
13. For first-strand cDNA synthesis, the concentration of poly (A^+) mRNA should be 50–200 μg/mL; otherwise, the yield will be reduced.
14. After *Rsa* I digestion, the cDNA size should be 0.1–2 kb and 0.5–10 kb. If size reduction is inadequate, repeat phenol/chloroform extraction, digestion, and ethanol precipitation steps.
15. Complete digestion of tester cDNA with *Rsa* I must be confirmed before ligation, as efficient adaptor ligation requires blunt-ended DNA fragments.

16. Adaptor ligation is performed only with tester cDNA (Adaptor 1 and Adaptor 2R separately), while driver cDNA remains without adaptors to enable selective amplification of differentially expressed sequences during suppression PCR.
17. To verify proper experimental performance, a positive control must be included.
18. To assess subtraction efficiency, gel band patterns should differ from those of the unsubtracted sample.
19. During cDNA insert amplification, increasing the number of cycles enhances amplification efficiency.
20. To prevent G/C degradation, avoid exposing the sequence reaction to light and oxygen prior to loading it into the instrument.
21. To avoid the low signal, ensure the sufficiency of primers and template prior to sequencing. Prior to cDNA synthesis, the purity of RNA is ensured for sufficient cDNA synthesis.
22. Ensure the purity of the PCR product to avoid a noisy signal during sequencing the samples.
23. Ensure the primer is appropriately designed to obtain good primer annealing.
24. Excessive PCR cycles can cause amplification bias and reduce subtraction efficiency. Use the recommended cycle number (primary and nested PCR) to maintain representation of rare transcripts.
25. Ensure efficient probe labeling and proper DNA denaturation (NaOH treatment) before spotting onto the membrane.
26. Prepare fresh hybridization and blocking solutions, and perform stringent post-hybridization washes to reduce nonspecific binding.
27. Use purified plasmid or PCR products, avoid mixed colonies, and optimize template concentration (50–100 ng per reaction).
28. When using [α-^{32}P] dCTP, follow radiation safety guidelines, ensure proper shielding, and remove unincorporated nucleotides thoroughly to reduce background noise.
29. Avoid the secondary structure of the template to get rid of early termination of the sequence reaction.
30. Ensure the transcript redundancy in order to detect even low-abundance transcripts.

References

1. Zhou B, Zhang L, Ullah A, Jin X, Yang X, Zhang X (2016) Identification of multiple stress responsive genes by sequencing a normalized cDNA library from sea-land cotton (*Gossypium barbadense* L.). PLoS ONE 11(3):e0152927
2. Sivakumar JS, Gurusamy D, Chellamuthu S, Rathnasamy SA, Krishnan V (2025) Development of transgenic cotton plants for insect resistance. Methods Mol Biol 2973:137–148.
3. Lv Y, Zhao L, Xu X, Wang L, Wang C, Zhang T, Guo W (2013) Characterization of expressed sequence tags from developing fibers of *Gossypium barbadense* and evaluation of insertion-deletion variation in tetraploid cultivated cotton species. BMC Genomics 14(1):170
4. Wang QQ, Liu F, Chen XS, Ma XJ, Zeng HQ, Yang ZM (2010) Transcriptome profiling of early developing cotton fiber by deep-sequencing reveals significantly differential expression of genes in a fuzzless/lintless mutant. Genomics 96(6):369–376
5. Raghavendra KP, Phanindra MLV, Kumar BK, Dhandapani G, Kumar PA (2011) Identification of differentially expressed genes during bud stage of cotton boll development using suppression subtractive hybridization and cDNA macroarray. J Plant Biochem Biotechnol 20(1):12–19
6. Parkinson J, Blaxter M (2009) Expressed sequence tags an overview. Methods Mol Biol 533:1–12.
7. Sivakumar JS, Gurusamy D, Chellamuthu S, Rathnasamy SA, Krishnan V (2025) Characterization of transgenic cotton plants by PCR. Methods Mol Biol 2973:157–170.
8. Chen YG (2003) Construction of a normalized cDNA library by mRNA-cDNA hybridization and subtraction. Generation of cDNA libraries: methods and protocols. Humana Press, Totowa NJ, pp 33-40
9. Gurusamy D (2014) Transcriptome analysis of cotton (*Gossypium hirsutum* L.) during boll development under drought condition and functional characterization of drought tolerance genes. Ph.D thesis. https://hdl.handle.net/10603/215568
10. Zhang L, Li FG, Liu CL, Zhang CJ, Zhang XY (2009) Construction and analysis of cotton (*Gossypium arboreum* L.) drought-related cDNA library. BMC Res Notes 2(1):120
11. Xie Z, Zhang Z, Cao Z, Chen M, Li P, Liu W, Chen Y (2017) An external substrate-free blue/white screening system in *Escherichia coli*. Appl Microbiol Biotechnol 101(9):3811–3820
12. Zeng F, Zhang X, Zhu L, Tu L, Guo X, Nie Y (2006) Isolation and characterization of genes associated to cotton somatic embryogenesis by suppression subtractive hybridization and macroarray. Plant Mol Biol 60(2):167–183
13. Conesa A, Götz S, García-Gómez JM, Terol J, Talón M, Robles M (2005) Blast2GO: a universal tool for annotation, visualization and analysis in functional genomics research. Bioinformatics 21(18):3674–3676
14. Aparicio G, Gotz S, Conesa A, Segrelles D, Blanquer I, Garcia JM, Talon M (2006) Blast2GO goes grid: developing a grid-enabled prototype for functional genomics analysis. Stud Health Technol Inform 120:194
15. Morrissy AS, Morin RD, Delaney A, Zeng T, McDonald H, Jones S, Marra MA (2009) Next-generation tag sequencing for cancer gene expression profiling. Genome Res 19(10):1825–1835
16. Sasaki K, Mitsuda N, Nashima K, Kishimoto K, Katayose Y, Kanamori H, Ohmiya A (2017) Generation of expressed sequence tags for discovery of genes responsible for floral traits of *Chrysanthemum morifolium* by next-generation sequencing technology. BMC Genomics 18(1):683
17. Reusch TB, Veron AS, Preuss C, Weiner J, Wissler L, Beck A, Bornberg-Bauer E (2008) Comparative analysis of expressed sequence tag (EST) libraries in the seagrass *Zostera marina* subjected to temperature stress. Mar Biotechnol 10(3):297–309

Chapter 15

Development of Single-Nucleotide Polymorphism (SNP) Markers in Sesame (*Sesamum indicum* L.)

Mohamed Kouighat and Abdelghani Nabloussi

Abstract

Sesame (*Sesamum indicum* L.) is an important oilseed crop with considerable agronomic potential, widely cultivated across tropical and subtropical regions. Despite its substantial economic value, genomic research in sesame has progressed more slowly than in major staple crops. The advent of high-throughput single-nucleotide polymorphism (SNP) markers has markedly advanced sesame breeding by enabling the development of dense genetic linkage maps, genome-wide association studies (GWAS), and marker-assisted selection (MAS). This chapter presents detailed protocols for SNP discovery and validation in sesame using next-generation sequencing (NGS) approaches, specifically genotyping-by-sequencing (GBS) and double-digest restriction-site associated DNA sequencing (ddRAD-seq). It provides step-by-step methods for extracting high-quality DNA from tissues rich in polysaccharides and polyphenols, constructing sequencing libraries, performing bioinformatic filtering, and validating SNPs using Kompetitive allele specific PCR (KASP) and high-resolution melting (HRM) analysis. Particular emphasis is placed on troubleshooting common challenges related to DNA quality, restriction enzyme digestion efficiency, and the optimization of bioinformatic parameters to achieve a sufficient number of reliable SNP calls across the sesame genome.

Keywords SNP discovery, GBS, ddRAD-seq, KASP assay, Molecular breeding, Genotyping

1 Introduction

Sesame (*Sesamum indicum* L., $2n=26$) is one of the species that belong to the Pedaliaceae family and is considered to be one of the first oleaginous crops ever to be recorded in the history of humankind [1]. Its main value is its high content of oil, varying between 50% and 60%, and the high quality of this oil, its properties owing to large part to the endogenous antioxidant elements sesamin and sesamolin [2–4]. Regardless of this nutritional and economic importance, sesame has long been an orphan crop, mainly because of the lack of extensive genomic resources, thus hindering specific genetic amelioration efforts, especially with regard to yield stability and resistance to biotic and abiotic stresses [5].

Nitish Kumar (ed.), *Molecular Markers in Plants*, Methods in Molecular Biology, vol. 3097,
https://doi.org/10.1007/978-1-0716-5663-1_15,

Simple sequence repeats (SSRs), amplified fragment length polymorphism (AFLP), and random amplified polymorphic DNA (RAPD), are the most commonly used conventional systems of molecular markers to measure genetic variation in sesame [6, 7]. Nonetheless, they are often constrained in use due to poor throughput, relatively high per-sample prices, and low marker densities to accurately map quantitative trait loci (QTLs) quantitatively. Conversely, single-nucleotide polymorphisms (SNPs) are the marker system of choice in contemporary breeding of plants, due to their abundance all over the genome, biallelic nature, and ability to be genotyped in large-scale, automated high-throughput systems [8]. The recent progress in the next-generation sequencing technologies has resulted in the design of high-density SNP materials in sesame. Such datasets have been useful in explaining the genetic basis of major agronomic objectives, which include determinacy in the growth of plants, shattering of capsules, and the molecular events that govern the production of oils [9–11].

SNP construction in sesame is typically performed through two major ways. To begin with, SNPs may be identified de novo by reduced-representation sequencing techniques, including double-digest Restriction-site Associated DNA sequencing (ddRAD-seq) and Genotyping-by-Sequencing (GBS) [12]. Secondly, SNP genotyping can be performed with specific target SNPs by high-resolution melting (HRM) and Kompetitive allele-specific PCR (KASP) analysis [13]. The reduced-representation methods are also relatively cost-effective to identify thousands of SNPs in large populations simultaneously, and KASP and HRM systems are especially useful in validating existing SNPs, as well as screening breeding groups of domesticated species using each particular marker [14].

The present chapter of the book describes strict rules that include discovery and validation procedures that have been specially developed to overcome the intrinsic limitations that are presented by the processing of sesame genomic DNA.

2 Materials

Every solution must be prepared using ultrapure water with a resistivity of 18 MΩcm at 25°C and in combination with analytical grade reagents. Unless it is mentioned otherwise, the reagents should be kept at ambient temperature. All the applicable waste-disposal regulations binding during the disposal of waste materials must be adhered to.

2.1 Plant Materials and DNA Extraction

1. Plant material: Fresh, young apical leaves of sesame (approx. 100–200 mg) collected from 2- to 3-week-old seedlings. Samples should be flash-frozen in liquid nitrogen immediately upon collection (*see* **Note 1**).

2. CTAB extraction buffer (2%): Prepared with 2% (w/v) cetyltrimethylammonium bromide (CTAB), 100 mM Tris–HCl (pH 8.0), 20 mM EDTA (pH 8.0), and 1.4 M NaCl. Autoclave the solution and store it at room temperature. Just before use, add 2% (v/v) β-mercaptoethanol [15] (*see* **Note 2**).
3. Wash buffer: 76% ethanol supplemented with 10 mM ammonium acetate.
4. Chloroform:isoamyl alcohol (24:1): Either freshly prepared or stored in a dark glass container at 4°C.
5. Isopropanol: Keep prechilled at −20°C.
6. RNase A: Prepare a 10 mg/mL stock solution and store at −20°C.
7. TE buffer: Comprising 10 mM Tris–HCl (pH 8.0) and 1 mM EDTA (pH 8.0).
8. Equipment: Includes a mortar and pestle, water bath set to 65°C, refrigerated centrifuge, NanoDrop spectrophotometer, and Qubit fluorometer.

2.2 SNP Discovery by Next-Generation Sequencing (GBS/ddRAD-seq)

1. Restriction enzymes: High-fidelity enzymes commonly used for sesame: *Ape*KI (for GBS) or the combination of *Eco*RI and *Mse*I (for ddRAD-seq) [10, 16] (*see* **Note 4**).
2. Adapters: Barcoded adapters (P1) and common adapters (P2) specific to the sequencing platform (e.g., Illumina).
3. Ligase: T4 DNA Ligase and corresponding 10× buffer.
4. PCR reagents: High-fidelity DNA polymerase (e.g., Phusion or NEBNext), dNTP mix (10 mM each) [17] (*see* **Note 5**).
5. Size selection: AMPure XP beads or Pippin Prep system for selecting fragments (typically 300–500 bp) [18].

2.3 SNP Validation and Genotyping Methods

1. KASP assay components [19]:
   - KASP Master Mix (2×) containing *Taq* polymerase, dNTPs, $MgCl_2$, and FRET cassettes (FAM and HEX).
   - Primer Mix: Two allele-specific forward primers and one common reverse primer.
   - Real-time PCR system compatible with FRET detection.
2. HRM analysis components [13]:
   - High-resolution melting dye (e.g., LCGreen, EvaGreen, or proprietary HRM Master Mix).
   - Specific primers flanking the SNP site (amplicon size 80–120 bp).
   - Real-time PCR instrument with high thermal uniformity and HRM analysis software.

2.4 Data Analysis Tools

1. Reference genome: *Sesamum indicum* reference genome (e.g., Zhongzhi No. 13 v2.0) [20] (*see* **Note 10**).
2. Software: TASSEL-GBS pipeline, STACKS, BWA-MEM, SAMtools, GATK, and R statistical environment [21].

3 Methods

3.1 Plant Material Collection and DNA Extraction

Sesame leaves are rich in polysaccharides and polyphenols, which have the ability to inhibit a downstream enzymatic reaction. The CTAB protocol proposed below is modified to guarantee the recovery of high-quality DNA to be used in next-generation sequencing (NGS) [12, 15].

1. Preheat the CTAB extraction buffer supplemented with 2% β-mercaptoethanol to 65°C.
2. Using liquid nitrogen, grind 200 mg of frozen young leaf tissue to a fine powder with a mortar and pestle, ensuring that the material does not thaw.
3. Transfer the powdered tissue into a 2 mL microcentrifuge tube containing 800 μL of pre-warmed CTAB buffer and mix thoroughly by vortexing.
4. Incubate the mixture at 65°C for 60 min, gently inverting the tube every 10 min to ensure uniform mixing.
5. Add 800 μL of chloroform:isoamyl alcohol (24:1) and mix carefully by inversion for 10 min to allow phase separation.
6. Centrifuge the sample at 12,000 × *g* for 15 min at room temperature.
7. Carefully transfer the upper aqueous phase to a new tube without disturbing the interphase.
8. Treat the extract with 5 μL of RNase A (10 mg/mL) and incubate at 37°C for 30 min.
9. Perform a second chloroform:isoamyl alcohol extraction (repeat Steps 5–7) to eliminate residual proteins and RNase.
10. Precipitate the DNA by adding 0.6 volumes of cold isopropanol and gently inverting the tube until the DNA becomes visible as white filaments.
11. Incubate the mixture at −20°C for 1 h, or overnight to maximize DNA yield.
12. Pellet the DNA by centrifugation at 12,000 × *g* for 10 min at 4°C, then carefully discard the supernatant.

13. Wash the DNA pellet twice with 500 µL of wash buffer (76% ethanol, 10 mM ammonium acetate) and air-dry for 10–15 min.
14. Dissolve the DNA in 50–100 µL of TE buffer and allow it to fully resuspend overnight at 4°C.

3.2 DNA Quality Assessment

1. *Evaluate DNA purity* using a NanoDrop spectrophotometer, aiming for an A260/A280 ratio between 1.8 and 2.0 and an A260/A230 ratio greater than 2.0 [22] (*see* **Note 3**).
2. *Determine DNA concentration* with a Qubit fluorometer using the dsDNA BR Assay Kit. Fluorometric measurement is critical for NGS library preparation because it specifically quantifies double-stranded DNA [23].
3. *Check DNA integrity* by running the sample on a 0.8% agarose gel. Intact DNA should appear as high-molecular-weight bands (>20 kb) without visible smearing.

3.3 Genotyping-by-Sequencing (GBS) Library Preparation

This protocol follows the method described by Elshire et al. [24], adapted for sesame [25, 26].

1. Digestion: In a 96-well plate, digest 100 ng of genomic DNA with 3.6 Units of *Ape*KI in a 20 µL reaction volume. Incubate at 75°C for 2 h.
2. Ligation: Add a ligation master mix containing T4 DNA ligase, ATP, and adapters. The reaction includes a unique barcoded adapter (0.1–0.6 pmol, depending on barcode) and a common adapter (common adapter concentration usually in excess). Incubate at 22°C for 1 h, followed by heat inactivation at 65°C for 30 min.
3. Pooling: Pool 5 µL from each well into a single tube (multiplexing).
4. PCR amplification: Amplify the pooled library to enrich for fragments with adapters. Use primers complementary to the adapters (*see* **Note 5**). Initial denaturation at 98°C for 30 s; followed by 18 cycles of 98°C for 10 s, 65°C for 30 s, and 72°C for 30 s; with a final extension at 72°C for 5 min.
5. Cleanup: Purify the PCR product using AMPure XP beads (0.8× ratio) to remove primers and small fragments (<150 bp).
6. Validation: Check the library distribution on an Agilent Bioanalyzer 2100. A typical GBS library shows a smear between 170 and 1000 bp.

3.4 ddRAD-seq Library Preparation

Alternative protocol using double digestion for reduced genome complexity [12].

1. Double digestion: Digest 200–500 ng of DNA with *Eco*RI (rare cutter) and *Mse*I (frequent cutter) at 37°C for 3 h.
2. Ligation: Ligate P1 adapter (barcoded, matching *Eco*RI overhang) and P2 adapter (matching *Mse*I overhang) using T4 DNA ligase.
3. Size selection: Pool samples and run on a 2% agarose gel. Excise the range of 300–500 bp and purify using a Gel Extraction Kit. Alternatively, use a Pippin Prep system.
4. Amplification: Perform PCR amplification similar to Step 3.3.4 but using primers specific to the ddRAD adapters.
5. Sequencing: Sequence the library on an Illumina platform (e.g., HiSeq or NovaSeq) to generate paired-end reads (PE150).

3.5 SNP Discovery and Filtering

Raw data processing is critical for reliable SNP calling [26].

1. Demultiplexing: Use the process_radtags component of STACKS or TASSEL to separate reads based on barcodes. Discard reads with low quality scores (Phred <20) or missing restriction sites.
2. Alignment: Align clean reads to the sesame reference genome (e.g., *Sesamum indicum* v2.0) using BWA–MEM with default parameters.
3. Variant calling: Use GATK HaplotypeCaller or SAMtools mpileup to identify variants.
4. Filtering: Apply hard filtering to remove false positives. Standard filters for sesame populations:
    1. Minor Allele Frequency (MAF) > 0.05.
    2. Missing data rate < 20% (*see* **Note 6**).
    3. Read depth (DP) > 5× per individual.
    4. Quality score (QUAL) > 30.
    5. Heterozygosity < 10% (for inbred lines) (*see* **Note 7**).

3.6 SNP Validation by KASP Assay

KASP is used to validate key SNPs discovered via GBS/ddRAD-seq [10, 17].

1. Primer design: Submit the SNP sequence (approx. 50 bp flanking regions) to the primer design software (e.g., LGC Genomics). This generates two Allele-Specific Primers (ASP1 and ASP2) and one Common Reverse Primer.
2. Reaction setup: Prepare a 10 μL reaction in a 96- or 384-well plate:
    - 5 μL 2× KASP Master Mix.
    - 0.14 μL Primer Mix.

- 5 μL Genomic DNA (10–20 ng total).

3. Thermal cycling:
   - Hot start: 94°C for 15 min.
   - Touchdown: Ten cycles of 94°C for 20 s and 61–55°C for 60 s, decreasing the annealing temperature by 0.6°C per cycle.
   - Amplification: 26 cycles of 94°C for 20 s and 55°C for 60 s.
4. Reading: Read fluorescence (FAM and HEX) on a compatible plate reader. Plot cluster data (*x*-axis: FAM, *y*-axis: HEX) to call genotypes (Fig. 1) (*see* **Note 8**).

3.7 SNP Validation by HRM Analysis

1. Primer design: Design primers to amplify a small amplicon (80–120 bp) containing the SNP. Ensure the melting domains are stable.
2. Reaction setup: Prepare 20 μL reaction [27]:
   1. 10 μL HRM Master Mix (2×) (*see* **Note 9**).

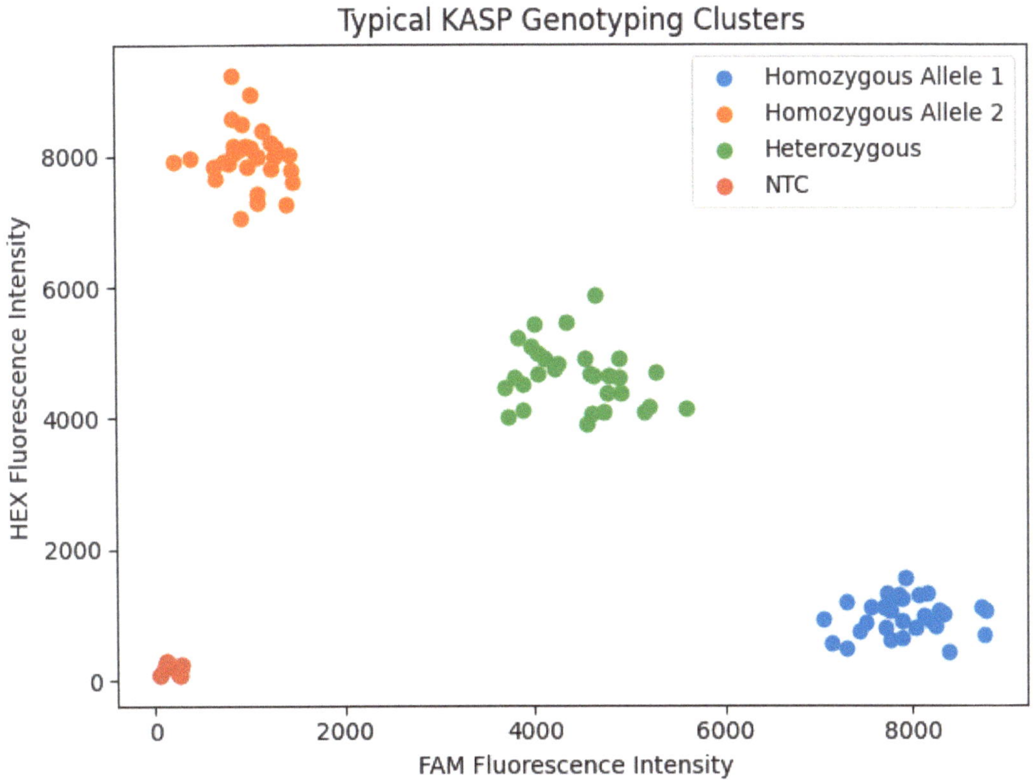

Fig. 1 Typical KASP genotyping results plotted on Cartesian coordinates. Three distinct clusters are expected: Homozygous Allele 1 (FAM high/HEX low), Homozygous Allele 2 (FAM low/HEX high), and Heterozygous (Intermediate fluorescence for both). NTC (No Template Controls) cluster near the origin

2. 1.0 μL Forward Primer (10 μM).
3. 1.0 μL Reverse Primer (10 μM).
4. 2.0 μL Genomic DNA (20 ng).
5. 6.0 μL ddH_2O.

3. PCR and melting:
   1. Amplification: 95°C for 2 min; 40 cycles of [95°C for 10 s, 60°C for 15 s, 72°C for 15 s].
   2. Melting curve: Heat from 65°C to 95°C at 0.1°C/s increments with continuous fluorescence acquisition.
4. Analysis: Use the difference plot software to normalize melting curves. Distinct genotypes will show divergent melting profiles due to the T_m shift caused by the SNP.

4 Notes

1. Tissue age: Older sesame leaves accumulate secondary metabolites that interfere with DNA extraction. Always prioritize 2–3 week old seedlings. If using older tissue, increase the PVP (polyvinylpyrrolidone) concentration in the extraction buffer [28].
2. Mucilage management: Sesame produces significant mucilage. If the DNA pellet is gelatinous and difficult to dissolve, it indicates polysaccharide contamination. Repeat the wash step with 70% ethanol or perform a high-salt precipitation (1.2 M NaCl) prior to isopropanol precipitation [29].
3. Spectrophotometry vs. fluorometry: NanoDrop often overestimates DNA concentration in sesame samples due to RNA or phenolic contamination. Always use Qubit (fluorometry) for normalizing DNA concentrations before library preparation to ensure equimolar pooling [30].
4. Restriction enzyme choice: For sesame GBS, *Ape*KI is generally effective. However, perform an in silico digest of the reference genome first to predict fragment distribution. If the genome coverage is insufficient, consider a two-enzyme system (*Pst*I/*Msp*I) [23].
5. PCR duplicates: In GBS/RAD-seq, PCR duplicates can skew allele frequency estimates. Include a deduplication step in the bioinformatic pipeline (e.g., using PICARD tools) or use unique molecular identifiers (UMIs) in adapters during library preparation [24].

6. Missing data: Sesame GBS datasets often have high missing data rates (up to 40%–50%). Imputation (using Beagle or LinkImpute) is recommended before using the data for GWAS, but only if the initial marker density is sufficient to infer haplotypes [31].
7. SNP filtering for heterozygosity: Sesame is predominantly self-pollinating. High heterozygosity in an inbred line sample usually indicates paralogous mapping (two similar genomic regions mapping to one reference locus). Filter out SNPs with heterozygosity >10% in inbred lines [25].
8. KASP cluster separation: If KASP clusters are tight or ambiguous, the DNA concentration may be too high or variable. Dilute DNA to a uniform 5–10 ng/μL. Ensure the "No Template Control" (NTC) is included to define the background fluorescence [19].
9. HRM specificity: HRM is sensitive to nonspecific amplification. Always run the PCR product on a gel during optimization to ensure a single band. Primer dimers can produce misleading melting curves at lower temperatures [27].
10. Reference genome selection: Several sesame genomes are available (Zhongzhi No. 13, Yuzhi 11). Use the reference genome that is most genetically similar to your breeding population to maximize alignment rates and minimize false-positive SNPs caused by structural variations [32].

References

1. Andrade PB, Freitas BM, Rocha EEM et al (2014) Floral biology and pollination requirements of sesame (*Sesamum indicum* L.). Acta Sci Anim Sci 36:93–99
2. Kouighat M, Nabloussi A, Adiba A et al (2022) First study of improved nutritional properties and antioxidant activity in novel sesame mutant lines as compared to their wild types. Plants 11:1099
3. Vasanthi E, Rajavel DS (2021) Biopesticidal effect of sesamin on subterranean termite *Odontotermes obesus* (Termitidae, Isoptera). J Entomol Res 45:444–446
4. Abbas S, Sharif MK, Sibt-e-Abbas M et al (2022) Nutritional and therapeutic potential of sesame seeds. J Food Qual 2022:1–9
5. Kouighat M, El Harfi M, Hanine H et al (2022) Moroccan sesame: current situation, challenges, and recommended actions for its development. OCL 29:27
6. Wang Z, Zhou F, Tang X et al (2023) Morphology and SSR marker-based genetic diversity analysis of sesame (*Sesamum indicum* L.) cultivars released in China. Agriculture 13:1885
7. Ali Al-Somain BH, Migdadi HM, Al-Faifi SA et al (2017) Assessment of genetic diversity of sesame accessions collected from different ecological regions using sequence-related amplified polymorphism markers. 3 Biotech 7: 82
8. Poplin R, Chang PC, Alexander D et al (2018) A universal SNP and small-indel variant caller using deep neural networks. Nat Biotechnol 36:983–987
9. Wei L, Li C, Duan Y et al (2019) A SNP mutation of SiCRC regulates seed number per capsule and capsule length of cs1 mutant in sesame. Int J Mol Sci 20:4056
10. Yol E, Basak M, Kızıl S et al (2021) A high-density SNP genetic map construction using ddRAD-seq and mapping of capsule shattering trait in sesame. Front Plant Sci 12:679659
11. Elsafy M, Badawi W, Ibrahim A et al (2026) Candidate genomic regions underlying capsule shattering in sesame revealed by multi-model

GWAS and field-based phenotyping. Theor Appl Genet 139:42
12. Aguirre NC, Filippi CV, Vera PA et al (2023) Double digest restriction-site associated DNA sequencing (ddRADseq) technology. In: Shavrukov Y (ed) Plant genotyping. Springer, New York, pp 37–57
13. Raizada A, Souframanien J (2021) SNP genotyping and diversity analysis based on genic SNPs through high-resolution melting analysis in blackgram (*Vigna mungo*). Genet Resour Crop Evol 68:1331–1343
14. Sahoo J, Mishra R, Joshi RK (2025) PCR-based single nucleotide polymorphism genotyping for crop improvement: current status and future prospects. Discov Plants 2:172
15. Turaki AA, Ahmad B, Magaji UF et al (2017) Optimised CTAB DNA extraction method for plant leaves rich in polysaccharides and polyphenolic compounds for reliable downstream molecular analyses. Afr J Biotechnol 16:1354–1365
16. Ruperao P, Bajaj P, Subramani R et al (2023) A pilot-scale comparison between single- and double-digest RAD markers generated using a GBS strategy in sesame (*Sesamum indicum* L.). PLoS ONE 18:e0286599
17. Kizil S, Basak M, Guden B et al (2020) Genome-wide discovery of InDel markers in sesame (*Sesamum indicum* L.) using ddRAD-seq. Plants 9:1262
18. Basak M, Uzun B, Yol E (2019) Genetic diversity and population structure of the Mediterranean sesame core collection using genome-wide SNPs developed by double-digest RAD-seq. PLoS ONE 14:e0223757
19. Rahman MZ, Hasan MT, Rahman J (2023) Kompetitive allele-specific PCR (KASP): an efficient high-throughput genotyping platform and its applications in crop variety development. In: Kumar N (ed) Molecular marker techniques. Springer, Singapore, pp 25–54
20. Wang L, Yu J, Zhang Y et al (2021) Sinbase 2.0: an updated database to study multi-omics in Sesamum indicum. Plants 10:272
21. Rachappanavar V, Padiyal A, Sharma JK, Negi N (2022) Analytical pipelines for GBS analysis. In: Sonah H, Goyal V, Shivaraj SM, Deshmukh RK (eds) Genotyping by sequencing for crop improvement. Wiley, pp 161–187
22. Kutateladze T, Karchkhadze K, Bitskinashvili K et al (2024) Novel PCR-based technology for the detection of sunflower in edible and used cooking oils. Foods 13:3760
23. Baloyi NH, Voster AA, Swanepoel P (2025) Bambara groundnut (Vigna subterranea) genetic diversity and population structure assessed through restriction-site associated DNA sequencing. Crop Sci 65:e21444
24. Elshire RJ, Glaubitz JC, Sun Q et al (2011) A robust, simple genotyping-by-sequencing approach for high diversity species. PLoS ONE 6:e19379
25. Uncu AO, Frary A, Karlovsky P, Doganlar S (2016) High-throughput SNP identification and mapping in the sesame genome using genotyping by sequencing analysis. Mol Breed 36:173
26. Seay D, Szczepanek A, De La Fuente GN et al (2024) Genetic diversity and population structure of a large USDA sesame collection. Plants 13:1765
27. Kim N, Kwon JS, Kang WH, Yeom SI (2023) High-resolution melting genotyping. In: Shavrukov Y (ed) Plant genotyping. Springer, New York, pp 337–349
28. Tyagi R, Sharma V (2013) Optimization of DNA extraction protocol for PCR-based methods in phenolic- and polysaccharide-rich Sesamum indicum L. Gene Conserve 12:56–64
29. Rieder J, Jemmi E, Adrian-Kalchhauser I (2023) A guide to DNA extraction protocols for ecologists, conservation managers, and environmental DNA researchers. Methods Ecol Evol 14:185–199
30. Sudan J, Raina M, Singh R et al (2017) A modified protocol for high-quality DNA extraction from seeds rich in secondary compounds. J Crop Improv 31:637–647
31. Zhao H, MacLeod IM, Keeble-Gagnere G et al (2025) Using genotype imputation to integrate canola populations for genome-wide association and genomic prediction of blackleg resistance. BMC Genomics 26:215
32. Wei X, Zhu X, Yu J et al (2016) Identification of sesame genomic variations from genome comparison of landrace and variety. Front Plant Sci 7:1169

Check for updates

Chapter 16

Application of Single Nucleotide Polymorphisms (SNPs) in Sesame (*Sesamum indicum* L.) for Genetic Diversity, Mapping, and Marker-Assisted Selection

Mohamed Kouighat and Abdelghani Nabloussi

Abstract

Sesamum indicum L. is one of the oldest oilseed, cultivated oleaginous species, which is highly valued due to the excellent quality of the oils and owing to its antioxidative effects. Even though it is a major economic crop, there is a historical gap in breeding of sesame compared to the major staple crops, mainly because of a lack of genomic resources. With the emergence of next-generation sequencing methods, single-nucleotide polymorphisms (SNPs) with their noted benefits, compared to older marker systems, such as dense marker coverage, whole-genome coverage, and high-throughput automation capability, have been identified. This chapter outlines elaborate procedures for using SNP markers in the sesame improvement schemes. It describes the use of high-quality DNA extraction, SNP discovery using genotyping-by-sequencing (GBS), and downstream uses, including the determination of high-density linkage maps, genetic diversity, and marker-assisted selection (MAS). Using these protocols, breeders will be able to deconstruct complex traits with ease—such as yield, oil content, and biotic stress, thus speeding up the production of better sesame resilient varieties.

Keywords SNP markers, Marker-assisted selection, Genetic diversity, Molecular breeding, Linkage mapping

1 Introduction

Sesame (*Sesamum indicum* L. $2n=26$) is a diploid species mainly propagated by self-pollination [1]. Also known as the Queen of Oilseeds, it is planted both in tropical and subtropical regions [2, 3]. Seeds are rich in oil and protein, and represent unique lignans including sesamin and sesamolin, which provide oxidative stability and have health benefits [4]. Despite its international importance, sesame production is limited by a continuum of factors, such as low yield capacity, breaking of the capsules, indeterminate growth pattern, vulnerability to biotic stresses (phyldody and fusarium wilt), and vulnerability to abiotic stresses (waterlogging and drought) [5–7].

Nitish Kumar (ed.), *Molecular Markers in Plants*, Methods in Molecular Biology, vol. 3097,
https://doi.org/10.1007/978-1-0716-5663-1_16,

Traditional breeding in sesame has been accomplished at a sluggish rate due to the limited genetic variation in the cultivated genetic pools and the economically complicated mode of inheritance of characteristics of yield. In order to overcome these obstacles, molecular breeding techniques are being applied more and more. The history of marker development in sessions is characterized by the shift in the technology of marker development in these utilizing hybridization methods, like restriction fragment length polymorphism (RFLP), to PCR-based technologies, like restriction fragment length polymorphism (RFLP) and simple sequence repeats (SSRs) [8, 9]. Although these markers are useful, they are often impeded by either low density or high cost per data point [10]. The SNP marker system has become the favorite of modern genomics studies [11].

The most common type of sequence variation used in plant genomes is represented by SNPs. They are the best to polarize the quantitative trait loci (QTL) fine-map, to perform genome-wide association studies (GWAS), and to perform marker-assisted selection (MAS) due to their biallelic nature, amenability to high-throughput automation, and general genome-wide [12, 13]. The completion of the sesame genome sequencing project has transformed the genetic makeup of this crop, making it easier to identify thousands of SNPs [14]. Genetic resources and molecular breeding workflow in sesame are described in Fig. 1. The chapter outlines detailed protocols to be used in the implementation of SNP markers to measure genetic diversity, build linkage maps, and MAS with the aim of improving breeding efficiency in sesame.

2 Materials

2.1 Plant Material and DNA Sources

1. Germplasm: A diverse panel of sesame accessions, including landraces, breeding lines, and wild relatives (e.g., *Syzygium malabaricum*), should be selected based on the specific research objective.
2. Tissue source: The appropriate material is young, healthy, non-expanded leaves (100–200 mg) collected 3–4 weeks after planting.
3. Storage: Leaf samples are to be collected in liquid nitrogen or silica gel and stored at −80°C. Tissue freeze-dried (lyophilized) should be used with high-throughput extraction.

2.2 DNA Extraction Reagents

Using molecular biology-grade water and autoclaved glassware, prepare all solutions. Store stock solutions as indicated [15].

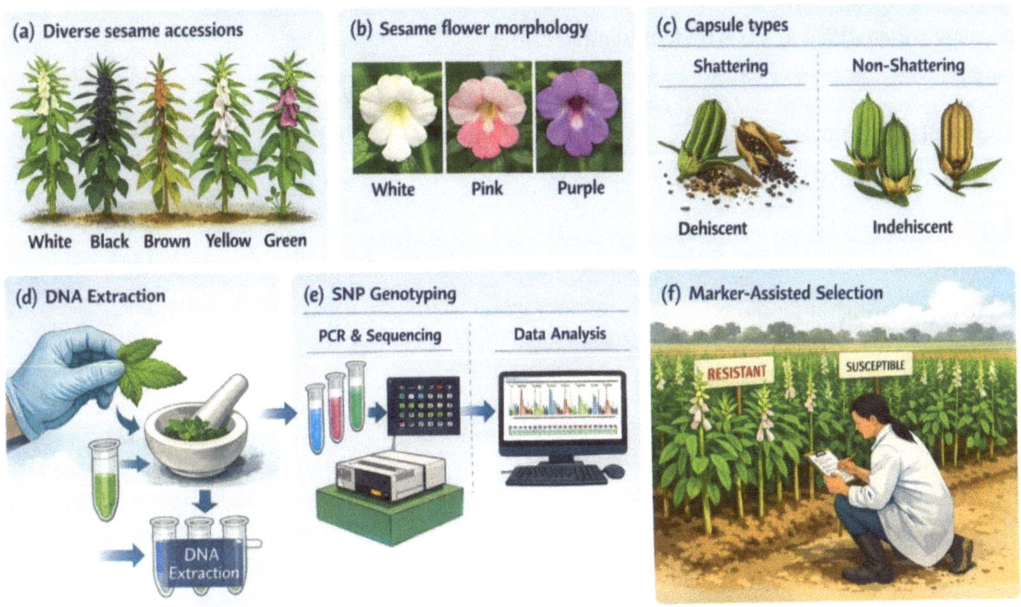

Fig. 1 Sesame germplasm and breeding pipeline. Genetic resources and molecular breeding workflow in sesame. (**a**) Diverse sesame accessions showing morphological variation, (**b**) flower morphology of sesame, (**c**) capsule types—shattering vs. non-shattering, (**d**) DNA extraction from young leaves, (**e**) SNP genotyping workflow, and (**f**) marker-assisted selection in breeding nursery

1. Chloroform:isoamyl alcohol (C:I): Isoamyl alcohol and Chloroform were mixed in a volumetric proportion of 24:1 (v/v). The solution obtained was kept in a dark bottle at 4°C.
2. Isopropanol: Cold isopropanol, kept at −20°C, was used in the process of DNA precipitation.
3. Ethanol: 70% (v/v) ethanol for washing DNA pellets.
4. TE buffer (1×): Composed of 1 mM EDTA and 10 mM Tris–HCl (pH 8.0) (*see* **Note 1**).
5. RNase: 10 mg/mL solution in 10 mL made and kept in −20°C.
6. Chemical composition of the CTAB DNA extraction buffer is tabulated in Table 1.

2.3 SNP Discovery and Genotyping

1. Library preparation: Restriction enzymes (e.g., *Ape*KI and *Pst*I) for GBS, adapters, and ligase enzymes [16].
2. Genotyping assays: validation is performed on a KASP (Kompetitive Allele Specific PCR) master mix, which contains FAM and HEX-labeled FRET cassettes, Taq DNA polymerase, a ROX passive reference dye, deoxynucleotide triphosphates, and $MgCl_2$) [17].

Table 1
Chemical composition of CTAB DNA extraction buffer

Chemical constituents	Stock concentration	Volume per 100 mL	Final concentration
Tris–HCl (pH 8.0)	1 M	10 mL	100 mM
NaCl	5 M	28 mL	1.4 M
EDTA (pH 8.0)	0.5 M	4 mL	20 mM
CTAB (Cetyltrimethylammonium bromide)	Solid	2 g	2% (w/v)
PVP-40 (Polyvinylpyrrolidone)	Solid	1 g	1% (w/v)
β-mercaptoethanol (immediately before use)	Liquid (14.3 M)	0.2 mL	0.2% (v/v)

3. Primers: allele-specific forward primers in combination with a universal reverse primer are developed to match certain single-nucleotide polymorphisms (SNPs) [18].

2.4 Data Analysis Software

1. Population genetics: The population structure can be analyzed using software tools such as TASSEL (v5.0), STRUCTURE (v2.3.4), or (v6.5) [19].
2. Mapping: Using JoinMap 4.1 or 5.0, MapDisto 2.0, and remaining markers can be used to construct a genetic linkage map [20].
3. Statistics: R statistical software (packages: *ape*, *poppr*, *qtl*).

3 Methods

3.1 DNA Extraction Protocol

The modified CTAB method yields high-quality genomic DNA suitable for sequencing and SNP genotyping [16, 21, 22].

1. Preheat the CTAB extraction buffer (Table 1) to 65°C in a water bath. Add β-mercaptoethanol just prior to use (*see* **Note 2**).
2. Using liquid nitrogen, pulverize 100 mg of fresh or frozen leaf tissue into a fine powder with either a mortar and pestle or a tissue lyser.
3. Transfer the powder to a 2 mL microcentrifuge tube containing 800 μL of pre-warmed CTAB buffer. Vortex vigorously to mix.

4. Incubate samples at 65°C for 60 min, inverting the tubes gently every 10–15 min.
5. Let the tubes cool to room temperature, then add 800 µL of chloroform:isoamyl alcohol (24:1, v/v) and mix gently by inverting for 10 min to form an emulsion.
6. Centrifuge at 12,000 × *g* for 15 min at room temperature.
7. Taking care not to disturb the interphase, gently transfer approximately 600 µL of the upper aqueous layer to a fresh 1.5 mL tube.
8. Add 0.6 volumes of cold isopropanol to the supernatant. Mix gently by inversion until DNA strands precipitate.
9. DNA Precipitation: Centrifuge at 12,000 × *g* for 10 min to pellet the DNA and carefully discard the supernatant.
10. Pellet Washing: Using 500 µL of 70% ethanol, wash the DNA pellet and centrifuge at 7000 × *g* for 5 min. Repeat this wash step twice.
11. Drying: Air-dry the pellet for 15–20 min until the ethanol smell is gone, taking care not to overdry.
12. Resuspension and RNA removal: Before adding 2 µL of RNase A, dissolve the pellet in 50–100 µL of TE buffer, and incubate at 37°C/30 min to degrade any RNA.
13. Quantify DNA concentration using a NanoDrop spectrophotometer or Qubit fluorometer. Check purity; the A260/280 ratio should be between 1.8 and 2.0 (*see* **Note 3**).

3.2 SNP Discovery

1. Library construction: Digest genomic DNA (100–200 ng) with restriction enzymes (e.g., *Ape*KI) to reduce genome complexity. Ligate barcoded adapters to the sticky ends [16].
2. Sequencing: Pool libraries and sequence on the NovaSeq platform or an Illumina HiSeq to generate 150 bp paired-end reads [23].

3.3 Bioinformatics Pipeline

1. Quality control: Use FastQC and trimmomatic to remove low-quality reads and adapters [24].
2. Alignment: Using reference genome (e.g., *Zhongzhi No. 13* v2.0) and BWA-MEM or Bowtie2, align clean reads to the sesame [25].
3. Variant calling: Use GATK (Genome Analysis Toolkit) or SAMtools to call SNPs [26].
4. Filtering: Filter raw SNPs based on: Missing data < 20%, minor allele frequency (MAF) > 0.05, and read depth > 5× [23] (*see* **Notes 4 and 5**).

3.4 SNP Genotyping

For validation or routine screening of specific markers (e.g., KASP assay) [27, 28]:

1. Prepare the PCR master mix containing 2× KASP Master Mix and the assay mix (two allele-specific forward primers and one common reverse primer) (*see* **Note 6**).
2. Dispense 5 μL of reaction mixture (containing ~10 ng of template DNA) into 96-well or 384-well PCR plates.
3. Thermal Cycling: for 15 min perform the initial activation of PCR at 94°C; followed by 10 touchdown cycles of 94°C/20 s and annealing from 61°C to 55°C/60 s; then complete 26 amplification cycles of 94°C/20 s and 55°C/60 s.
4. Read endpoint fluorescence using a FRET-capable plate reader. Plot allelic discrimination data (FAM vs. HEX intensity) (see Fig. 2).

3.5 Genetic Diversity Assessment

1. Construct a genotyping matrix (rows = samples, columns = SNP markers) (Table 2).

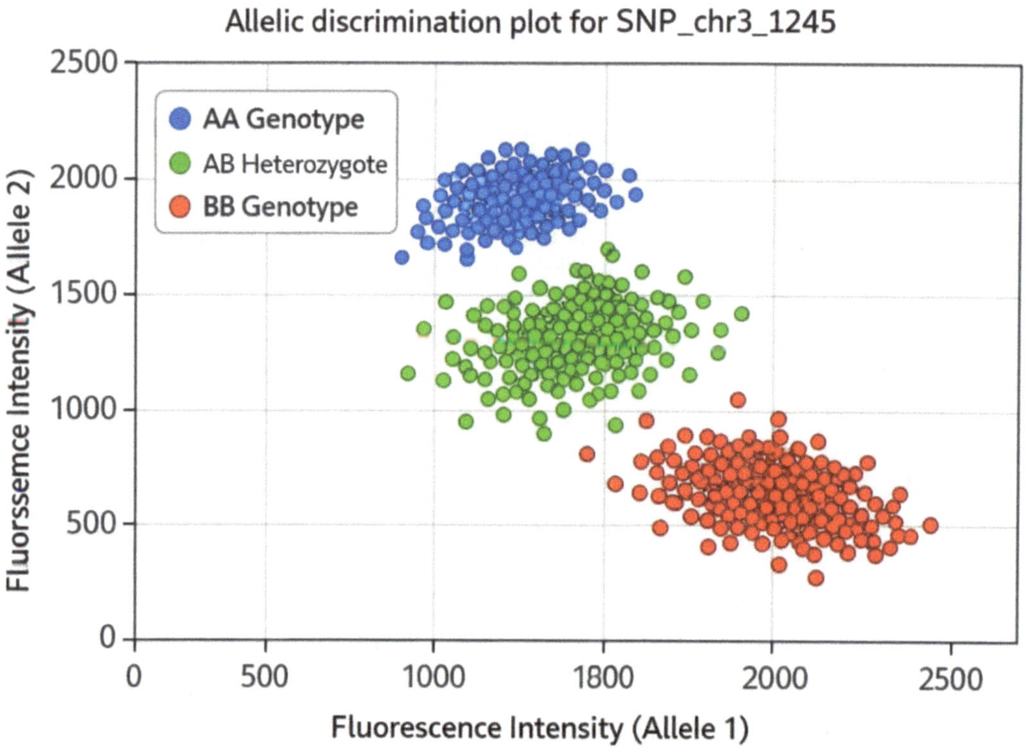

Fig. 2 Typical SNP genotyping results and polymorphism detection. Allelic discrimination plot showing SNP genotyping results for marker SNP_chr3_1245. Blue clusters represent the AA genotype, red clusters represent the BB genotype, and green clusters represent the AB heterozygote. DNA samples from 96 sesame accessions. The axis shows fluorescence intensity for allele 1 (*x*-axis) and allele 2 (*y*-axis)

Table 2
Selected SNP markers for sesame genetic analysis [29]

SNP ID	Chromosome	Position (Mb)	Alleles	Gene annotation	Application
SiSNP_01_4523	Chr 1	4.52	A/G	FAD2 desaturase	Oil Quality
SiSNP_02_1102	Chr 2	11.02	C/T	TIR–NBS–LRR	Disease Resistance
SiSNP_03_8890	Chr 3	8.89	A/T	Unknown	Diversity
SiSNP_05_2341	Chr 5	2.34	G/C	Cellulose synthase	Capsule Shattering
SiSNP_06_6712	Chr 6	6.71	T/A	Cytochrome P450	Sesamin Content
SiSNP_08_0921	Chr 8	0.92	C/G	WRKY transcription factor	Drought Tolerance
SiSNP_11_1450	Chr 11	14.50	A/C	Auxin response factor	Plant Height

2. Use GenAlEx or PowerMarker to calculate diversity indices: number of alleles (Na), observed heterozygosity (Ho), effective number of alleles (Ne), fixation index (Fst), expected heterozygosity (He), and polymorphism information content (PIC) [30].
3. Conduct the analysis using STRUCTURE software, evaluating *K* values ranging from one to ten with five to ten replicates per K. Using the Evanno method to determine the most likely number of clusters [19].
4. Construction of the neighbor-joining (NJ) tree by Nei's genetic distance, employing software such as TASSEL or MEGA [23] (*see* Fig. 3).

3.6 Linkage Map Construction

1. Develop a mapping population (F_2, RIL, or BC_1F_1) derived from two parents with contrasting traits [31] (*see* **Note 7**).
2. Genotype the population (minimum 150 individuals) using genome-wide SNPs [14] (*see* **Note 8**).
3. Filter markers for segregation distortion using chi-square tests (1:2:1 for F_2 or 1:1 for RIL) [32].
4. Import data into JoinMap. Group markers into Linkage Groups (LGs) using a minimum LOD score of 3.0–5.0 [20].
5. Order markers within groups using regression mapping or maximum likelihood algorithms [23].
6. Convert recombination frequencies to map distances (cM) using the Kosambi mapping function [33].

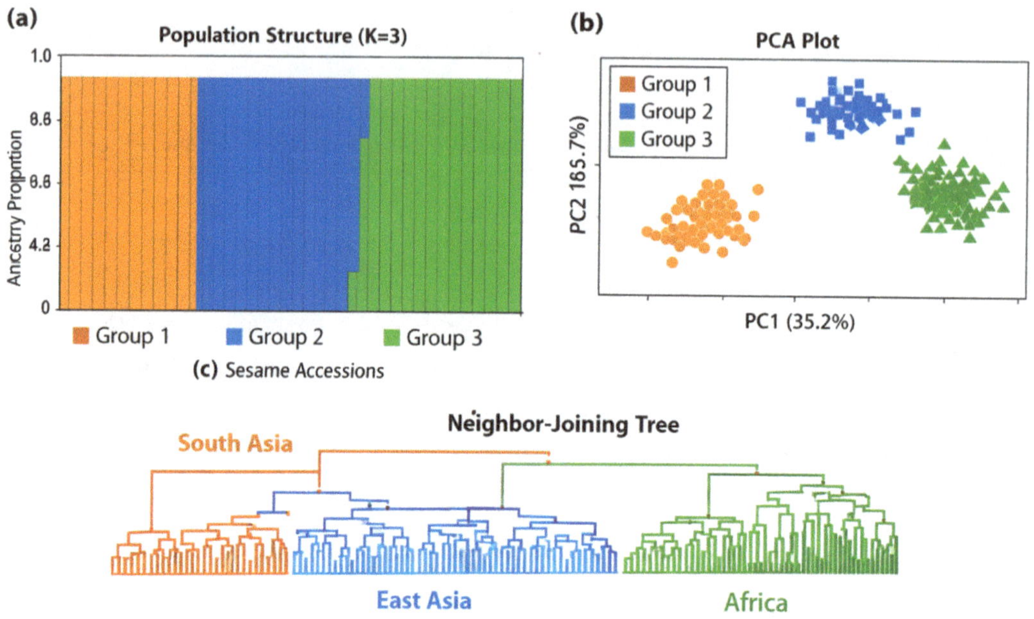

Fig. 3 Typical population structure and genetic relationships. Genetic diversity analysis of 120 sesame accessions using 2500 SNP markers. (**a**) Population structure analysis ($K=3$) showing three major genetic groups, (**b**) principal component analysis (PCA) plot with PC1 and PC2 explaining 35.2% and 18.7% of variation, respectively, (**c**) neighbor-joining dendrogram based on genetic distance showing clustering of accessions by geographic origin

3.7 Marker-Assisted Selection (MAS) Protocol

1. Marker validation: Confirm the association between a SNP and the trait of interest (e.g., *SiSNP_05_2341* for non-shattering) in a distinct validation population.
2. Foreground selection: Screen breeding lines to select those carrying the favorable allele (homozygous donor genotype).
3. Background selection: Use genome-wide SNPs to select individuals with the highest percentage of the recurrent parent genome to reduce linkage drag.
4. Advance selected lines to yield trials.

4 Notes

1. TE buffer: Ensure the pH of the TE buffer is exactly 8.0. DNA dissolves best in slightly alkaline conditions. Acidic pH can cause hydrolysis of DNA.
2. Buffer preparation: CTAB dissolves slowly. Heat the buffer to 65°C on a magnetic stirrer to ensure complete dissolution. Do not autoclave the buffer after adding β-mercaptoethanol.

3. DNA quality: For SNP genotyping, particularly GBS, high molecular weight DNA is crucial. The A260/280 ratio must be strict (1.8–2.0). A260/230 < 1.8 indicates polysaccharide or phenol contamination, which inhibits enzymatic digestion.
4. SNP calling: High depth (>10×) is preferred for accurate heterozygote calling. For diversity studies, lower depth (5×) may suffice if imputation is used, but for MAS, high confidence is required.
5. MAF threshold: When filtering SNPs, setting the Minor Allele Frequency (MAF) too high (>0.10) may discard rare alleles valuable for specific traits. A standard threshold is 0.05.
6. Genotyping failure: Low call rates in KASP assays are often due to poor DNA quality or incorrect primer concentration. Repurify DNA or optimize primer annealing temperatures.
7. Software: Linkage mapping algorithms (e.g., maximum likelihood vs. regression) can yield different maps. Compare outputs from different software (JoinMap vs. R/qtl) for robustness.
8. Population size: For linkage mapping, small populations (<100) result in inflated map distances and poor marker ordering. A minimum of 150–200 RILs is recommended for high-resolution mapping.

References

1. Andrade PB de, Freitas BM, Rocha EE, de M et al (2014) Floral biology and pollination requirements of sesame (*Sesamum indicum* L.). Acta Sci Anim Sci 36:93–99
2. Kouighat M, Channaoui S, Labhilili M et al (2021) Novel genetic variability in sesame induced via ethyl methane sulfonate. J Crop Improv 35:654–665. https://doi.org/10.1080/15427528.2020.1861155
3. Aadi ON, Almarie AA (2024) Yield and quality of sesame under various weed removal periods and predicting the optimum weed control time. In: IOP conference series: earth and environmental science. IOP Publishing, p 052061
4. Abbas S, Sharif MK, Sibt-e-Abbas M et al (2022) Nutritional and therapeutic potential of sesame seeds. J Food Qual 2022:1–9. https://doi.org/10.1155/2022/6163753
5. Abbas AA, Berhe M, Kefale H et al (2025) Physiological and biochemical responses to cold stress in sesame (*Sesamum indicum* L.) during the early growth stage. Plant Soil Environ 71(5):293–307. https://doi.org/10.17221/79/2025-PSE
6. kouighat M, Kettani R, El Fechtali M, Nabloussi A (2024) Exploring mechanisms of drought-tolerance and adaptation of selected sesame mutant lines. J Agric Food Res 15:100911. https://doi.org/10.1016/j.jafr.2023.100911
7. Farhaoui A, Kouighat M, Taoussi M, Lahlali R (2025) Soil-borne fungal pathogens of sesame: insights into pathogenesis and control. Physiol Mol Plant Pathol 138:102726
8. Wang Z, Zhou F, Tang X et al (2023) Morphology and SSR markers-based genetic diversity analysis of sesame (*Sesamum indicum* L.) cultivars released in China. Agriculture 13:1885
9. Ali Al- Somain BH, Migdadi HM, Al-Faifi SA et al (2017) Assessment of genetic diversity of sesame accessions collected from different ecological regions using sequence-related amplified polymorphism markers. 3 Biotech 7:82. https://doi.org/10.1007/s13205-017-0680-2
10. Ekanayake W, Phuong T, Lee J, Lee S (2024) Selection of informative markers using machine learning approaches and genome-wide association studies to improve genomic prediction in Hanwoo cattle: a simulation study. J Anim Breed Genom 8(1):17–32s
11. Sahoo J, Mishra R, Joshi RK (2025) PCR-based single nucleotide polymorphism (SNP)

genotyping for crop improvement–current status and future prospects. Discov Plants 2:172. https://doi.org/10.1007/s44372-025-00262-9
12. Kefale H, Wang L (2022) Discovering favorable genes, QTLs, and genotypes as a genetic resource for sesame (*Sesamum indicum* L.) improvement. Front Genet 13:1002182
13. Li D, Dossa K, Zhang Y et al (2018) GWAS uncovers differential genetic bases for drought and salt tolerances in sesame at the germination stage. Genes 9:87
14. Basak M, Uzun B, Yol E (2019) Genetic diversity and population structure of the Mediterranean sesame core collection with use of genome-wide SNPs developed by double digest RAD-Seq. PLoS ONE 14:e0223757
15. Cho Y-S, Choi Y-L, Chung C-H (1997) An effective method for isolating genomic DNA from leaves of sesame and perilla. Appl Biol Chem 40:262–265
16. Aguirre NC, Filippi CV, Vera PA et al (2023) Double digest restriction-site associated DNA sequencing (ddRADseq) technology. In: Shavrukov Y (Ed) Plant Genotyping. Springer, US, New York, NY, pp37–57
17. Kizil S, Basak M, Guden B et al (2020) Genome-wide discovery of InDel markers in sesame (*Sesamum indicum* L.) using ddRAD-Seq. Plants 9:1262
18. Purru S, Sahu S, Rai S et al (2018) GinMicrosatDb: a genome-wide microsatellite markers database for sesame (*Sesamum indicum* L.). Physiol Mol Biol Plants 24:929–937. https://doi.org/10.1007/s12298-018-0558-8
19. Ali F, Nadeem MA, Barut M et al (2020) Genetic diversity, population structure and marker-trait association for 100 seed weight in international safflower panel using silicoDArT marker information. Plants 9:652
20. Yol E, Basak M, Kızıl S et al (2021) A high-density SNP genetic map construction using ddRAD-seq and mapping of capsule shattering trait in sesame. Front Plant Sci 12:679659
21. Turaki AA, Ahmad B, Magaji UF et al (2017) Optimised cetyltrimethylammonium bromide (CTAB) DNA extraction method of plant leaf with high polysaccharide and polyphenolic compounds for downstream reliable molecular analyses. Afr J Biotechnol 16:1354–1365
22. Sudan J, Raina M, Singh R et al (2017) A modified protocol for high-quality DNA extraction from seeds rich in secondary compounds. J Crop Improv 31:637–647. https://doi.org/10.1080/15427528.2017.1345028
23. Seay D, Szczepanek A, De La Fuente GN et al (2024) Genetic diversity and population structure of a large USDA sesame collection. Plants 13:1765
24. Bonfiglio F, Legati A, Lasorsa VA et al (2024) Best practices for germline variant and DNA methylation analysis of second- and third-generation sequencing data. Hum Genomics 18:120. https://doi.org/10.1186/s40246-024-00684-8
25. Wu X, Heffelfinger C, Zhao H, Dellaporta SL (2019) Benchmarking variant identification tools for plant diversity discovery. BMC Genomics 20:701. https://doi.org/10.1186/s12864-019-6057-7
26. Wang L, Han X, Zhang Y et al (2014) Deep resequencing reveals allelic variation in *Sesamum indicum*. BMC Plant Biol 14:225. https://doi.org/10.1186/s12870-014-0225-3
27. Rahman MdZ, Hasan MdT, Rahman J (2023) Kompetitive allele-specAllele-Specific PCR (KASP): an efficient high-throughput genotyping platform and its applications in crop variety development. In: Kumar N (ed) Molecular marker techniques. Springer Nature Singapore, Singapore, pp 25–54
28. Li Y, Yu T, Li J et al (2021) A novel fluorescent FRET hairpin probe switch for aflD gene detection in real fermented soybean paste. Food Anal Methods 14:2469–2477. https://doi.org/10.1007/s12161-021-02080-7
29. sesame-bioinfo.org/SesameFG/. http://www.sesame-bioinfo.org/SesameFG/. Accessed 23 Jan 2026
30. Azon CF, Fassinou Hotegni NV, Adjé CO et al (2024) Molecular diversity and agronomic performance of sesame (*Sesamum indicum*) cultivars in Benin: local cultivars and lines introduced from China. Plant Environ Interact 5:e70024. https://doi.org/10.1002/pei3.70024
31. Wu K, Liu H, Yang M et al (2014) High-density genetic map construction and QTLs analysis of grain yield-related traits in Sesame (*Sesamum indicum* L.) based on RAD-Seq techonology. BMC Plant Biol 14:274. https://doi.org/10.1186/s12870-014-0274-7
32. Zhang Z, Jiang Y, Zhu H et al (2025) Segregation distortion and linkage analysis in F2 populations of pumpkin (*C. Moschata* Duch). Plant Gene 43:100508
33. Xu G, Cui Y, Li S et al (2024) High-density genetic map construction and QTL mapping to identify genes for blight defense-and yield-related traits in sesame (*Sesamum indicum* L.). Front Plant Sci 15:1446062

Check for updates

Chapter 17

ISSR-PCR for Genetic Diversity Analysis in Faba Bean

Oumaima Chetto, Loubna Belqadi, Mounia Ennami, Etienne Bucher and Abdelghani Nabloussi

Abstract

Molecular marker-based diversity assessment is essential for crop improvement strategies in faba bean (*Vicia faba* L.), a diploid grain legume with limited genomic resources relative to model species such as *Medicago truncatula* or major crops like soybean (*Glycine max*). ISSR markers provide a PCR-based, multilocus fingerprinting approach requiring no prior sequence information, making them particularly suitable for orphan crops. This protocol provides comprehensive, step-by-step procedures covering the complete analytical pipeline from genomic DNA isolation to statistical interpretation. This chapter includes: (1) CTAB DNA extraction optimized for polyphenol removal, (2) ISSR-PCR with optimization guidelines, (3) gel electrophoresis protocols, (4) manual and software-based fragment sizing with *Rf*-based interpolation methods, (5) binary matrix construction and band scoring, and (6) statistical analysis for diversity parameters calculation. Comprehensive notes provide troubleshooting guidance for common problems and data interpretation standards. While optimized for faba bean, the protocol approach allows ready adaptation to other plant species through tissue-specific modifications to extraction procedures and PCR conditions.

Keywords *Vicia faba*, Genetic diversity, Molecular markers, PCR optimization, DNA extraction, Legumes

1 Introduction

Faba bean (*Vicia faba* L., $2n = 12$) is a diploid, cool-season grain legume cultivated globally for human consumption, animal feed, and soil improvement through biological nitrogen fixation [1]. Despite its agronomic importance and nutritional value as a protein-rich crop, faba bean breeding programs face challenges including narrow genetic diversity in cultivated germplasm, susceptibility to biotic stresses (fungal pathogens, parasitic weeds) and abiotic stresses (drought, heat), and limited genomic resources compared to model legumes such as *Medicago truncatula* or *Glycine max* [2]. Assessment of genetic diversity is fundamental to crop improvement strategies, enabling characterization of germplasm collections, identification of

Nitish Kumar (ed.), *Molecular Markers in Plants*, Methods in Molecular Biology, vol. 3097,
https://doi.org/10.1007/978-1-0716-5663-1_17,

diverse parental lines for breeding [3], monitoring genetic erosion in cultivated varieties, establishing core collections for efficient germplasm management [4], and understanding population structure and evolutionary relationships [5].

Molecular markers provide objective, environment-independent tools for diversity assessment, overcoming limitations of morphological characterization, which is time-consuming, influenced by environmental factors, and limited by the number of distinguishable traits.

1.1 ISSR Markers: Advantages and Principles

Inter-simple sequence repeat (ISSR) markers, first described by [6], are PCR-based markers that amplify DNA segments located between adjacent, oppositely oriented microsatellite regions. ISSR primers are typically 16–25 nucleotides in length, consisting of a di or trinucleotide repeat motif (e.g., $(GA)_8$, $(CAC)_5$) anchored at the 3′ or 5′ end by 1–4 arbitrary nucleotides to increase specificity [7]. The arbitrary nucleotides allow annealing to microsatellite-adjacent regions, generating multilocus amplification patterns across the genome. ISSR markers offer several advantages for genetic diversity studies in faba bean and other crops:

- No prior sequence information required: Unlike SSR (simple sequence repeat) markers, which require species-specific primer design based on genomic sequences, ISSR primers can be applied across diverse taxa without prior knowledge of the target genome [2]. This is particularly valuable for orphan crops like faba bean, where genomic resources have historically been limited.
- High reproducibility: ISSR-PCR employs longer primers (16–25 bp) and higher annealing temperatures (45°C–55°C) compared to RAPD (random amplified polymorphic DNA) markers, resulting in superior reproducibility within and between laboratories [7]. Studies report >90% reproducibility for well-optimized ISSR protocols [8].
- High polymorphism detection: ISSR markers typically detect 60%–95% polymorphism in diverse germplasm collections, providing sufficient resolution for population genetics and diversity studies [9–11]. The multilocus nature of ISSR (5–15 bands per primer) enables efficient screening compared to single-locus markers.
- Cost-effectiveness: ISSR-PCR requires minimal infrastructure (standard thermal cycler, gel electrophoresis system) and no expensive sequencing or genotyping platforms, making it accessible to laboratories with limited resources.
- Technical simplicity: ISSR protocols are straightforward to implement, requiring basic molecular biology skills and equipment available in most plant genetics laboratories.

1.2 Applications in Faba Bean Genetics and Breeding

ISSR markers have been successfully employed in faba bean for diverse applications. Multiple studies have utilized ISSR markers to characterize genetic variation in faba bean germplasm from different geographic origins, revealing population structure and identifying diverse accessions for breeding programs [3, 9, 10, 12, 13]. Additionally, ISSR fingerprinting enables discrimination of morphologically similar varieties and detection of genetic contamination in seed lots, supporting variety registration and seed certification programs [14]. ISSR markers have also been linked to agronomically important traits [15, 16], although their application in marker-assisted selection remains limited compared to SSR or SNP markers, due to their dominant inheritance, which prevents heterozygote identification [17].

2 Materials

All materials should be prepared fresh or stored according to the manufacturer's recommendations. Use molecular biology grade reagents throughout the protocol.

2.1 Plant Material and Sample Collection

- Fresh young leaves from faba bean genotypes.
- Liquid nitrogen for snap freezing.
- Silica gel desiccant (alternative for field collection).
- Zip-lock plastic bags (labeled).
- 70% ethanol for leaf surface sterilization.

2.2 DNA Extraction Reagents

2.2.1 CTAB Extraction Buffer

- 2%–3% CTAB (cetyltrimethylammonium bromide).
- 100 mM Tris–HCl, pH 8.0.
- 20–30 mM EDTA, pH 8.0.
- 1.4 M NaCl.
- 1.5%–2% Polyvinylpyrrolidone (PVP-40).
- 0.2%–0.3% β-mercaptoethanol (add fresh before use).

2.2.2 Other DNA Extraction Materials

- Chloroform:isoamyl alcohol (24:1).
- Isopropanol (ice-cold, −20°C).
- 70% and 95% ethanol (ice-cold).
- 3 M Sodium acetate, pH 4.8.
- TE buffer (10 mM Tris–HCl, 1 mM EDTA, pH 8.0).
- RNase A (10 mg/ml stock solution)

2.3 PCR Amplification Reagents

- Taq DNA polymerase (5 U/μl) with supplied buffer.
- 10× PCR buffer (typically supplied with Taq).
- $MgCl_2$ solution (25 mM stock).
- dNTPs mix (10 mM each of dATP, dTTP, dGTP, dCTP).
- ISSR primers (10 μM working stock).
- Molecular biology grade water (sterile, nuclease-free).

2.4 ISSR Primer Selection

Select primers from prescreened universal ISSR primer sets. Commonly used primers for faba bean can be seen in Table 1.

2.5 Gel Electrophoresis Materials

- Agarose (molecular biology grade).
- 1× TBE buffer (89 mM Tris–borate, 2 mM EDTA, pH 8.3).
- DNA staining solution (GelRed, SYBR Green, or ethidium bromide).
- DNA ladder (100 bp or/and 1 kb).
- 6× DNA loading dye.

2.6 Equipment and Supplies

- Mortar and pestle (prechilled) or tissue grinder.
- Microcentrifuge tubes (1.5 and 2.0 ml).
- PCR tubes (0.5 ml).
- Thermal cycler with programmable temperature control.
- Microcentrifuge (capable of 12,000–15,000 rpm).
- Water bath or heating block (37°C, 60°C, 65°C).
- Vortex mixer.
- UV spectrophotometer or NanoDrop.
- Horizontal gel electrophoresis system.
- Gel documentation system: Integrated imaging system (e.g., VisionCapt™, ChemiDoc™, Gel Doc™ EZ) OR standalone UV transilluminator (302 nm) with digital camera.
- UV-protective face shield or amber safety glasses (depending on system type).
- Gel spatula.
- Micropipettes (2–10, 10–100, 100–1000 μl) with tips.
- −20°C and −80°C freezers for sample storage.

Table 1
Commonly used primers for faba bean [3, 11, 18]

Primer name	Sequence (5′ → 3′)	Repeat type
UBC-807	$(AG)_8T$	Dinucleotide
UBC-808	$(AG)_8C$	Dinucleotide
UBC-817	$(CA)_8A$	Dinucleotide
UBC-826	$(AC)_8C$	Dinucleotide
UBC-827	$(AC)_8G$	Dinucleotide
UBC-841	$(GA)_8YC$	Dinucleotide
UBC-843	$(CT)_8RA$	Dinucleotide
UBC-844	$(CT)_8RC$	Dinucleotide
ISSR-47	$(AC)_8YT$	Dinucleotide
IMA9Z	$(GA)_8CG$	Dinucleotide
IMA834	$(AG)_8YT$	Dinucleotide
F7	$(AC)_8CT$	Dinucleotide
F8	$(AG)_8CC$	Dinucleotide
F10	$(CA)_8AG$	Dinucleotide
ISSF1	$(AG)_8TA$	Dinucleotide
HB-12	$(CAC)_3GC$	Trinucleotide
F4	$(AGC)_4CT$	Trinucleotide
ISSR-01	$(GA)_8C$	Dinucleotide
ISSR-02	$(CA)_8C$	Dinucleotide
ISSR-03	$(GT)_8C$	Dinucleotide
ISSR-04	$(GA)_6CC$	Dinucleotide
ISSR-05	$(CA)_6GC$	Dinucleotide
ISSR-06	$(GATA)_4CGC$	Tetranucleotide
ISSR-07	$(ATA)_3GAC$	Trinucleotide
ISSR-08	$(AGAC)_4GC$	Tetranucleotide
ISSR-09	$(GATA)_4GC$	Tetranucleotide
ISSR-10	$(GACA)_4A$	Tetranucleotide

Note: Y = C or T; R = A or G (degenerate bases)

3 Methods

3.1 Sample Collection and Preparation

1. Collect young, healthy leaves from each faba bean genotype during the early vegetative stage (3–4 week old plants).
2. Wipe leaf surfaces with 70% ethanol to remove surface contaminants.
3. Immediately snap-freeze samples in liquid nitrogen or place in zip-lock bags with silica gel for field collections.
4. Store frozen samples at −80°C until DNA extraction. Silica-dried samples can be stored at room temperature in airtight containers.

3.2 Genomic DNA Extraction (Modified CTAB Protocol)

3.2.1 CTAB Extraction Buffer Preparation (Modified from [19])

1. Prepare CTAB extraction buffer: Mix 2%–3% CTAB, 100 mM Tris–HCl (pH 8.0), 20–30 mM EDTA (pH 8.0), 1.4 M NaCl, and 1.5%–2% PVP-40.
2. Preheat buffer to 60°C–65°C in a water bath.
3. Add 0.2%–0.3% β-mercaptoethanol immediately before use (perform in fume hood).

3.2.2 Tissue Grinding and Lysis

1. Grind 100–150 mg of fresh leaf tissue or 50–100 mg of dried tissue in liquid nitrogen using a prechilled mortar and pestle to a fine powder. Avoid thawing during grinding.
2. Immediately transfer ground tissue to a 2.0 ml microcentrifuge tube.
3. Add 800–1000 μl of preheated CTAB extraction buffer (with β -mercaptoethanol).
4. Vortex vigorously for 30 s to mix thoroughly.
5. Incubate at 60°C–65°C for 45 min with gentle mixing every 10 min.

3.2.3 Organic Extraction (Modified from [20])

1. Allow tubes to cool to room temperature.
2. Add an equal volume (800–1000 μl) of chloroform:isoamyl alcohol (24:1).
3. Mix gently by inversion for 5–10 min.
4. Centrifuge at 12,000–13,000 rpm for 10–15 min at 4°C.
5. Carefully transfer the upper aqueous phase to a fresh 1.5 ml tube using a micropipette (avoid interface).
6. Repeat chloroform extraction (Steps 7–10) if the aqueous phase is cloudy or dark-colored.

3.2.4 DNA Precipitation and Washing

1. Add 1/10 volume of 3 M sodium acetate (pH 4.8) to the aqueous phase.
2. Add two volumes of ice-cold isopropanol (−20°C).
3. Mix gently by inversion and incubate at 4°C overnight for a better yield.
4. Centrifuge at 12,000–13,000 rpm for 10 min at 4°C.
5. Carefully discard supernatant without disturbing the DNA pellet.
6. Wash pellet with 500 μl of ice-cold 70% ethanol.
7. Centrifuge at 12,000 rpm for 5 min at 4°C.
8. Discard supernatant and repeat wash with 500 μl of 70% ethanol.
9. Air-dry pellet at room temperature for 10–15 min (avoid overdrying).

3.2.5 DNA Resuspension and RNASE Treatment

1. Dissolve the DNA pellet in 50–100 μl of TE buffer (pH 8.0) or nuclease-free water.
2. Incubate at 37°C for 1–2 h or overnight at 4°C to ensure complete dissolution.
3. Add RNase A to a final concentration of 10 μg/ml.
4. Incubate at 37°C for 30–45 min.
5. Store DNA at −20°C for short-term storage or −80°C for long-term storage.

3.3 DNA Quality and Quantity Assessment

3.3.1 Gel Electrophoresis Quality Check

1. Prepare 1% agarose gel in 1× TBE buffer.
2. Load 5 μl of the DNA sample mixed with 1 μl of 6× loading dye.
3. Run electrophoresis at 90–100 V for 60 min.
4. Visualize under UV transilluminator to confirm high molecular weight DNA and absence of smearing.

3.3.2 Spectrophoto metric Quantification

1. Measure DNA concentration using a spectrophotometer or NanoDrop at 260 and 280 nm.
2. Assess DNA purity by calculating A_{260}/A_{280} ratio (should be 1.8–2.0).
3. Check A_{260}/A_{230} ratio (should be 2.0–2.2) to detect polysaccharide contamination.
4. Dilute DNA samples to a working concentration of 25–50 ng/μl for PCR.

3.4 ISSR-PCR Amplification (Modified from [6])

3.4.1 Primer Screening (Initial Optimization)

1. Screen 20–30 ISSR primers using 3–5 representative genotypes. Use DNA samples with sufficient quantity (>200 ng/µl concentration or >500 ng total) to allow multiple optimization tests without depleting working stocks.
2. Test different annealing temperatures (45°C–65°C) using gradient PCR if available.
3. Select primers that produce clear, reproducible, and polymorphic banding patterns.

3.4.2 PCR Reaction Mix Preparation

1. Prepare master mix for each primer on ice. To ensure consistency and minimize pipetting errors, prepare a master mix for all samples as a bulk solution, then aliquot to individual tubes before adding template DNA separately.
2. Calculate total volume needed for ($n+2$) reactions, where
3. n = number of samples. The two additional reactions compensate for pipetting loss and provide a safety margin.
4. Prepare master mix containing all components EXCEPT template DNA.
5. Vortex master mix gently and briefly centrifuge to collect contents.
6. Aliquot 18 µl of master mix into each PCR tube.
7. Add 2 µl of template DNA (25–50 ng/µl) individually to each tube.

Table 2
ISSR-PCR reaction mix composition (20 µl per reaction)

Component	Volume (µl)
10× PCR buffer	2.0
$MgCl_2$ (25 mM)	1.6–2.4[1]
dNTPs mix (10 mM each)	0.4–0.6
ISSR Primer (10 µM)	0.5–1.0
Taq DNA polymerase (5 U/µl)	0.15–0.30
Template DNA (25–50 ng/µl)	2.0
Nuclease-free water	Up to 20.0
Total volume	**20.0**

[1]Optimize $MgCl_2$ concentration (2.0–3.0 mM final concentration) for each primer during screening (*see* **Note 1**)

8. Cap tubes, mix by pipetting or brief vortexing, and centrifuge briefly before placing in the thermocycler.

   Standard 20 µl reaction composition is shown in Table 2.

3.4.3 PCR Cycling Conditions

1. Initial denaturation: 94°C for 3–5 min.
2. 35–40 cycles:
   a) Denaturation: 94°C for 40–60 s.
   b) Annealing: 45°C–65°C for 30–60 s (primer-dependent).
   c) Extension: 72°C for 1.5–2 min.
3. Final extension: 72°C for 7–10 min.
4. Hold: 4°C (indefinite).

3.4.4 PCR Controls

1. Include a negative control (no template DNA) in each amplification batch.
2. Include a positive control using a known template that amplifies well.
3. Run duplicate reactions for each genotype–primer combination to verify reproducibility.

3.5 Gel Electrophoresis and Band Visualization

3.5.1 Gel Preparation

1. Prepare 2.0% agarose gel in 1× TBE buffer for optimal band resolution.
2. Add DNA staining dye (GelRed or SYBR Green) according to manufacturer instructions.
3. Cast gel with appropriate comb size (15–20 wells for small studies).
4. Allow gel to solidify completely (45–60 min) before use.

3.5.2 Sample Loading and Electrophoresis

1. Mix 8–10 µl of PCR product with 1–2 µl of 6× loading dye.
2. Load DNA ladder (100 bp and 1 kb) in the first and last lanes for size estimation.
3. Load samples systematically with consistent order across all gels.
4. Run electrophoresis at 100–120 V for 2–3 h, or until the dye front reaches approximately 75%–80% of the gel length.

100 bp Ladder	S1	S2	S3	S4	S5	S6	…	S12	S13	1 kb Ladder

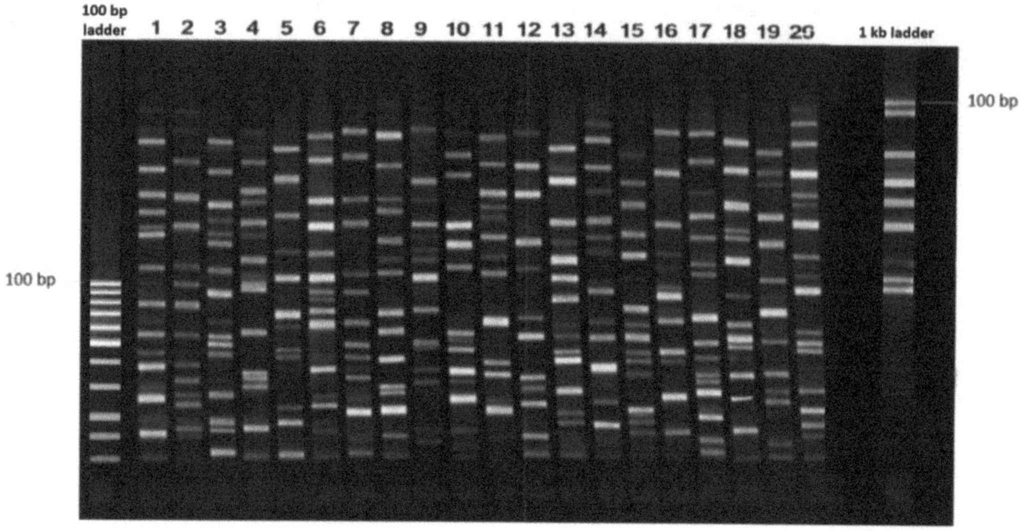

Fig. 1 Representative ISSR-PCR amplification profile showing polymorphic banding. The gel includes molecular weight markers in the first and last lanes for size determination. Samples (Lanes 1–20) are labeled numerically to ensure unbiased scoring

3.5.3 Documentation

1. Visualize bands using a UV or blue light imaging system (*see* **Note 2** for troubleshooting weak, absent, or smeared bands).
2. Capture high-resolution images using a gel documentation system.
3. Label all lanes clearly in digital images.
4. Save images in multiple formats (TIFF for analysis, JPEG for documentation); record exposure settings, date, sample identifiers, and primer information (Fig. 1).

3.5.4 Molecular Weight Estimation and Band Sizing

1. Estimate molecular weights of ISSR bands using a DNA ladder as a reference.
2. For bands aligning exactly with ladder markers, directly assign the corresponding molecular weight (e.g., band at 500 bp marker = 500 bp).
3. For bands between ladder markers, use interpolation to calculate size:
   a) Measure migration distance from well to center of each band using ruler (mm precision).
   b) Calculate Rf (retardation factor) for all bands:
      Rf = Migration distance from well/total distance to dye front [21, 22].
      (Rf values range from 0 to 1; smaller fragments have higher Rf).
   c) Choose interpolation method:

Linear interpolation (for narrow ranges, <500 bp between markers)

$$\text{Size (unknown)} = \text{Size (lower)} + \left(\frac{Rf\ (\text{unknown}) - Rf\ (\text{lower})}{Rf\ (\text{upper}) - Rf\ (\text{lower})}\right) \times (\text{Size (upper)} - \text{Size (lower)})$$

Semilogarithmic interpolation (preferred for wide ranges, 200–2000 bp)

$$\log_{10}[\text{Size (unknown)}] = \log_{10}[\text{Size (lower)}] + \left(\frac{Rf\ (\text{unknown}) - Rf\ (\text{lower})}{Rf\ (\text{upper}) - Rf\ (\text{lower})}\right) \times (\log_{10}(\text{Size (upper)}) - \log_{10}(\text{Size (lower)}))$$

Then: Size (unknown) = $10^{[\text{calculated value}]}$
Where:

- Size (lower) = nearest smaller ladder band size
- Size (upper) = nearest larger ladder band size

This method accounts for the logarithmic relationship between DNA fragment size and electrophoretic mobility [21, 23].
Rf values calculated as in step 65b.
(*see* **Note 3** for step-by-step calculations and practical applications).

4. Software-assisted sizing (recommended for routine analysis):
   a) Load gel image (TIFF format) into analysis software (Image Lab™ [24], VisionCapt™ [25], ImageJ [26], GelAnalyzer [27], or similar).
   b) Define lanes and assign known sizes to ladder bands.
   c) Verify standard curve quality (R^2 should be >0.98); adjust if needed.
   d) Perform automated band detection; manually verify and correct.
   e) Export sizing data to a spreadsheet.
5. Manual sizing (when software unavailable):
   a) Create a table with ladder sizes, migration distances, and *Rf* values.

b) Plot semilog graph: x-axis = Rf (0–1.0), y-axis = $\log_{10}$(fragment size). Draw the best-fit line through ladder points.

c) For each unknown band: locate Rf on the x-axis, draw a vertical line to the curve, then a horizontal line to the y-axis to read the size.

d) Alternative: Use Excel scatter plot (Rf vs. $\log_{10}$ size), generate trendline equation ($y = mx + b$), calculate: Size = $10^{[m(Rf)+b]}$ [28, 29].

6. Report fragment sizes with appropriate rounding:
   - 200–1000 bp: round to nearest 10 bp (e.g., 597 bp → 600 bp).
   - 1000–2500 bp: round to nearest 50 bp (e.g., 1247 bp → 1250 bp).

7. Record all fragment sizes in the data table with primer name, sample identifier, and size for binary matrix construction.

3.5.5 Reproducibility Assessment

1. Re-amplify 10%–20% of samples to verify reproducibility.
2. Calculate reproducibility percentage: (matching bands/total bands) × 100.
3. Accept primers with >90% reproducibility (*see* **Note 4** for troubleshooting inconsistent results between replicates).

3.6 Binary Matrix Construction and Band Scoring

1. The binary matrix converts ISSR banding patterns into numerical data for genetic diversity analysis. Each band is scored as present (1) or absent (0) across all genotypes.
2. Create a spreadsheet with genotypes as rows and ISSR bands (loci) as columns.
3. Label each column systematically: Primer name + fragment size (e.g., UBC807_600 bp, UBC825_350 bp).
4. Score each band position independently:
   - Clear, distinct band = 1.
   - Absence of band = 0.
   - Faint but reproducible bands = 1.
   - Very faint/questionable bands = 0 (exclude from analysis).
   - Smeared regions or artifacts = 0.
5. Repeat for all primers, adding columns to the matrix.
6. Exclude monomorphic bands (all 1 s or all 0 s).
7. Export final matrix in Excel (.xlsx) and CSV formats for analysis software (POPGENE [30], GenAlEx [31], NTSYS-pc [32]). See Table 3.

Table 3
Example binary matrix structure for ISSR genetic diversity analysis

Genotype ID	UBC807_250	UBC807_400	UBC807_650	UBC825_300	...	Total bands
Genotype_1	1	0	1	0	...	8
Genotype_2	1	1	0	0	...	6
Genotype_3	0	0	1	1	...	7
...	...	...	...	...	...	

Note: Independent scoring by two people improves accuracy. Resolve discrepancies through discussion and gel image reexamination

3.7 Statistical Analysis of Genetic Diversity

3.7.1 Diversity Parameters

Calculate genetic diversity indices using specialized software (POPGENE v1.32 [30], GenAlEx v6.5 [31], or equivalent). These parameters quantify the level and distribution of genetic variation within and among populations.

1. Percentage of polymorphic bands (PPB)
   - Measures the proportion of variable loci in the dataset.
   - Formula: PPB = (Number of polymorphic loci/total loci) × 100 [33].
   - Higher values (>70%) indicate greater genetic diversity; values <50% suggest low variation or genetic bottlenecks.
2. Polymorphic information content (PIC)
   - Assesses the discriminatory power of each primer.
   - Formula: $PIC = 2pi(1 - pi)$ for dominant markers [34].
   - Range: 0 (monomorphic) to 0.5 (maximum informativeness).
   - Primers with PIC > 0.25 are considered highly informative.
3. Shannon's information index (*I*)
   - Quantifies genetic diversity accounting for both richness and evenness of allele distribution
   - Formula: $I = -\Sigma pi \ln(pi)$ [35].
   - Range: 0 to $\ln(2) \approx 0.693$ for dominant markers.
4. Marker index (MI)
   - Evaluates the overall utility of each primer.
   - MI = PIC × EMR (where EMR = effective multiplex ratio = number of polymorphic bands per primer) [33].

- Higher MI values indicate more efficient primers for diversity studies.

3.7.2 Genetic Similarity and Distance Matrices

Genetic relationships between genotypes are quantified using similarity or distance coefficients calculated from the binary matrix.

1. Choose an appropriate similarity coefficient:
   a) Jaccard coefficient: ($a/(a+b+c)$) [36]
      - Excludes double absences (0–0 matches).
      - Preferred when band absence may result from technical artifacts.
      - Range: 0 (no shared bands) to 1 (identical profiles).
      - Most commonly used for ISSR data.

   b) Dice coefficient: ($2a/(2a+b+c)$) [37]
      - Gives double weight to shared presences.
      - More sensitive to similarities.

   c) Simple matching coefficient: ($(a+d)/(a+b+c+d)$) [38]
      - Includes double absences.
      - Generally not recommended for ISSR due to homoplasy.

   where a = bands present in both genotypes, b = present in genotype 1 only, c = present in genotype 2 only, d = absent in both.
2. Generate pairwise similarity matrix using genetic analysis software (NTSYSpc [32], GenAlEx [31], PAST [39], R packages [40]: poppr, adegenet, vegan).
3. Convert similarity to distance: Distance = 1 – Similarity [41].

See **Note 5** for guidance on individual-level vs. population-level analysis selection.

3.7.3 Cluster Analysis

1. Perform cluster analysis to group genotypes based on genetic similarity, revealing population structure and phylogenetic relationships.
2. Construct UPGMA (unweighted pair group method with arithmetic mean) dendrogram:
   - Import the distance/similarity matrix into the clustering software.
   - Select UPGMA algorithm (assumes constant evolutionary rates; appropriate for within-species diversity studies).
   - Alternative methods: Neighbor-joining (NJ) for phylogenetic analysis, Ward's method for hierarchical grouping.

3. Generate a dendrogram using available software:
   - Genetics-specific: NTSYSpc [32], MEGA [42], DARwin [43].
   - General statistical: SPSS [44] (Hierarchical Cluster Analysis), XLSTAT [45] (Agglomerative Hierarchical Clustering), R [40] (hclust, ape).
   - Visualization: TreeView [46], FigTree [47], iTOL [48].
4. Assess dendrogram reliability:
   - Perform bootstrap analysis with ≥1000 replicates (10,000 recommended for publication).
   - Bootstrap support values interpretation [49]:

     >70% = strong support for grouping.
     50%–70% = moderate support.
     <50% = weak/unreliable clustering.

4 Notes

1. Magnesium optimization:
   $MgCl_2$ concentration strongly affects primer annealing specificity. Test range: 1.5, 2.0, 2.5, 3.0 mM final concentration. Too low (<1.5 mM): weak or no amplification. Optimal (2.0–2.5 mM): clear, reproducible bands. Too high (>3.5 mM): nonspecific amplification, smearing. Optimize for each primer individually during the screening phase. Record optimal concentration in primer database for future reference.
2. Troubleshooting PCR amplification failures
   Problem: No bands visible.
   Solutions:
   - Check DNA quality and concentration (use fresh dilution).
   - Verify primer integrity (reconstitute fresh working stocks).
   - Increase $MgCl_2$ concentration in 0.5 mM increments (test 2.0–3.5 mM)
   - Extend annealing time to 60 s.
   - Check Taq polymerase activity (use positive control).

   Problem: Weak or faint bands.
   Solutions:
   - Increase template DNA to 75–100 ng per reaction
   - Increase primer concentration to 1.2 μM.
   - Extend extension time by 30 s.

- Increase cycle number to 40–42 (but monitor for artifacts).

Problem: Smearing or excessive background.
Solutions:

- Decrease DNA concentration (try 1:20 dilution).
- Lower $MgCl_2$ concentration.
- Increase annealing temperature by 2°C–3°C.
- Reduce the cycle number to 35.
- Use fresh dNTPs (old dNTPs cause artifacts).

3. Step-by-step fragment size calculations with a practical example
The retardation factor (R*f*) is calculated as R*f* = *d*/*D*, where *d* is the migration distance from well to band center, and *D* is the total migration distance from well to dye front.

   - R*f* values range from 0 to 1, where R*f* = 0 indicates no migration (fragment at well origin) and R*f* = 1 represents migration to the dye front (not observed for DNA fragments under standard conditions).
   - DNA fragments typically show R*f* values between 0.3 and 0.9 in the 100–3000 bp size range.
   - Small fragments (100–500 bp) exhibit high mobility (R*f* = 0.7–0.9) due to efficient matrix penetration, while large fragments (1000–3000 bp) show reduced mobility (R*f* = 0.3–0.5) due to increased steric hindrance.
   - This inverse size-mobility relationship enables fragment size estimation through interpolation.

   Representative calculation example:

   - Consider a gel with a 100 bp DNA ladder (fragments in 100 bp increments) and ISSR-amplified DNA. If the dye front migrated 6.0 cm and ladder bands migrated as follows: 500 bp = 4.2 cm, 700 bp = 3.6 cm, 1000 bp = 3.0 cm, calculate R*f* by dividing each migration distance by 6.0 cm:
   - 500 bp: R*f* = 4.2/6.0 = 0.700; the 500 bp fragment migrated 70% of the total electrophoretic distance, consistent with expected high mobility for small DNA fragments.
   - 700 bp: R*f* = 3.6/6.0 = 0.600; the 700 bp fragment, being larger than the 500 bp marker, exhibited reduced mobility (60% of maximum migration), demonstrating the inverse size–R*f* relationship.
   - 1000 bp: R*f* = 3.0/6.0 = 0.500; the 1000 bp fragment showed further mobility reduction (50% of maximum), confirming progressive decrease in migration velocity with increasing molecular weight.

Table 4
Representative example of Rf determination in agarose gel electrophoresis

Fragment/band identity	Migration distance (cm)	Rf (calculated)
Dye front (blue line)	6.0	1.000
500 bp ladder band	4.2	0.700
700 bp ladder band	3.6	0.600
1000 bp ladder band	3.0	0.500
Your unknown ISSR band #1	3.5	0.583
Your unknown ISSR band #2	4.0	0.667

- For unknown ISSR bands migrating 3.5 cm and 4.0 cm, *Rf* values are 0.583 and 0.667 (Table 4), respectively. These fall between characterized markers and require interpolation for size estimation.

Linear interpolation approach:

- Identify bracketing markers for the unknown band. For a band with *Rf* = 0.583, the brackets are 500 bp (*Rf* = 0.700) and 700 bp (*Rf* = 0.600).
- Apply the formula:

$$\text{Size (unknown)} = \text{Size (lower)} + \left(\frac{Rf\ \text{(unknown)} - Rf\ \text{(lower)}}{Rf\ \text{(upper)} - Rf\ \text{(lower)}}\right) \times (\text{Size (upper)} - \text{Size (lower)}).$$

- Substituting values: Size (unknown) = 500 + [(0.583 − 0.700)/(0.600 − 0.700)] × [700 − 500] = 734 bp.
- Round to nearest 10 bp (accounting for gel resolution limits and inter-gel variability of ±5%–10%): 730 bp.

Semi-logarithmic interpolation (recommended approach):

- This method accounts for the logarithmic relationship between Rf and molecular weight, providing superior accuracy especially for fragments bracketed by widely-spaced markers (>500 bp separation), extreme size ranges (<300 bp or >1500 bp), or nonoptimal electrophoresis conditions.
- Convert fragment sizes to base-10 logarithms: $\log_{10}(500) = 2.699$ and $\log_{10}(700) = 2.845$.
- Apply the formula:

$$\log_{10}[\text{Size (unknown)}] = \log_{10}[\text{Size (lower)}] + \left(\frac{Rf\ \text{(unknown)} - Rf\ \text{(lower)}}{Rf\ \text{(upper)} - Rf\ \text{(lower)}}\right) \times (\log_{10}(\text{Size (upper)}) - \log_{10}(\text{Size (lower)})).$$

- Calculate: $\log_{10}$[Size (unknown)] = 2.699 + [(0.583 − 0.700)/(0.600 − 0.700)] × [2.845 − 2.699] = 2.870.
- Convert back using antilog: Size (unknown) = $10^{2.870}$ = 741 bp.
- Round to nearest 10 bp: 740 bp.

4. Inconsistent results between replicates
   - Prepare fresh master mix for each run.
   - Ensure thorough mixing of the master mix before aliquoting.
   - Use calibrated pipettes and check for bubbles.
   - Maintain consistent thermal cycler position (edge effects occur).

5. Guidelines for choosing individual-level vs. population-level analysis

   Individual-level analysis is used to:
   - Calculate pairwise genetic similarities/distances between all individual genotypes.
   - Generate a similarity matrix for downstream clustering and ordination analyses.

   Use individual-level analysis for:
   - Germplasm characterization and genetic diversity assessment in breeding collections.
   - Cultivar identification and distinctness determination.
   - Parent selection in breeding programs.
   - Duplicate detection in gene banks.
   - Assessment of genetic relationships among breeding lines or accessions.

   Individual-level analytical outputs:
   - Dendrograms (cluster analysis).
   - Principal Component Analysis (PCA).
   - Principal Coordinates Analysis (PCoA).

   Population-level analysis is used to:
   - Organize samples into biologically, geographically, or breeding-relevant population groups.
   - Quantify genetic variation within and among populations to determine population structure and differentiation levels.

   Use population-level analysis for:
   - Comparison of genetic diversity across geographic regions or agroecological zones.

- Assessment of genetic differentiation between wild and cultivated populations.
- Evaluation of breeding program structure (base population vs. elite lines).
- Gene flow and population structure studies.
- Conservation genetics prioritization.

Population-level analytical outputs:

- Analysis of molecular variance (AMOVA): partition variation within and among populations.
- Fixation indices (PhiPT): analogous to FST for dominant markers.
- Pairwise population genetic distances: Nei's distance for populations.
- Principal coordinates analysis (PCoA): with population centroids.

References

1. Khazaei H, O'Sullivan DM, Stoddard FL, Adhikari KN, Paull JG, Schulman AH, Andersen SU, Vandenberg A (2021) Recent advances in faba bean genetic and genomic tools for crop improvement. Legume Sci 3(3):e75. https://doi.org/10.1002/leg3.75
2. Chetto O, Belqadi L, Kouighat M, Nabloussi A (2025) Role of molecular markers in development of disease resistant varieties in crop plants: case studies of pulses. In: Sharma NK (ed.) Molecular markers and crop improvement. Springer, p 5. 10.1007/978-3-031-87401-7_5
3. Chetto O, Belqadi L, Fatemi ZEA, Charafi J, Kouighat M, Najmi A, El Fechtali M, Houmanat K, Nabloussi A (2024) Assessment of elite faba bean lines for enhanced productivity and resilience in contrasting challenging environments using phenotypic and molecular markers. J Agric Food Res 18:101488. https://doi.org/10.1016/j.jafr.2024.101488
4. Qin Q, Dong Y, He J, Chen J, Wu D, Zhang S (2025) Assessment of genetic diversity and construction of core germplasm in populations of *Acorus tatarinowii* based on SNP markers. J Appl Res Med Aromat Plants 44:100605. https://doi.org/10.1016/j.jarmap.2024.100605
5. Soltis DE, Moore MJ, Burleigh G, Soltis PS (2009) Molecular markers and concepts of plant evolutionary relationships: progress, promise, and future prospects. Crit Rev Plant Sci 28(1–2):1–15. https://doi.org/10.1080/07352680802665297
6. Zietkiewicz E, Rafalski A, Labuda D (1994) Genome fingerprinting by simple sequence repeat (SSR)-anchored polymerase chain reaction amplification. Genomics 20(2):176–183
7. Sarla N, Reddy MP, Siddiq EA (2001) Application of inter simple sequence repeat (ISSR) polymorphism in assessing diversity in crop plants. Indian J Plant Genet Resour 14(2):246–247
8. Moresco RM, Maniglia TC, De Oliveira C, Margarido VP (2013) The pioneering use of ISSR (inter simple sequence repeat) in Neotropical anurans: preliminary assessment of genetic diversity in populations of *Physalaemus cuvieri* (Amphibia, Leiuperidae). Biol Res 46(1):53–57. https://doi.org/10.4067/S0716-97602013000100008
9. Wang H, Zong X, Guan J, Yang T, Sun X, Ma Y, Redden R (2012) Genetic diversity and relationship of global faba bean (*Vicia faba* L.) germplasm revealed by ISSR markers. Theor Appl Genet 124(7):789–797
10. Salazar-Laureles ME, Pérez-López DJ, González-Huerta A, Vázquez-García LM, Valadez-Moctezuma E (2015) Genetic variability analysis of faba bean accessions using inter-simple sequence repeat (ISSR) markers. Chil J Agric Res 75:122–130
11. Neda B, Feyissa T, Dagne K, Assefa E (2021) Genetic diversity and population structure of

Ethiopian faba bean (*Vicia faba* L.) germplasm revealed by ISSR markers. Asia-Pac J Mol Biol Biotechnol 29:8–25
12. Asfaw BM, Dagne K, Wakayo GK, Kemal SA, Muleta KT (2018) Genetic diversity study of Ethiopian faba bean (*Vicia faba* L) varieties based on phenotypic traits and inter simple sequence repeat (ISSR) markers. Afr J Biotechnol 17:433–446
13. Qahtan AA, Al-Atar AA, Abdel-Salam EM, El-Sheikh MA, Gaafar ARZ, Faisal M (2021) Genetic diversity and structure analysis of a worldwide collection of faba bean (*Vicia faba*) genotypes using ISSR markers. Int J Agric Biol 25(3):683–691. https://doi.org/10.17957/IJAB/15.1717
14. Abo El-Kheir ZA, Abdel-Hady MS, El-Naggar HMH, Abd El-Hamed AR (2010) Molecular and biochemical markers of some *Vicia faba* L. cultivars in response to broomrape infestation. Nat Sci 8(11):252–260
15. Jugran A, Rawat S, Dauthal P, Mondal S, Bhatt ID, Rawal RS (2013) Association of ISSR markers with some biochemical traits of Valeriana jatamansi Jones. Ind Crops Prod 44:671–676. https://doi.org/10.1016/j.indcrop.2012.09.004
16. Hou W, Zhang X, Yan Q, Li P, Sha W, Tian Y, Liu Y (2018) Linkage map of a gene controlling zero tannins (zt-1) in faba bean (*Vicia faba* L.) with SSR and ISSR markers. Agronomy 8(6):80. https://doi.org/10.3390/agronomy8060080
17. Hammadi H, Abederrahim MA, Elbekkay M, Ferchichi A (2013) Comparison of the effectiveness of ISSR and SSR markers in determination of date palm (*Phoenix dactylifera* L.) agronomic traits. Aust J Crop Sci 7:763–769
18. Shebl G, Ahmed H, Kato A (2018) ISSR markers of some faba bean (*Vicia faba* L.) genotypes in relation to phenylalanine accumulation. Egypt J Exp Biol (Bot) 14(1):73–82. https://doi.org/10.5455/egyjebb.20180306054752
19. Doyle JJ, Doyle JL (1987) A rapid DNA isolation procedure for small quantities of fresh leaf tissue. Phytochem Bull 19:11–15
20. Murray MG, Thompson WF (1980) Rapid isolation of high molecular weight plant DNA. Nucleic Acids Res 8(19):4321–4326
21. Southern EM (1979) Measurement of DNA length by gel electrophoresis. Anal Biochem 100(2):319–323. https://doi.org/10.1016/0003-2697(79)90235-5
22. Sambrook J, Russell DW (2001) Molecular cloning: a laboratory manual, 3rd edn. Cold Spring Harbor Laboratory Press, New York
23. Raghava GPS (1994) Improved estimation of DNA fragment length from gel electrophoresis data using a graphical method. Biotechniques 17(1):100–104
24. Gough EJ, Gough NM (1984) Direct calculation of the sizes of DNA fragments separated by gel electrophoresis using programmes written for a pocket calculator. Nucleic Acids Res 12(1):845–853
25. Image Lab Software. Bio-Rad Laboratories, Hercules, CA, USA. Available at https://www.bio-rad.com/en-us/product/image-lab-software
26. VisionCapt Software. Vilber Lourmat, Collégien, France. Available at https://www.vilber.com/visioncapt/
27. Schneider CA, Rasband WS, Eliceiri KW (2012) NIH Image to ImageJ: 25 years of image analysis. Nat Methods 9(7):671–675. https://doi.org/10.1038/nmeth.2089
28. Heras J, Domínguez C, Mata E, Pascual V, Lozano C, Torres C, Zarazaga M (2015) GelJ—a tool for analyzing DNA fingerprint gel images. BMC Bioinform 16:270. https://doi.org/10.1186/s12859-015-0703-0
29. Bio-Rad Laboratories (n.d.) Standard curve determination for DNA band sizes using semi-log plots. Bio-Rad educational guide
30. Powell W, Morgante M, Andre C, Hanafey M, Vogel J, Tingey S, Rafalski A (1996) The comparison of RFLP, RAPD, AFLP and SSR (microsatellite) markers for germplasm analysis. Mol Breed 2(3):225–238
31. Yeh FC, Yang RC, Boyle T (1999) POPGENE Version 1.32: Microsoft Window-Based Freeware for Population Genetic Analysis. University of Alberta, Edmonton, AB, Canada. Available at https://sites.ualberta.ca/~fyeh/popgene.html
32. Peakall R, Smouse PE (2012) GenAlEx 6.5: genetic analysis in Excel. Population genetic software for teaching and research—an update. Bioinformatics 28(19):2537–2539. https://doi.org/10.1093/bioinformatics/bts460. http://biology-assets.anu.edu.au/GenAlEx/Welcome.html
33. Rohlf FJ (2000) NTSYS-pc: Numerical Taxonomy and Multivariate Analysis System, Version 2.1. Exeter Software, Setauket, NY, USA. Available at http://www.exetersoftware.com/cat/ntsyspc/ntsyspc.html
34. Roldan-Ruiz I, Dendauw J, Van Bockstaele E, Depicker A, De Loose M (2000) AFLP markers reveal high polymorphic rates in ryegrasses (Lolium spp.). Mol Breed 6(2):125–134
35. Shannon CE (1948) A mathematical theory of communication. Bell Syst Tech J 27:379–423
36. Jaccard P (1908) Nouvelles recherches sur la distribution florale. Bull Soc Vaudoise Sci Nat 44:223–270
37. Dice LR (1945) Measures of the amount of ecologic association between species. Ecology 26(3):297–302. https://doi.org/10.2307/1932409

38. Sokal RR, Michener CD (1958) A statistical method for evaluating systematic relationships. Univ Kans Sci Bull 38:1409–1438
39. Hammer Ø, Harper DAT, Ryan PD (2001) PAST: Paleontological Statistics Software Package for Education and Data Analysis. Palaeontol Electron 4(1):9. https://www.nhm.uio.no/english/research/resources/past/
40. R Core Team (2023) R: A language and environment for statistical computing. R Foundation for Statistical Computing, Vienna, Austria. Available at https://www.R-project.org/
41. Sneath PHA, Sokal RR (1973). Numerical taxonomy: the principles and practice of numerical classification. W.H. Freeman, San Francisco
42. Tamura K, Stecher G, Kumar S (2021) MEGA11: molecular evolutionary genetics analysis version 11. Mol Biol Evol 38(7):3022–3027. https://doi.org/10.1093/molbev/msab120. https://www.megasoftware.net/
43. Perrier X, Jacquemoud-Collet JP (2006) DARwin software. Available at http://darwin.cirad.fr/
44. IBM SPSS Statistics for Windows, Version 28.0. IBM Corp., Armonk, NY, USA. Availeble at https://www.ibm.com/products/spss-statistics
45. XLSTAT Statistical and Data Analysis Solution. Addinsoft, Paris, France. Available at https://www.xlstat.com
46. Page RDM (1996) TreeView: an application to display phylogenetic trees on personal computers. CABIOS 12(4):357–358. https://doi.org/10.1093/bioinformatics/12.4.357. http://taxonomy.zoology.gla.ac.uk/rod/treeview.html
47. Rambaut A (2009) FigTree v1.4.4. Available at http://tree.bio.ed.ac.uk/software/figtree/
48. Letunic I, Bork P (2021) Interactive Tree Of Life (iTOL) v5: an online tool for phylogenetic tree display and annotation. Nucleic Acids Res 49(W1):W293–W296. https://doi.org/10.1093/nar/gkab301. https://itol.embl.de/
49. Hillis DM, Bull JJ (1993) An empirical test of bootstrapping as a method for assessing confidence in phylogenetic analysis. Syst Biol 42(2):182–192

Chapter 18

Development of SNPs for Crop Improvement

Shouvik Gorai, Afridi Mondal, Md Imtiyaz, Bandela Chandana, Subhasis Samanta, V. Umesh Kumar, Soham Hazra and Anirban Maji

Abstract

Single-nucleotide polymorphisms (SNPs) represent the most abundant form of genetic variation in crop genomes and serve as highly informative molecular markers for modern breeding programs. Advances in next-generation sequencing technologies have enabled rapid and high-throughput SNP discovery across diverse crop species, including those with large and complex genomes. Reliable SNP development requires an integrated workflow encompassing germplasm selection, high-quality DNA isolation, library preparation, sequencing, sequence data processing, variant calling, stringent filtering, annotation, and experimental validation. This chapter presents a stepwise methodological framework for SNP discovery and validation tailored for crop improvement applications. The described protocols emphasize reproducibility, data quality, and scalability to support downstream analyses such as genetic diversity assessment, association mapping, marker-assisted selection, and genomic prediction.

Keywords Single nucleotide polymorphism, SNP discovery, Next-generation sequencing, Crop genomics, Molecular markers, Variant calling, Genomic selection

1 Introduction

The integration of molecular genomics into crop improvement has shifted breeding strategies toward DNA-guided selection. Among molecular markers, single-nucleotide polymorphisms (SNPs) are the most abundant and informative markers in plant genomes [1, 2]. Their high density and genome-wide distribution make them suitable for high-throughput genotyping and modern breeding applications [2, 3].

SNPs arise from naturally occurring mutations generated through replication errors, recombination, and environmental influences, contributing to genetic diversity within crop populations [4]. These variations occur in coding and regulatory regions where they may influence gene expression and phenotypic traits [5, 6]. Advancements in next-generation sequencing have enabled efficient genome-wide detection of nucleotide variation in diverse crop species [5]. Both whole-genome sequencing and reduced-representation approaches

Nitish Kumar (ed.), *Molecular Markers in Plants*, Methods in Molecular Biology, vol. 3097,
https://doi.org/10.1007/978-1-0716-5663-1_18,

support scalable SNP identification across genomes of varying sizes. Robust SNP development requires coordinated laboratory and computational workflows involving germplasm selection, DNA extraction, sequencing library preparation, sequence analysis, variant detection, filtering, annotation, and validation. Optimization of these steps is critical for generating reliable marker datasets. SNP markers support genetic diversity studies, population structure analysis, linkage mapping, genome-wide association studies, and genomic selection, thereby accelerating crop improvement programs [7–11]. An experimental and bioinformatics pipeline for SNP development for crop improvement is presented in Fig. 1. This chapter outlines a reproducible protocol for SNP discovery and validation applicable to a wide range of crop species.

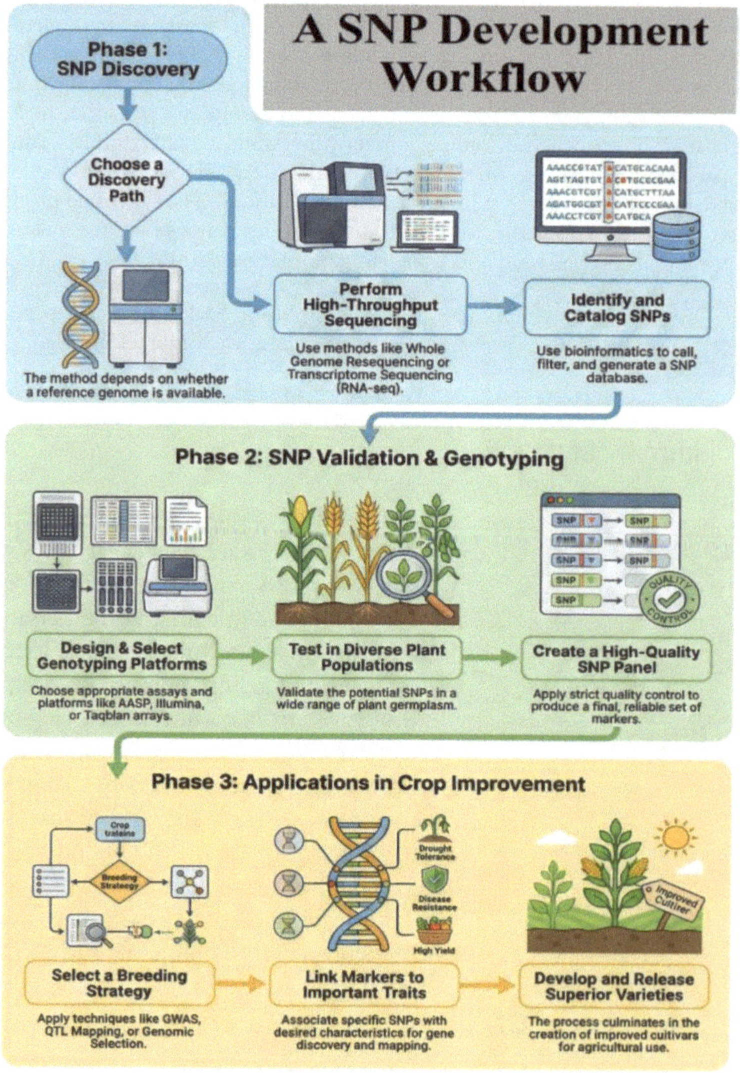

Fig. 1 Experimental and bioinformatics pipeline for SNP development

2 Materials

2.1 Biological Materials

1. Young and healthy leaf tissues were collected from selected germplasm accessions.
2. Molecular biology-grade reagents and ultrapure water.

2.2 Buffers and Solutions for Genomic DNA Isolation

1. CTAB extraction buffer: 2% CTAB, 100 mM Tris–HCl (pH 8.0), 20 mM EDTA, 1.4 M NaCl, 1% PVP-40, and 0.2% β-mercaptoethanol (added before use).
2. Chloroform:isoamyl alcohol (24:1).
3. 100% isopropanol.
4. 70% ethanol.
5. TE buffer: 10 mM Tris–HCl and 1 mM EDTA (pH 8.0).
6. RNase A (10 mg mL^{-1}).

2.3 Library Preparation Reagents

1. Restriction digestion buffer containing Tris–HCl, $MgCl_2$, and NaCl.
2. End-repair buffer with dNTPs and ATP.
3. Adapter ligation buffer containing PEG-8000.
4. PCR amplification buffer with $MgCl_2$ and dNTPs.

2.4 Quality Control Reagents

1. 1× TAE electrophoresis buffer.
2. DNA loading dye.
3. Library dilution buffer (10 mM Tris–HCl with 0.1% Tween-20).

2.5 Equipment and Computational Resources

1. High-throughput sequencing platform supporting multiplex libraries.
2. Thermal cycler with heated lid.
3. Real-time PCR system.
4. Gel electrophoresis apparatus with visualization system.
5. High-throughput SNP genotyping platform [12–14].
6. High-performance computing infrastructure for sequence analysis [15, 16].
7. Quality control software such as FastQC and MultiQC.
8. Read aligners, including BWA and Bowtie2.
9. Alignment tools such as SAMtools and Picard.
10. Variant callers including GATK, FreeBayes, and SAMtools/BCFtools.
11. Variant filtering and annotation tools such as VCFtools and SnpEff.

12. Visualization software, including Integrative Genomics Viewer (IGV).
13. Bioinformatics pipelines such as Stacks, TASSEL, and PLINK [17–19].

3 Methods

3.1 Selection of Germplasm for SNP Discovery

1. Select genetically diverse accessions representing breeding lines, landraces, and wild relatives.
2. Define the objective of SNP discovery (diversity analysis, GWAS, or breeding).
3. Assess genetic diversity using SSR markers or existing SNP arrays [20, 21].
4. Remove redundant accessions.
5. Construct a core collection using PCA or clustering methods [20, 22].
6. Select representative samples from each cluster (*see* **Note 1**).
7. Determine sample size based on genome characteristics and study objectives [23, 24].
8. Validate accession identity prior to sequencing.
9. Evaluate DNA quality to ensure reliable library preparation (*see* **Note 2**).

3.2 Genomic DNA Isolation and Quality Assessment

3.2.1 DNA Isolation

1. Collect 100–200 mg of young leaf tissue [25, 26].
2. Freeze tissue in liquid nitrogen and grind it into a fine powder.
3. Transfer powder into preheated CTAB extraction buffer.
4. Incubate at 60°C–65°C for 30–60 min.
5. Add chloroform:isoamyl alcohol (24:1) and mix gently.
6. Centrifuge and transfer the aqueous phase.
7. Precipitate DNA with cold isopropanol and incubate at −20°C.
8. Centrifuge to obtain the DNA pellet.
9. Wash pellet with 70% ethanol and air-dry.
10. Resuspend DNA in TE buffer.
11. Treat with RNase A at 37°C for 30 min.
12. Perform secondary purification if required.

3.2.2 DNA Quality Assessment

1. Measure absorbance at 260 and 280 nm; confirm A260/A280 ≈1.8 [27].
2. Confirm A260/A230 between 2.0 and 2.2.
3. Assess integrity using 0.8%–1.0% agarose gel electrophoresis.
4. Verify the presence of a high-molecular-weight band.

3.3 SNP Discovery Approaches

3.3.1 Whole Genome Sequencing

1. Fragment high-quality genomic DNA.
2. Ligate sequencing adapters.
3. Amplify libraries via PCR.
4. Sequence libraries to obtain genome-wide coverage.

3.3.2 Reduced-Representation Sequencing

1. Digest genomic DNA using restriction enzymes such as ApeKI (*see* **Note 3**).
2. Ligate adapters containing sequencing primers and barcodes.
3. Amplify fragments using high-fidelity polymerase.
4. Perform size selection.
5. Purify amplified libraries.
6. Assess library quality.
7. Denature pooled libraries and dilute to recommended concentrations.
8. Load libraries onto a sequencing flow cell.
9. Generate clusters through bridge amplification.
10. Perform single-end or paired-end sequencing using NGS platforms.
11. Convert fluorescent signals into nucleotide sequences.
12. Generate FASTQ files and demultiplex reads [28, 29] (*see* **Notes 4 and 5**).

3.4 Sequence Data Processing

1. Assess raw FASTQ files using FastQC.
2. Remove adapters using Trimmomatic or Cutadapt.
3. Trim low-quality bases (Phred < 20–30).
4. Demultiplex reads based on index sequences.
5. Align reads to a reference genome using BWA or Bowtie2 when available.
6. Convert SAM files to sorted BAM files.
7. Perform de novo assembly if no reference genome exists.
8. Generate consensus sequences and identify candidate SNPs (*see* **Note 6**).

3.5 SNP Calling

1. Process aligned BAM files using variant callers such as GATK, BCFtools, or FreeBayes.
2. Perform joint variant calling across samples.
3. Export variants in VCF format (*see* **Note 7**).

3.6 SNP Filtering and Quality Control

1. Evaluate SNP datasets for depth, quality scores, and missing data [30, 31].
2. Apply minimum read depth thresholds (≥5–10).
3. Remove SNPs with low Phred quality.
4. Apply mapping quality threshold (MQ ≥ 30).
5. Exclude SNPs with excessive missing genotypes (>10%–20%).
6. Filter variants using minor allele frequency ≥0.01–0.05 [30].
7. Remove multiallelic and monomorphic loci.
8. Inspect selected variants using IGV.

3.7 SNP Annotation and Validation

1. Annotate SNPs using SnpEff, ANNOVAR, or Ensembl VEP.
2. Classify variants based on genomic position.
3. Map genes harboring SNPs to Gene Ontology pathways.
4. Select high-quality SNPs with uniform genome distribution [32].
5. Validate variants using Sanger sequencing (*see* **Note 8**).
6. Perform allele-specific PCR where applicable.
7. Convert validated SNPs into KASP or TaqMan assays (*see* **Note 9 and 10**).

4 Notes

1. Include wild relatives to maximize allelic diversity [33, 34].
2. Ensure equal DNA contribution during pooling to prevent allele bias [12, 35, 36].
3. Reduced-representation sequencing is recommended for large genomes to minimize cost [35].
4. Use high-molecular-weight DNA free from contaminants [13, 14].
5. Avoid PCR overamplification to reduce duplicate reads.
6. Conduct pilot sequencing to optimize marker density.
7. Archive raw data with metadata to ensure reproducibility.
8. Validate SNPs across multiple populations to improve transferability [37].

9. Adjust filtering thresholds according to population structure [38, 39].
10. Ensure high-quality flanking sequences for assay development in polyploid crops [37, 40].

References

1. Kage U, Kumar A, Dhokane D, Karre S, Kushalappa AC (2016) Functional molecular markers for crop improvement. Crit Rev Biotechnol 36(5):917–930. https://doi.org/10.3109/07388551.2015.1062743
2. Thomson MJ (2014) High-throughput SNP genotyping to accelerate crop improvement. Plant Breed Biotechnol 2(3):195–212. https://doi.org/10.9787/PBB.2014.2.3.195
3. Varshney RK, Dubey A (2009) Novel genomic tools and modern genetic and breeding approaches for crop improvement. J Plant Biochem Biotechnol 18(2):127–138. https://link.springer.com/article/10.1007/BF03263311
4. Gou X, Shao Y, Wang X, Shi H, Yu J, Li X, Guo T (2024) Evolutionary patterns of DNA base composition at polymorphic sites highlight the role of the environment in shaping barley and rice genomes. Plant Genome 17(2):e20456. https://doi.org/10.1002/tpg2.20456
5. Kumar, S, Banks, TW, Cloutier, S (2016) SNP discovery through next-generation sequencing and its applications. In: Crop breed, pp 209–244. https://doi.org/10.1155/2012/831460
6. Jehan T, Lakhanpaul S (2006) Single nucleotide polymorphism (SNP)-methods and applications in plant genetics: a review. Indian J Biotechnol 5(4):435
7. Dwiningsih Y, Rahmaningsih M, Alkahtani J (2020) Development of single nucleotide polymorphism (SNP) markers in tropical crops. Adv Sustain Sci Eng Technol 2(2):343558. https://journal.upgris.ac.id/index.php/asset/article/view/6279
8. Bhandari HR, Bhanu AN, Srivastava K, Singh MN, Shreya HA (2017) Assessment of genetic diversity in crop plants-an overview. Adv Plants Agric Res 7(3):279–286. https://doi.org/10.15406/apar.2017.07.00255
9. Singh N, Choudhury DR, Singh AK, Kumar S, Srinivasan K, Tyagi RK, Singh R (2013) Comparison of SSR and SNP markers in estimation of genetic diversity and population structure of Indian rice varieties. PLoS ONE 8(12):e84136. https://doi.org/10.1371/journal.pone.0084136
10. Yasir M, He S, Sun G, Geng X, Pan Z, Gong W, Du X (2019) A genome-wide association study revealed key SNPs/genes associated with salinity stress tolerance in upland cotton. Genes 10(10):829. https://doi.org/10.3390/genes10100829
11. Challa S, Neelapu NR (2018) Genome-wide association studies (GWAS) for abiotic stress tolerance in plants. In: Biochemical, physiological and molecular avenues for combating abiotic stress tolerance in plants. Academic Press, 135–150. https://doi.org/10.1016/B978-0-12-813066-7.00009-7pp
12. Fan JB, Oliphant A, Shen R, Kermani BG, Garcia F, Gunderson KL, Hansen M, Steemers F, Butler SL, Deloukas P, Galver L, Hunt S, McBride C, Bibikova M, Rubano T, Chen J, Wickham E, Doucet D, Chang W, Campbell D, Zhang B, Kruglyak S, Bentley DR, Haas J, Rigault P, Zhou L, Stuelpnagel J, Chee MS (2003) Highly parallel SNP genotyping. Cold Spring Harb Symp Quant Biol 68:69–78. https://doi.org/10.1101/sqb.2003.68.69
13. Doyle JJ, Doyle JL (1987) A rapid DNA isolation procedure for small quantities of fresh leaf tissue. Phytochem Bull 19:11–15. https://doi.org/10.1007/978-1-4615-0277-4_12
14. Porebski S, Bailey LG, Baum BR (1997) Modification of a CTAB DNA extraction protocol for plants containing high polysaccharide and polyphenol components. Plant Mol Biol Rep 15:8–15. https://doi.org/10.1007/BF02772108
15. Mardis ER (2008) Next-generation DNA sequencing methods. Annu Rev Genomics Hum Genet 9:387–402. https://doi.org/10.1146/annurev.genom.9.081307.164359
16. Schatz MC, Delcher AL, Salzberg SL (2010) Assembly of large genomes using second-generation sequencing. Genome Res 20:1165–1173. https://doi.org/10.1101/gr.101360.109
17. Andrews KR, Good JM, Miller MR, Luikart G, Hohenlohe PA (2016) Harnessing the power of RADseq for ecological and evolutionary genomics. Nat Rev Genet 17:81–92. https://doi.org/10.1038/nrg.2015.28
18. DeAngelis MM, Wang DG, Hawkins TL (1995) Solid-phase reversible immobilization for the isolation of PCR products. Nucleic Acids Res 23:4742–4743. https://doi.org/10.1093/nar/23.22.4742

19. Nielsen R, Paul JS, Albrechtsen A, Song YS (2011) Genotype and SNP calling from next-generation sequencing data. Nat Rev Genet 12:443–451. https://doi.org/10.1038/nrg2986
20. McCouch SR et al. (2012) Agriculture: feeding the future. Nature 499:23–24. https://doi.org/10.1038/499023a
21. Hamblin MT et al. (2011) Population genetics of genomics-based crop improvement methods. TIG 27:98–106. https://doi.org/10.1016/j.tig.2010.12.003
22. Frankel OH, Brown AHD (1984) Current plant genetic resources—a critical appraisal. In: Genetic resources of crop plants. https://www.researchgate.net/profile/Anthony-Brown-17/publication/279524317_Plant_genetic_resources_today_a_critical_appraisal/links/5594c68508ae21086d1eead6/Plant-genetic-resources-today-a-critical-appraisal.pdf
23. Nielsen R et al. (2011) Genotype and SNP calling from next-generation sequencing data. Nat Rev Genet 12:443–451. https://doi.org/10.1038/nrg2986
24. Davey JW, Hohenlohe PA, Etter PD, Boone JQ, Catchen JM, Blaxter ML (2011) Genome-wide genetic marker discovery and genotyping using next-generation sequencing. Nat Rev Genet 12:499–510. https://doi.org/10.1038/nrg3012
25. Doyle JJ, Doyle JL (1987) A rapid DNA isolation procedure for small quantities of fresh leaf tissue. Phytochem Bull 19:11–15
26. Doyle JJ, Doyle JL (1990) Isolation of plant DNA from fresh tissue. Focus 12:13–15
27. Russell DW, Sambrook J (2001) Molecular cloning: a laboratory manual, vol 1. Cold Spring Harbor Laboratory, Cold Spring Harbor, NY, p 112. https://www.cshlpress.com/pdf/sample/2013/MC4/MC4FM.pdf
28. Good JM (2011) Reduced representation methods for subgenomic enrichment and next-generation sequencing. Mol Methods Evolut Genet 772:85–103. https://doi.org/10.1007/978-1-61779-228-1_5
29. Barabaschi D, Guerra D, Lacrima K, Laino P, Michelotti V, Urso S, Cattivelli L (2012) Emerging knowledge from genome sequencing of crop species. Mol Biotechnol 50(3):250–266. https://doi.org/10.1007/s12033-011-9443-1
30. Mammadov J, Aggarwal R, Buyyarapu R, Kumpatla S (2012) SNP markers and their impact on plant breeding. Int J Plant Genomics 2012(1):728398. https://doi.org/10.1155/2012/728398
31. Stahl K, Papiol S, Budde M, Heilbronner M, Oraki Kohshour M, Falkai P, Bickeböller H (2025) Aggregating single nucleotide polymorphisms improves filtering for false-positive associations postimputation. G3 15(5):jkaf043. https://doi.org/10.1093/g3journal/jkaf043
32. Wang Y, Wu X, Li Y, Feng Z, Mu Z, Wang J, Li G (2021) Identification and validation of a core single-nucleotide polymorphism marker set for genetic diversity assessment, fingerprinting identification, and core collection development in bottle gourd. Front Plant Sci 12:747940. https://doi.org/10.3389/fpls.2021.747940
33. Jannink JL, Lorenz AJ, Iwata H (2010) Genomic selection in plant breeding: from theory to practice. Brief Funct Genomics 9(2):166–177. https://doi.org/10.1093/bfgp/elq001
34. Hufford MB et al. (2012) Crop biodiversity: an unfinished magnum opus of nature. Annu Rev Plant Biol 63:727–751. https://doi.org/10.1146/annurev-arplant-042817-040240
35. Metzker ML (2010) Sequencing technologies-the next generation. Nat Rev Genet 11:31–46. https://doi.org/10.1038/nrg2626
36. Eid J, Fehr A, Gray J, Luong K, Lyle J, Otto G, Peluso P, Rank D, Baybayan P, Bettman B, Bibillo A, Bjornson K, Chaudhuri B, Christians F, Cicero R, Clark S, Dalal R, deWinter A, Dixon J, Foquet M, Gaertner A, Hardenbol P, Heiner C, Hester K, Holden D, Kearns G, Kong X, Kuse R, Lacroix Y, Lin S, Lundquist P, Ma C, Marks P, Maxham M, Murphy D, Park I, Pham T, Phillips M, Roy J, Sebra R, Shen G, Sorenson J, Tomaney A, Travers K, Trulson M, Vieceli J, Wegener J, Wu D, Yang A, Zaccarin D, Zhao P, Zhong F, Korlach J, Turner S (2009) Real-time DNA sequencing from single polymerase molecules. Science 323:133–138. https://doi.org/10.1126/science.1162986
37. Varshney RK, Terauchi R, McCouch SR (2014) Harvesting the promising fruits of genomics: applying genome sequencing technologies to crop breeding. PLoS Biol 12:e1001883. https://doi.org/10.1371/journal.pbio.1001883a
38. Garrison E, Marth G (2012) Haplotype-based variant detection from short-read sequencing. Preprint at arXiv https://doi.org/10.48550/arXiv.1207.3907
39. Li H, Handsaker B, Wysoker A, Fennell T, Ruan J, Homer N, Marth G, Abecasis G, Durbin R (2009) The sequence alignment/map format and SAMtools. Bioinformatics 25:2078–2079. https://doi.org/10.1093/bioinformatics/btp352
40. Makhoul M, Rambla C, Voss-Fels KP, Hickey LT, Snowdon RJ, Obermeier C (2020) Overcoming polyploidy pitfalls: a user guide for effective SNP conversion into KASP markers in wheat. Theor Appl Genet 133(8):2413–2430. https://doi.org/10.1007/s00122-020-03608-x

Chapter 19

Non-Sequencing-Based Microsatellite Development in Pomegranate (*Punica granatum*): Enrichment and Screening Protocol

Mehrana Koohi-Dehkordi and Mehran E. Shariatpanahi

Abstract

The chapter reports an inexpensive, non-sequencing-based technique to develop microsatellite markers in pomegranate (*Punica granatum*). Microsatellites are also known as the polymorphic genetic markers that are usually used in genetic diversity, breeding, and population genetics. Even though next-generation sequencing (NGS) has quickly demonstrated speedy advancement in the development of markers, the given protocol provides a viable option in species with limited genomic information or in cases when sequencing is not a viable option. The strategy uses the enrichment strategy as a way of selectively isolating the DNA fragments having the microsatellite repeats, followed by a detailed screening procedure to identify the polymorphic loci. This is also geared towards providing an economical form of high-quality genetic markers production that can be utilized in genetic diversity, fingerprinting, and linkage mapping studies. The approach can be particularly applicable to the laboratories that do not have significant access to advanced sequencing technology.

Keywords *Punica granatum*, Microsatellites, SSR enrichment, Hybridization-based capture, DNA markers, Pomegranate biotechnology, Molecular screening

1 Introduction

Simple sequence repeats (SSRs) or microsatellites are very polymorphic genetic markers that are widely applicable in plant genetics to determine genetic diversity, linkage mapping, and cultivar identification [1]. These are the markers that have been successfully utilized to measure genetic variation and determine the phylogenetic relationship between pomegranate genotypes [2]. New SSR markers have been developed using non-sequencing-based techniques that are essential to species with limited genomic resources [3]. They can effectively identify novel loci of microsatellites, without requiring a large amount of sequencing, and turn out to be an affordable way to conduct genetic research on pomegranate [1].

Nitish Kumar (ed.), *Molecular Markers in Plants*, Methods in Molecular Biology, vol. 3097,
https://doi.org/10.1007/978-1-0716-5663-1_19,

Non-sequencing-based methods of developing SSR markers have resulted in the discovery of highly polymorphic markers, which are essential to marker-assisted selection and breeding programs [4]. Although the non-sequencing-based technique provides a viable alternative to the development of the microsatellite markers, the combination of the sequencing technologies can be used to make the genetic studies more conclusive and accurate. This protocol describes an extensive enrichment and screening strategy that will effectively find new microsatellite loci without necessarily requiring large-scale sequencing. This is a cheaper and more accessible alternative that can increase genetic analysis tools in species with limited sequencing resources.

2 Materials

2.1 Genomic DNA Extraction and Quality Assessment

- CTAB extraction buffer: 1.4 M NaCl, 20 mM EDTA (pH 8), 100 mM Tris-HCl (pH 8), 2% CTAB.
- Chloroform–isoamyl alcohol: 24 ml Chloroform, 1 ml isoamyl alcohol.
- RNase-A solution: 10 mg/ml RNase A, 10 mM Tris–HCl (pH 7.5), 15 mM NaCl.
- Tris–HCl 1 M: 121.1 g Tris base in 1 L dH2O, pH adjusted to 7.5 with HCl.
- TE (T10E1) buffer: 10 mM Tris–HCl (pH 7.8), 1 mM EDTA.

2.2 Enzymatic Digestion and Purification

- Denaturation solution: 0.25 M NaOH and 0.25 M NaCl.

2.3 Cloning and Bacterial Transformation

- LB broth (per liter): 10 g Tryptone, 5 g Yeast Extract, 5 g NaCl.
- LB selection agar: 100 ml LB agar, 500 μl X-gal, 500 μl IPTG, 100 μl ampicillin.
- SOC medium (per liter): 10 g Tryptone, 5 g yeast extract, 5 g NaCl, 10 mM $MgSO_4$, 10 mM $MgCl_2$, 20 mM glucose, pH adjusted to 7.0.
- Trituration buffer: 100 mM $CaCl_2$, 70 mM $MgCl_2$, 40 mM sodium acetate (pH 5.5).
- IPTG: 238 mg/ml H_2O.
- Xgal: 20 mg/ml Dimethyl formamide.

2.4 Plasmid Extraction and Rapid Screening

- Washing solution: 10 mM Tris (pH 8), 100 mM NaCl, 10 mM EDTA.

- Solution I: 40 ml 250 mM EDTA, 1 ml 25 mM Tris (pH 8), 500 ml 100 mM glucose, 435 ml ddH_2O.
- Solution II (freshly prepared): 5 ml 2 N NaOH, 5 ml 10% SDS, 90 ml ddH_2O.
- Solution III: 143 ml Glacial Acetic Acid, 100 ml ddH_2O, pH adjusted to 4.8 with KOH, full up to 500 ml with ddH_2O.
- 2× Cracking buffer (500 ml): 5 ml 1 M NaOH, 1 ml 0.5 M EDTA, 0.5 g SDS, 0.025 g bromocresol green, 5 ml glycerol.
- Lysis buffer: 10 mM Tris–HCl (pH 8), 1 mM EDTA, 15 W/V sucrose, 2 mg/ml lysosyme, 0/02 mg/ml RNase A.

2.5 Hybridization and Detection (Enrichment/Dot Blot)

- Hybridization buffer: 5× SSC, 0.1% N-lauryl sarcosine, 0.02% SDS, 1% blocking reagent.
- Hybridization buffer with formamide: 5× SSC, 0.1% N-lauryl sarcosine, 0.02% SDS, 2% blocking reagent, 50% deionized formamide.
- Blocking sock reagent (10×): 10% blocking reagent in maleic acid.
- 10× SSC: 0.15 M NaCitrate, 1.5 M NaCl, adjust to pH 7.
- 2× washing solution: 2× SSC, containing 0.1% SDS.
- Blocking solution: 1 g blocking reagent in 100 ml maleic acid buffer.
- Blocking stock reagent (10×): 10% blocking reagent in maleic acid.

2.6 Electrophoresis and Visualization

- Acrylamide gel (40%): 19 g Acrylamide, 1 g bisacrylamide, volume to 50 ml with dH_2O.
- Tris-acetate EDTA (TAE) buffer: 0.04 M Tris–acetate, 0.001 M EDTA.
- Tris–borate EDTA (TBE) buffer: 0.089 M Tris, 0.089 M boric acid, 0.002 M EDTA.
- Gel loading buffer: 25 mgr Bromophenol blue, 25 mgr Xylene cyanol, 3 ml glycerol.
- Gel loading buffer (with formamide): 96 ml Formamide, 2 ml 0.5 M EDTA, 0.25% bromophenol blue, 0.25% xylene cyanol.

3 Methods

3.1 Genomic DNA Extraction from Plant Samples

DNA is extracted according to the Murray and Thompson method [5] with minor modifications as follows.

1. Grind 250 mg (small scale) or 800 mg (large scale) of young pomegranate leaves in liquid nitrogen to a fine powder.

2. Mix the powder with 800 μl (small) or 10 ml (large) of CTAB extraction buffer preheated to 65°C, adding beta-mercaptoethanol.
3. Incubate at 65°C for 30 min, mixing gently by inversion.
4. Add an equal volume of chloroform/isoamyl alcohol (24:1), centrifuge for 15 min, and transfer the upper aqueous phase to a new tube.
5. Precipitate DNA by adding 0.6 volumes of cold Isopropanol and storing at −20°C for 10 min.
6. Centrifuge to pellet the DNA, wash with 70% ethanol, dry the pellet, and resuspend in sterile ddH_2O.
7. Add 3 μl of RNase A (10 mg/μl) to the sample and incubate at 37°C for 3 h, and repeat Steps 4–6.

3.2 Quantification and Quality Assessment

1. Measure DNA concentration and purity using spectrophotometry; ensure the A260/A280 ratio is between 1.7 and 1.9.
2. Run samples on a 0.7% agarose gel to verify DNA integrity; high-quality DNA appears as a sharp band, while smearing indicates degradation (*see* **Note 1**).

3.3 Enzymatic Digestion and Gel Purification

1. Blunt-end digestion: Digest 10 μg of genomic DNA with *Alu*I, *Hae*III, or *Rsa*I (4-base cutters) overnight at 37°C [6] (*see* **Note 2**).
2. Sticky-end digestion: Digest 500 ng of DNA with *Eco*RI (6-base cutter) and *Tru*9I (4-base cutter) for 3 h at 37°C using OPA buffer (*see* **Note 3**).
3. Heat the samples at 65°C for 10 min to inactivate enzymes.
4. Run digested DNA on a 0.7% agarose gel and excise the 200–1000 bp range with a sterile scalpel [7].
5. Determine the weight of the excised gel slice. Add sodium iodide (NaI) at a 3:1 ratio by gel weight. Incubate the tubes at 50°C for 5 min until the gel has fully dissolved in the sodium iodide.
6. Add 5–10 μl of resuspended glass milk (silica suspension) to the sample (the amount depends on the DNA concentration observed in the gel). Mix thoroughly by vortexing.
7. Shake the samples for 10 min to bind the DNA to the silica particles.
8. Centrifuge the mixture at 14,000 rpm for 5 s. Carefully discard the supernatant.
9. Resuspend the pellet in 500 μl of Wash solution. Centrifuge at 14,000 rpm for 5 s and discard the supernatant. Repeat this washing step three times.

10. Dry the pellet thoroughly to eliminate any remaining ethanol from the wash.
11. Resuspend the pellet in 20 μl of sterile distilled water to elute the DNA from the silica
12. Let the suspension sit at room temperature for 10 min, then spin at 14,000 rpm for 5 s.
13. Transfer approximately 18 μl of the supernatant (containing the purified DNA) to a new sterile tube.
14. To estimate the concentration of the purified DNA, electrophorese 3 μl of the final product. Store the remaining purified DNA at 4°C for use in subsequent stages.

3.4 Adapter Ligation

1. Prepare the MluI adapter by mixing equal molar concentrations of its two oligonucleotides, heating to 80°C for 5 min, and cooling at room temperature for 1 h to anneal.

MluI Oligonucleotide 21 mer	5′-CTC TTG CTT ACG CGT GGA CTA-3
MluI Oligonucleotide 25 mer	5′-PO4 - TAG TCC ACG CGT AAG CAA GAG CAC A-3′
MluI Adaptor	3′-ATC AGG TGC GCA TTC GTT CTC-5′ 5′-PO_4-TAG TCC ACG CGT AAG CAA GAG CAC A-3′

2. Ligate the adapter to purified fragments using T4 DNA Ligase and PEG (*see* **Note 4**) at 4°C for 10–12 h [8].
3. For sticky-ended fragments, ligate *Eco*RI and *Tru*9I adapters in the presence of ATP and OPA buffer.
   EcoRI adaptor: 5′-CTC GTA GAC TGC GTA CC-3′
   3′-CAT CTG ACG CAT GGT TAA-5′
   Tru 91 adaptor: 5′-GAC GAT GAG TCC TGA G-3′
   3′-TA CTC AGG ACT CAT-5′
4. Validate ligation by PCR using adapter-specific primers (MluI, E000, T000, for EcoRI and Tru91 adapters) under programmed thermal cycles.
   E_{000}: 5′-GAC TGC GTA CCA ATT C-3′
   T_{000}: 5′-GAT GAG TCC TGA GTA A-3′
5. Electrophoresis at 60 V for 2.5 h. Marker III can be used as a reference to determine the size and estimated concentration of the DNA bands.
6. Hold samples at 4°C for the hybridization step.

3.5 Microsatellite Capture Using Hybridization

3.5.1 Preparation of Microsatellite Sequences

1. Dilute the synthesized oligonucleotide sequences ((AG)15, (GT)15, and (ATT)10) with sterile double-distilled water to reach a concentration of 50 pmol/µl (*see* **Note 13**).
2. To ensure the DNA is fully dissolved, store the diluted samples at 4°C for 10–12 h.
3. Take 5 pmol (50 ng) of each sequence as a template based on Edwards et al. [9].
4. Add the denaturation solution to each sample, seal the tubes with parafilm, and incubate at room temperature for 10 min (*see* **Note 5**).
5. Heat the tubes at 95°C for 5 min, then immediately place them on ice.
6. Spot the denatured sequences onto 1 × 0.5 cm nylon membranes.
7. Let the membranes dry completely at room temperature.
8. Incubate the membranes at 120°C for 30 min to fix the single-stranded DNA onto the surface.

3.5.2 Pre-Hybridization

1. Add 1500 µl of hybridization buffer containing 50% deionized formamide to the hybridization tubes containing the nylon membrane (*see* **Note 8**).
2. Set the hybridization oven (Biometra Compact line OV4) to 42°C and the rotor speed to 1.
3. Incubate for 1 h to block the membrane.

3.5.3 Preparation of a DNA Sample as a Probe

1. Use the adapter-ligated digested DNA as the probe at a concentration of 100 ng per 500 µl of hybridization buffer.
2. Denature the genomic DNA using the same procedure described for the template (treatment with NaOH/NaCl followed by heating at 95°C and rapid cooling on ice).
3. For the hybridization step, add the denatured DNA to the hybridization buffer.

3.5.4 Selective Hybridization (Capture Procedure)

1. Once the pre-hybridization period has elapsed, add the required amount of single-stranded DNA (300 ng of DNA for every 1500 µl of hybridization buffer) to the tubes.
2. Return the tubes to the oven and incubate them for 10–12 h under the same conditions as previously set (42°C with rotor speed 1).

3.5.5 Stringency Washes of the Nylon Membrane

1. Wash the membrane twice for 15 min per wash in 2× SSC/0.1% SDS at room temperature.
2. Wash twice with 0.5× SSC containing 0.1% SDS at 65°C for 15 min each (*see* **Note 9**).
3. Perform a final wash with 0.2× SSC at room temperature for 15 min (*see* **Note 9**).

3.5.6 DNA Elution and Reamplification

1. Cut the membrane into small pieces (approx. 3 mm in size) with a sterile scalpel.
2. Resuspend the pieces in 100 μl of sterile ddH_2O in a 1.5 ml tube.
3. Heat the samples to 100°C for 10 min to elute the captured DNA.
4. Using the eluted DNA as a template, perform PCR with adapter-specific primers (MluI or E000/T000). Verify the amplification on a 1.2% TBE agarose gel.

3.6 Cloning and Transformation

1. To achieve a high frequency of recombinant clones, ligate the enriched DNA into the pGEM-T Easy Vector and incubate the reaction overnight at 4°C [8].
2. Preparation of competent cells: Grow *E. coli* strain MC1061 in LB medium until reaching the appropriate growth phase. Centrifuge the cells and resuspend the pellet in ice-cold Trituration buffer (containing $CaCl_2$, $MgCl_2$, and sodium acetate). Incubate on ice for 30 min before concentrating the cells for transformation
3. Transformation and selection:
   - Mix 150 μl of competent cells with the ligation product and incubate on ice for 30 min.
   - Place the sample at 42°C for 2 min, then immediately move it to ice for another 2 min.
   - Add 500 μl of pre-warmed SOC medium and shake at 37°C for 30–60 min.
   - Plate the mixture on LB agar containing ampicillin, X-gal, and IPTG; incubate for 12–16 h at 37°C to identify recombinants through blue–white screening; white colonies contain potential SSR inserts (Fig. 1).

3.7 Screening for Recombinant Clones

3.7.1 Rapid Cracking Screening

To quickly screen white colonies, mix a loop of the colony with 25 μl ddH_2O and 25 μl 2× cracking buffer. Electrophorese 15 μl of this suspension on a 1.2% agarose gel; clones with inserts will migrate more slowly than the blue colony negative control.

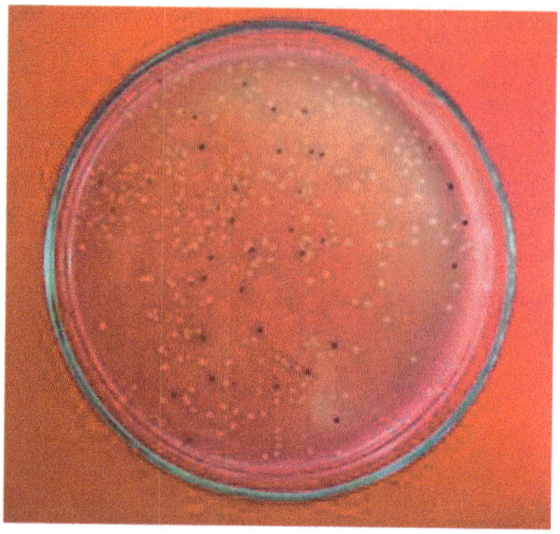

Fig. 1 A sample of clones was obtained via TA cloning. White clones were selected based on blue–white screening [10]

3.7.2 Plasmid Extraction (Minipreparation)

- Inoculate a single loop of the desired clones into 3 ml of liquid LB broth containing ampicillin (100 mg/μl) [4].
- Shake the cultures at 37°C for 10–12 h.
- Pellet the bacterial cells by centrifugation at 14,000 rpm for 2 min.
- Resuspend the pellet in 50 μl of Wash Buffer, then centrifuge at 14,000 rpm for 2 min.
- After discarding the supernatant, resuspend the pellet in 100 μl of Solution I, and incubate for 5 min at room temperature.
- Add 10 μl of Lysozyme solution (50 mg/ml) and incubate the suspension for 10 minutes at room temperature.
- Add 150 μl of freshly prepared Solution II. Mix the contents gently by inverting the tube and place it on ice for 5 min.
- Add 6 μl of RNase-A (10 mg/ml) to the mixture.
- Add 150 μl of Solution III, mix gently, and incubate on ice for 5 min.
- Centrifuge at 14,000 rpm for 10 min. Transfer the supernatant to a new, sterile tube.
- Add an equal volume of chloroform/isoamyl alcohol (24:1) to the supernatant. Centrifuge at 14,000 rpm for 5 min and transfer the upper phase to a new tube.
- Add 0.6 volumes of cold Isopropanol and store at −20°C for 10 min.
- Centrifuge for 15 min at 14,000 rpm.
- Discard the supernatant and wash the pellet with 70% ethanol.

- Dissolve the dried pellet in 25 μl of sterile dH_2O.
- To evaluate the quality of the extracted plasmids, electrophorese 5 μl of the sample on a 1.2% TBE agarose gel.
- To confirm the presence and correct size of the insert, perform a restriction digest with EcoRI.

3.7.3 Screening of Clones Using PCR

- Designing PCR primers:
  Two specific primers, pGEM1 and pGEM2, were designed using Oligo software from the flanking regions of the insertion site in the pGEM-T vector.
  - Primer sequences:
    - pGEM1: 5′-GTA ATA CGA CTC ACT ATA GGG CGA AT-3′.
    - pGEM2: 5′- AGA ATA CTC AAG CTA TGC ATC CAA CG-3′.
  - SSR Motif Primers: (AG)15, (GT)15, and (ATT)10.
- Screening: To screen for the presence of the target SSR sequence, separate PCR reactions are carried out. Each reaction uses a pair of primers: one vector-specific primer (either pGEM1 or pGEM2) and one SSR motif primer.
- Specific annealing temperatures:
  - For tracking (AG), (GT), and (ATT) sequences, the recommended annealing temperatures are 65.3°C, 65°C, and 55.1°C, respectively (*see* **Note 7**).

3.8 Dot Blot Hybridization

1. DNA preparation
   - Use 100 ng plasmid DNA per sample for spotting (lower concentrations are ineffective).
   - Mix samples with the denaturation solution.
   - Incubate at room temperature for 10 min.
   - Add 2× SSC to denatured samples to achieve the final volume before loading.
2. Membrane and apparatus setup
   - Clean the dot blot apparatus thoroughly.
   - Soak a layer of filter paper and a nylon membrane in distilled water.
   - Place layers in the device, tighten screws, and connect to the vacuum suction.
3. Sample loading:
   - Load denatured samples into the 96-well dot blotter according to a predefined sample map.

- Apply vacuum suction to transfer DNA onto the membrane.
- Dry membrane at room temperature, then bake at 120°C for 30 min to fix DNA.

4. Probe labeling:
   - Label SSR probes ($(AG)_{15}$, $(GT)_{15}$, $(ATT)_{10}$) using the DIG Oligonucleotide 3′-End Labeling Kit.
   - Incubate labeling reaction at 37°C for 1 h.
   - Stop the reaction by adding 0.2 M EDTA (pH 8.0).
5. Hybridization:
   - Pre-hybridize membrane in 40 ml hybridization buffer for 1 h:
     - 65°C for $(AG)_{15}/(GT)_{15}$ motifs.
     - 50°C for $(ATT)_{10}$ motif, (see **Note 6**).
   - Prepare hybridization probe mixture:
     - Dilute 50 pmol labeled probe in 10 ml hybridization buffer.
     - Boil to remove secondary structures.
   - Replace the pre-hybridization buffer with a fresh buffer containing the probe.
   - Hybridize for 10–12 h at the appropriate temperature.
6. Stringency washes:
   - Wash three times with 2× SSC + 0.1% SDS, 10 min each, at room temperature.
   - Wash two times with 0.2× SSC + 0.1% SDS, 15 min each, at hybridization temperature.
   - Wash once with 0.2× SSC, 5 min, at room temperature.

3.8.1 Detection and Coloring

1. Incubate the membrane in blocking buffer for 1 h.
2. Transfer membrane to blocking buffer containing antibody (1:1000 dilution) and incubate for 1 h.
3. Wash membrane twice in washing buffer, 15 min per wash.
4. Incubate in detection buffer for 5 min.
5. Cover the membrane with NBT/BCIP solution. Incubate in the dark without agitation to prevent background staining.
6. Stop the reaction by rinsing with water once colored spots appear, (see **Note 10**) (Fig. 2).

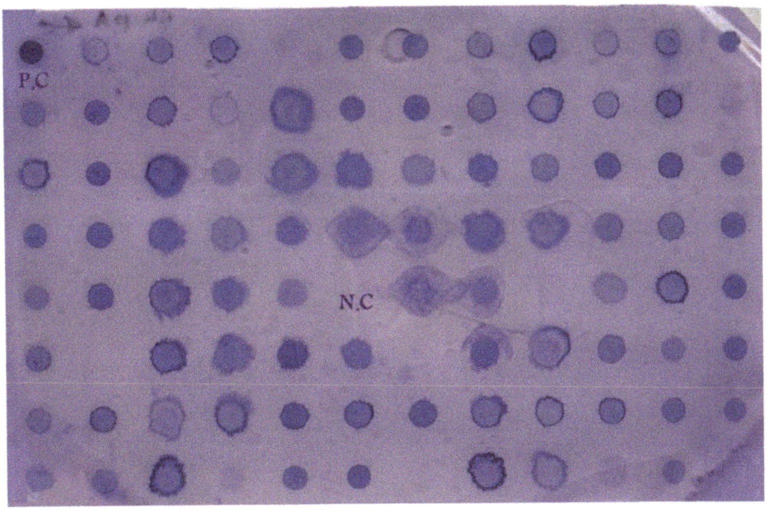

Fig. 2 Dot blot hybridization results for AG sequence-enriched clones. *N.C.* negative control, *P.C.* positive control (Source: [10])

3.9 Stripping and Reusing the Membrane

- Decoloring: To remove the blue color, incubate the membrane in dimethylformamide at 55°C–60°C until white.
- De-probing: Remove the probe by washing twice (20 min each) in a solution of 0.2 M NaOH and 0.1% SDS at 37°C, followed by a final 2× SSC wash.

3.10 Plasmid Extraction for DNA Sequencing (SEQLAB Kit Method)

To achieve maximum efficiency in sequencing pomegranate DNA inserts within the pGEM-T vector, plasmids are extracted using the **SEQLAB Kit**.

3.11 Nucleotide Sequencing and Analysis

Provide 3 µg of plasmid DNA for sequencing in the forward direction.

Check the electrogram's quality in Chromas, use BLAST to find and match any parts of the sequence that come from the pGEM-T vector. Delete those vector sequences to get the clean pomegranate insert data.

3.12 Designing Primers for Microsatellite (SSR) Marker Development

The primers should be designed to flank microsatellite loci containing a sufficient quantity of tandem repeats (Table 1). Design primers with the following specifications:

- Length: 20–25 nucleotides.
- Annealing Temperature: ≥45°C.
- GC Content: Balanced.
- Amplicon Size: 100–400 base pairs.
- Secondary Structures: Must be free of hairpins and primer–dimers.

Table 1
Primer information for the designed primers.

Primer name	Sequences	Amplicon (bp)
A10 1–83 A10 2–83	5′-GCT CTC CGA TAG TGA TTC A-3′ 5′-CCG CAT ATA GAC AAA GAA AG-3′	210
A18 1–83 A18 2–83	5′-GGG TTT TGT GTG GGG AAG ATG GA-3′ 5′-ATG CCA TCG TTG CCT CCC TGT-3′	141
A16 1–83 A16 2–83	5′-GCG TCC ACT TTA CTG CTT TC-3′ 5′-CAG TAA CCG GAT GCA AAG TC-3	409
A9 1–83 A9 2–83	5′-GCG TGC ATG ATG GGA GAA G-3′ 5′-CGA AGT TAG TCA AGT GGG TG-3′	245

3.13 Optimization of PCR Reactions

- Dilute synthesized primers with sterile double-distilled water to a working concentration of 20 pmol/μl.
- PCR reaction composition (15 μl): Template DNA: 2 μl (50 ng), 10× Buffer, dNTPs (10 mM), forward/reverse primers (20 pmol/μl), *Taq* DNA polymerase (4 u/μl)
- Thermal cycling steps: The thermal cycling protocol consists of:
  - An initial denaturation step at 94°C for 5 min.
  - 35 cycles of (Denaturation at 94°C for 60 s, Annealing at a target-specific temperature for 60 s, Extension at 72°C for 45 s).
  - A final extension at 72°C for 10 min.
- Optimal annealing temperatures: A18 Primer: 70.1°C, A10 Primer: 62.5°C, A16 Primer: 54.0°C.

3.14 Polymorphism Analysis of Designed Primers

3.14.1 Electrophoresis on Denaturing Sequencing Gel

- Apply a binding solution to the surface of the large plate.
- Coat the small plate with a repellent solution.
- Prepare a 6% denaturing gel containing urea (as a denaturant), TBE buffer, 40% acrylamide/bisacrylamide (19:1), APS, and TEMED.
- For optimal performance, conduct an initial 30-min warm-up at 100 W without any samples to achieve a temperature range of 45°C–55°C.
- Mix PCR products with an equal volume of formamide loading buffer. Denature at 95°C for 5 min, then immediately transfer to ice.
- Proceed with electrophoresis by loading the samples and running the gel for 90 min, using the identical parameters established during the pre-run.

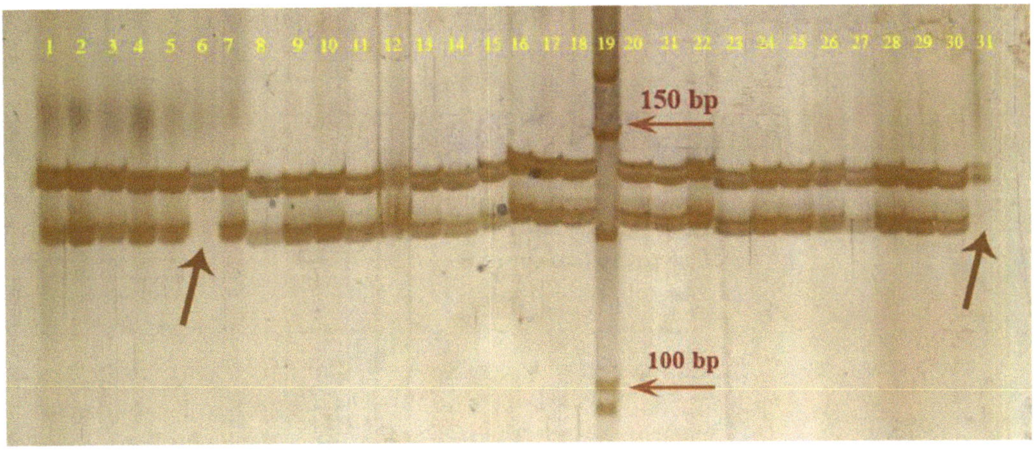

Fig. 3 DNA amplification results from multiple pomegranate cultivars using primer A18 (141 bp amplicon) [10]. The samples are, from left to right: 1—Golnar Fars, 2—Khafri Red Skin, 3—Shahvar Shirin, 4—Gurur Kerrez, 5—Zhenjeh Arous Khafr Soursh, 6—Nabati Ardakan, 7—Hasibi Mehriz, 8—Besed Sangan, 9—Ardestani Red Skin, 10—Tab and Ler, 11—Ardestani White Skin, 12—Besed Ladiz, 13—Asali Sarvestan, 14—Wahshi Turk Sarvestan, 15—Soursh Moemol Zabol, 16—Debhai Sarjangal, 17—Dokhtar Hamoumi Varamin, 18—Golu Gandeh Ney Riz, 19—Marker XIII, 20—Barg Rosti Ardel, 21—Ameneh Khatuni, 22—Toq Gernd, 23—Ardestani Soursh, 24—Sabi Bam Kerman, 25—Kalle Gawi (Khorasan), 26—Melas Sour Yazd, 27—Kalle Gawi (Sistan and Baluchestan), 28—Bi-Khafar Jahrom, 29—Pesht Siah Yazdi, 30—Khoshe Nar Mazandaran, 31—Gerch Shahvar Yazd. Arrows indicate polymorphic varieties

3.14.2 Silver Nitrate Staining Protocol

1. Submerge the gel in *10% acetic acid* for 30 min (for sequencing gel).
2. Incubate in *0.1% or 0.2% silver nitrate solution* for 10–30 min.
3. Briefly rinse in cold distilled water for 10 s (*see* **Note 11**).
4. Submerge in a cold *Developer solution* (sodium carbonate/sodium hydroxide and formaldehyde) until bands appear (1–10 min) (*see* **Note 12**).
5. Stop the reaction by placing the gel in 10% acetic acid for 2–5 min.
6. Rinse with distilled water, then dry thoroughly before imaging (Fig. 3).

4 Notes

1. DNA quality and degradation: A smeared DNA band on an agarose gel is one of the signs of degradation, which can be caused by thawing of tissues during grinding or improper utilization of liquid nitrogen.
2. Enzyme combinations: Multiple restriction enzymes aid in the capturing of a larger number of SSR loci, and avoid marker clustering, making future genetic maps better.

3. Ligation inefficiency: Blunt-end ligation is a less efficient reaction compared to sticky-end ligation due to the absence of complementary overhangs to direct the reaction.
4. Enhancing the reaction: To improve the efficiency of MluI adapter ligation, the protocol recommends 10% polyethylene glycol (PEG) in the protocol and 10 mM ATP. PEG is a molecular crowding agent and facilitates ligation since it ensures that the ends of the DNA stay close to each other. At the same time, ATP is the necessary source of energy to mediate the enzyme activity of T4 DNA Ligase.
5. To eliminate secondary structures, combine 50 ng of each sequence with a solution that contains a denaturing agent. Warm the mixture to 95 °C within 5 min and then pour into ice.
6. Probe length challenges: Short 30-bp oligonucleotide probes (e.g., $(AG)_{15}$, $(GT)_{15}$) form unstable hybrids, which demand stringent temperature regulation to ensure fragments remain immobilized on the membrane.
7. Motif-specific temperatures: The best temperatures of hybridization are motif specific: AG and GT repeats have the best hybridization temperature at 65°C, and ATT motives require 50°C–55°C to stabilize annealing.
8. Formamide utility: Formamide (30–50) in the hybridization buffer facilitates the hybridization to be done at a lower temperature, as the weak and nonspecific interactions are destabilized.
9. Sensitivity of washing: Stringency washing with 0.2× SSC is more efficient at eliminating nonspecific hybrids than 0.5× SSC, and this enhances the specificity of the process of enriching the samples [5].
10. Dot blot vs. PCR: PCR is quicker as the initial screening method, but may provide ambiguous or false results. The dot blot technique is the most reliable and precise method of verifying the positive clones before sequencing.
11. Precise timing: The duration of the water wash between silver nitrate staining and development must be strictly restricted to 10 s; otherwise, silver ions will wash away the DNA, resulting in weak bands.
12. Background noise: To avoid high backgrounds on polyacrylamide gels, the developer solution should be chilled to a temperature less than 10 °C.
13. Motif success rates: AG motifs are generally more frequent and more effective in the capture attempts than the GT or ATT motifs in pomegranate.

References

1. Mahajan Sagar R, Vaishali M, Bhosale S (2018) Molecular characterization of cultivated and wild genotypes of *Punica granatum* L.(Pomegranate) by using SSR marker. Int J Life Sci Sci Res 4:1786–1794
2. Liu C, Li J, Qin G (2020) Genome-wide distribution of simple sequence repeats in pomegranate and their application to the analysis of genetic diversity. Tree Genet Genomes 16(2):36
3. Ebrahimi S, Sayed-Tabatabaei BE, Sharifnabi B (2010)). Microsatellite isolation and characterization in pomegranate (*Punica granatum* L.). Iran J Biotechnol 8(3):156–163
4. Ravishankar KV, Chaturvedi K, Puttaraju N, Gupta S, Pamu S (2015) Mining and characterization of SSRs from pomegranate (*Punica granatum* L.) by pyrosequencing. Plant Breed 134(2):247–254
5. Murray MG, Thompson WF (1980) Rapid isolation of high molecular weight plant DNA. Nucleic Acids Res 8:4321
6. Bloor PA, Barker FS, Watts PC, Noyes HA, Kemp SJ (2001) Microsatellite libraries by enrichment. Anim Genomics Lab http://www.genomics.liv.ac.uk/animal/research/MicrosatelliteEnrichment
7. Zane L, Bargelloni L, Patarnello T (2002) Strategies for microsatellite isolation: a review. Mol Ecol 11:1–16
8. Sambrook J, Fritsch EF, Maniatis T (1989). Molecular cloning: a laboratory manual, 2nd ed. Cold Spring Harbour Laboratory Press, New York, USA
9. Edwards K, Johnstone C, Thompson C (1991) A simple and rapid method for the preparation of plant genomic DNA for PCR analysis. Nucleic Acids Res 19(6):1349
10. Koohi-Dehkordi M, Sayed-Tabatabaei BE, Yamchi A, Danesh-Shahraki A (2006) Microsatellite markers in pomegranate. In: XXVII International Horticultural Congress-IHC2006: II International Symposium on Plant Genetic Resources of Horticultural, vol 760, pp 179–184

Chapter 20

Development of Genome-Wide Simple Sequence Repeat Markers from Whole-Genome Sequence of Mungbean (*Vigna radiata*)

Rakshana Palaniswamy, Vignesh Mohanavel, Veera Ranjani Rajagopalan, Raghu Rajasekaran, Raveendran Muthurajan and Sudha Manickam

Abstract

Mungbean (*Vigna radiata* L.) is a major pulse crop cultivated extensively across Asia and is considered an important model species among legumes because of its diploid genome and relatively small genome size. The availability of reliable, economical, and highly polymorphic molecular markers is critical for assessing genetic diversity, constructing linkage maps, and implementing functional genomics-based breeding programs. Among the available marker systems, simple sequence repeat (SSR) markers are highly polymorphic, abundant, co-dominant, and extensively used for allelic characterization across diverse crop species. This chapter describes the methodology for developing SSR markers and their use in genetic diversity studies in mungbean. Genome-wide SSR screening is performed to identify SSR motifs in the mungbean whole-genome sequence using the Phobos tandem repeat detection tool. The identified SSR motifs are annotated and categorized based on motif length, repeat type, and chromosomal location to facilitate the selection of informative loci for marker development. Primers are designed for selected SSR loci using appropriate flanking sequences and subsequently validated through PCR amplification across a panel of diverse mungbean genotypes. The polymorphism of these SSR markers is assessed by analyzing distinct and reproducible amplification patterns. The developed SSR markers constitute a valuable genomic resource for genetic diversity studies, molecular characterization, and genomics-assisted breeding applications in mungbean.

Keywords Genetic diversity, Mungbean, Simple sequence repeats, Whole-genome sequence

1 Introduction

Mungbean (*Vigna radiata* L.) is a short-duration grain legume cultivated on more than six million hectares and plays a crucial role in cereal-based cropping systems due to its adaptability, nitrogen-fixing ability, and suitability for intercropping and dryland agriculture [1]. Conventional breeding approaches have contributed to varietal development; however, their efficiency is limited when dealing with quantitatively inherited traits and low levels of genetic variability. Recent

Nitish Kumar (ed.), *Molecular Markers in Plants*, Methods in Molecular Biology, vol. 3097,
https://doi.org/10.1007/978-1-0716-5663-1_20,

advances in high-throughput sequencing technologies have enabled the application of genomics-assisted breeding strategies, wherein molecular markers are used to complement phenotypic selection [2]. Availability of a high-quality reference genome for mungbean has further expanded genomic resources for precise marker development and trait-association studies.

Among the various classes of molecular markers, simple sequence repeats (SSRs) have been widely adopted for genetic diversity analysis, linkage mapping, quantitative trait locus (QTL) identification, and marker-assisted breeding due to their abundance, high polymorphism, co-dominant inheritance, reproducibility, and cost-effectiveness [3]. SSRs consist of tandemly repeated short nucleotide motifs ranging from one to six base pairs and are classified into di-, tri-, tetra-, penta-, and hexanucleotide repeats. These repeats may occur as perfect, imperfect, or compound motifs and are ubiquitously distributed throughout eukaryotic genomes, predominantly in non-coding regions, though they also occur within genes and regulatory sequences [4]. High mutation rates at SSR loci, primarily caused by DNA polymerase slippage during replication, generate extensive allelic variation among genotypes [5].

The informativeness of SSR markers is commonly assessed using polymorphic information content (PIC), which reflects both the number of alleles and the allele frequency distribution. SSR markers with high PIC values are particularly useful for germplasm characterization, assessment of genetic diversity, parent selection, and hybrid purity testing in mungbean. Furthermore, polymorphic SSRs tightly linked to genes or QTLs can be effectively utilized for linkage mapping, association mapping, fine mapping, map-based cloning, and marker-assisted introgression of target traits. Although SSR markers have been extensively used in plant breeding, their application in mungbean has historically been constrained by the limited availability of genome-wide, evenly distributed, and highly polymorphic markers [6–8]. Recent studies have addressed these gaps by developing and validating WGS-derived SSR markers, demonstrating their effectiveness across diverse *Vigna* species and their strong cross-species transferability, making them well-suited for genetic diversity assessment and population structure analyses [9, 10].

The availability of the mungbean whole genome sequence (WGS) has overcome many of these limitations by enabling systematic, genome-wide identification and characterization of SSRs across all chromosomes [11]. Genome-wide SSR (gSSR) mining facilitates comprehensive analysis of repeat abundance, motif composition, repeat length variation, and chromosomal distribution, and allows efficient primer design from conserved flanking regions. SSR markers designed on perfect di-, tri-, and tetranucleotide SSR motifs with PIC > 0.5 are highly useful for germplasm fingerprinting, genetic diversity analysis, hybrid purity testing, and marker-assisted trait introgression. High-density gSSR markers are particularly valuable for precise genetic mapping, QTL identification, and broadening the genetic base of cultivated mungbean by incorporating diverse germplasm and wild relatives. Moreover, a comprehensive microsatellite database of *Vigna* species called "VigSatDB" database harbors over 800,000 putative SSR loci developed from multiple *Vigna* genome assemblies, including *Vigna radiata*. The user interface in this database facilitates the selection of motif regions by chromosome location and tools for marker development for breeding applications [12].

Previous genome-wide SSR studies in mungbean have reported low to moderate genetic diversity, with an average of 2–6 alleles per locus and polymorphism levels ranging from 60% to 86%, often clustering genotypes into two to four major groups using UPGMA analysis. For example, [6] identified 1493 SSRs using 454 sequencing technology and validated 60 polymorphic markers with an average of 2.68 alleles per locus, highlighting both the potential and limitations of early SSR resources. The incorporation of sequence-based SSR resources with broad germplasm collections continues to reveal important genetic diversity that can be exploited for crop improvement, while ongoing advances in high-density genotyping and genome assembly are enabling more precise resolution of trait-associated loci. Owing to their robustness, reproducibility, and cost-effectiveness, SSR markers offer practical advantages over SNP-based platforms in resource-limited mungbean breeding programs and contribute to the development of consensus genetic maps that integrate molecular and morphological traits for accelerated varietal improvement.

This chapter describes the methodology involved in SSR marker development, encompassing genome-wide SSR mining from the mungbean whole-genome sequence, primer design, PCR validation of SSR primers across mungbean genotypes, polymorphism scoring, and data analysis for genetic diversity studies.

2 Materials

2.1 Plant Material

Seeds of diverse mungbean genotypes collected from the National Pulses Research Center, Tamil Nadu Agricultural University (TNAU), Vamban, Tamil Nadu, India.

2.2 DNA Extraction

1. CTAB extraction buffer: 100 mM Tris-HCl (pH 8.0), 20 mM EDTA, 1.4 M NaCl and 2% CTAB (*see* **Note 1**).
2. β-mercaptoethanol.
3. Chloroform/isoamyl alcohol: Mix 24 parts of chloroform with one part of isoamyl alcohol to obtain 24:1 (v/v) ratio.
4. Isopropanol.

5. 70% ethanol.
6. TE buffer (1×):10 mM Tris-HCl (pH 8.0) and 1 mM EDTA (*see* **Note 2**).
7. UV Bio-spectrophotometer.

2.3 Polymerase Chain Reaction and Agarose Gel Electrophoresis

1. PCR Master Mix (2×): Taq DNA polymerase, dNTPs (dATP, dCTP, dGTP, dTTP), Taq reaction buffer, magnesium chloride, and nuclease-free water (purchased as ready-to-use premix).
2. Primer solution: Dilute 100 μM standard stock to a final concentration of 10 μM working stock with nuclease-free water.
3. Thermal Cycler (PCR machine).
4. TBE buffer (1×): 89 mM Tris-base, 89 mM Boric acid, and 2 mM EDTA (*see* **Note 3**).
5. Gel electrophoresis tank.
6. Agarose gel (0.8% or 3%) contains 0.2 μg/mL ethidium bromide (*see* **Note 4**).
7. DNA ladder.
8. UV transilluminator.

3 Methods

3.1 Mining Genomic SSR Motifs from the Mung Bean Genome

1. Retrieve the mungbean whole genome sequence (WGS; Vr 1.0) from the Legume Information System (https://legumeinfo.org/gbrowsevigra1.0).
2. Screen genome-wide SSR motifs using the Phobos tandem repeat finder plugin available in Geneious software [13]. Identify SSRs based on the following parameters:
   - Unit length ranging from 2 to 8 bp; exclude mononucleotide (1 bp) repeats to avoid sequencing or assembly errors.
   - Minimum repeat threshold: ≥8 for dinucleotide, ≥5 for trinucleotide, ≥4 for tetranucleotide, ≥3 for pentanucleotide, and ≥2 for hexanucleotide motif to ensure all identified SSRs are above 18 bp length (repeat number correlates with polymorphism).
   - The algorithm detected both perfect and imperfect (compound) SSRs across forward/reverse strands (*see* **Note 5**).
3. Examine the motif composition, abundance, repeat length, and chromosome distribution of perfect and imperfect SSRs across all eleven mungbean chromosomes.

4. Export the results as a tab-delimited feature table, and filter the dataset to retain only perfect di-, tri-, and tetra-nucleotide SSRs with a minimum of five repeat units for primer design. Discard imperfect repeats and longer motifs to prioritize highly polymorphic markers (*see* **Note 6**).

The overall workflow for SSR development and validation is illustrated in Fig. 1.

3.2 SSR Primer Design

Design primer pairs from genomic regions flanking selected perfect SSR motifs using Primer Premier 3.0 as described by [14]. Follow the design criteria mentioned below,

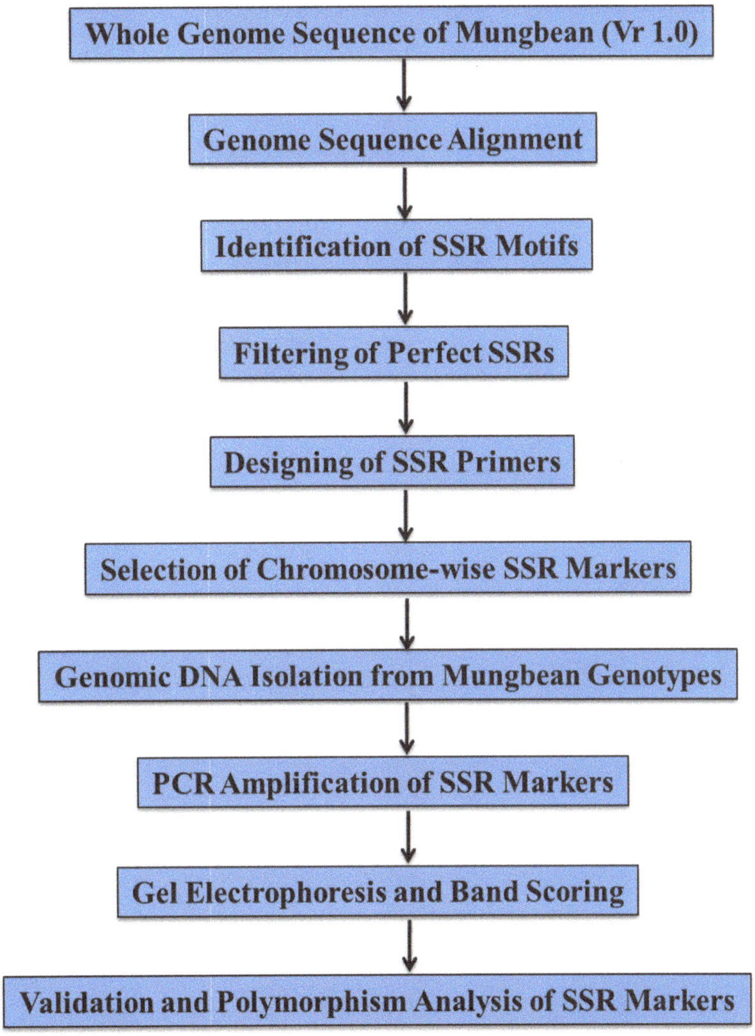

Fig. 1 Schematic representation for the development and validation of mungbean SSR markers

1. Amplicon size 100–500 bp.
2. Primer length 15–30 bp.
3. GC content 40%–60%.
4. Melting temperature (Tm) 45°C–60°C.
5. Maximum Tm difference between primer pairs 3°C–5°C.
6. Exclude stretch of mononucleotide repeats.
7. Avoid direct or inverted repeats between forward and reverse primers.
8. Avoid secondary structures such as hairpins, homo-dimers and hetero-dimers within primer pairs.

3.3 Genomic DNA Isolation (CTAB Method)

1. Isolate genomic DNA from the mungbean genotypes using the modified CTAB protocol [15].
2. Collect approximately 0.5 g leaf sample from 10-day-old seedlings and grind to a fine powder in liquid nitrogen using a pre-chilled mortar and pestle.
3. Transfer the powdered sample to a micro-centrifuge tube with pre-warmed CTAB extraction buffer supplemented with β-mercaptoethanol (freshly added before use).
4. Incubate the sample mix at 65°C for 30–45 min with intermittent mixing.
5. Centrifuge at 10,000 rpm for 10 min at 4°C and transfer the aqueous phase to a fresh tube.
6. Add equal volume of chloroform/isoamyl alcohol (24:1), mix gently, and centrifuge at 10,000 rpm for 15 min.
7. Take aqueous phase in a fresh tube and add 0.6–0.7 volume of ice-cold isopropanol.
8. Incubate the samples at −20°C for 3–4 h and centrifuge to recover the DNA.
9. Wash the DNA pellet with 70% ethanol, air-dry, and resuspend in TE buffer (pH 8.0).
10. Assess quality and quantity of the genomic DNA using a UV Bio-spectrophotometer and agarose gel electrophoresis (0.8%).
11. Adjust DNA concentration to 50 ng μL^{-1} for PCR analysis.

3.4 Validation of SSR Primers and Polymerase Chain Reaction

1. Select 50–100 SSR markers for validation using PCR analysis. To ensure uniform genome coverage, select a minimum of five SSR markers per chromosome across all 11 mung bean chromosomes.
2. Prepare 20 μL reaction volume that contains 100 ng of genomic DNA, 0.5–1.0 μM each of forward and reverse primers, 10 μL

of 2× PCR master mix, and make up to the appropriate volume with nuclease-free water.

3. Perform PCR amplifications in a thermal cycler using the program in Table 1.
4. Load the PCR-amplified products on a 3% agarose gel placed in a gel electrophoresis tank containing 1× TBE running buffer, along with a suitable DNA ladder for size reference. Resolve the SSR fragments at 80 V for 2–3 h and then visualize the resulting banding patterns using a UV transilluminator system.

3.5 Polymorphism Scoring and Genetic Diversity Analysis

An SSR primer is considered polymorphic when it produces two alleles per locus in diploid species, whereas polyploid species may exhibit more than two alleles per locus. Markers that produce a single allele across all genotypes are considered monomorphic/noninformative and thus excluded from analysis. Count distinct band sizes across the population and determine the number of SSR alleles per locus (*see* **Note 7**). As mungbean has a diploid genome, it is classified as an SSR marker that produces two distinct bands for a polymorphic marker (Fig. 2).

Score polymorphism of the SSR marker and use it for genetic diversity analysis as described below.

1. SSR allele scoring: Examine the gel banding pattern obtained for each SSR primer and identify alleles based on their amplicon size (bp). For each genotype and each primer, score the presence of a specific band as "1" and its absence as "0." Include only those bands that appear distinct, clear and reproducible for scoring.
2. Construct a binary data matrix {0,1} for polymorphism analysis.

Table 1
PCR thermal cycling conditions were followed for SSR genotyping

Steps	Stages	Temperature	Time
1	Initial denaturation	94°C	3 min
2	Denaturation	94°C	1 min
3	Annealing	55°C–60°C	1 min
4	Extension	72°C	1 min
5	Repeat steps 2–4 for 35 cycles		
6	Final extension	72°C	10 min
7	Hold	12°C	∞

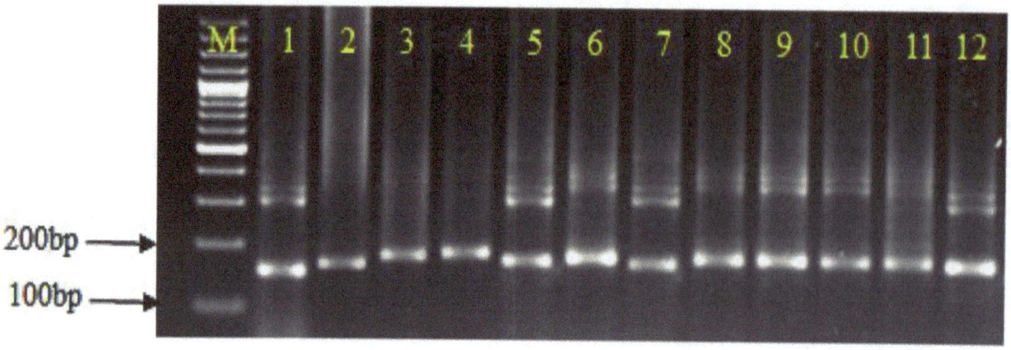

M- Marker 100 bp; 1-12: VRM (Gg) 1, HG 22, CO (Gg) 6, CO (Gg) 7, CO (Gg) 8, VBN (Gg) 2, VBN (Gg) 3, ADT3, Tenkasi, Paiyur-1, Srivilliputhur and Kovilpatti.

Fig. 2 SSR marker screening in mungbean genotypes (lanes 1–12). The presence of two distinct amplicons (between 100 and 200 bp) indicates polymorphism/allelic variations among the genotypes tested (Source: [16])

3. Calculate polymorphic information content (PIC) value for each SSR primer(locus) to measure marker informativeness and discriminatory power using the formula proposed by Botstein et al. [5]:

$$PIC = 1 - \sum_{i=1}^{n} (p_i^2)$$

Where

p_i represents the frequency of the ith allele
n represents the total number of alleles at that locus.

Higher PIC values indicate greater polymorphism and higher discriminatory power, while lower PIC values indicate poorly informative or nearly monomorphic markers (*see* **Note 7**).

4. For genetic diversity analysis, use the binary data matrix to calculate genetic similarity among genotypes using appropriate correlation, association or genetic distance coefficients such as Jaccard, Dice, or Nei. Subject the resulting similarity matrix to hierarchical cluster methods, including "Unweighted Pair Group Method with Arithmetic Mean (UPGMA)", "Neighbor-Joining (NJ)," or Ward's method to visualize genetic relationships among the genotypes in the form of dendrograms/phylogenetic trees. All analyses can be performed using statistical software such as "NTSYS-pc," or "DARwin."

4 Notes

1. CTAB extraction buffer (2%): To prepare 500 mL of CTAB buffer, mix 50 mL of 1 M Tris-HCl (pH 8.0), 20 mL of 0.5 M EDTA, 140 mL of 5 M NaCl, and 10 g CTAB.Mix thoroughly and adjust the volume with distilled water.
2. TE buffer (1×): To prepare 200 mL of 1× TE buffer, mix 2 mL of 1 M Tris–HCl (pH 8.0) and 0.4 mL of 0.5 M EDTA. Mix thoroughly and adjust the volume with distilled water.
3. TBE buffer (1×): To prepare 1000 mL of 10× TBE buffer, dissolve 108 g of Tris-base, 55 g of boric acid, and 40 mL of 0.5 M EDTA in distilled water. Mix thoroughly and adjust the volume with distilled water. Dilute one part of 10× TBE buffer with nine parts of distilled water to obtain 1× TBE running buffer.
4. Agarose gel (0.8% or 3%): To prepare 100 mL of agarose gel, dissolve an appropriate amount of agarose (0.8 g for 0.8% and 3 g for 3%) in 100 mL of 1× TBE buffer. Heat the solution until the agarose dissolves completely, then cool to 50–60°C. Add 4 μL of 5 mg/mL ethidium bromide solution to the gel and mix thoroughly. Pour the gel mix into a casting tray with a comb and allow it to solidify. After solidification, remove the comb and use the gel for electrophoresis.
5. A perfect SSR repeat refers to a DNA sequence in which a single motif is repeated continuously without any interruptions, while an imperfect SSR repeat contains base mismatches/interruptions such as base substitutions, additions, or deletions within the repeat regions.
6. Perfect SSR motifs are more susceptible to errors during DNA replication, as DNA polymerase slippage frequently leads to variation in the number of repeat units. This results in a higher mutation rate and greater polymorphism among genotypes. On the contrary, imperfect SSR motifs contain pre-existing base interruptions that stabilize the repeat region and reduce ligation errors, resulting in lower variability and lesser polymorphism. In addition, perfect SSRs with short motif lengths (di-, tri-, and tetra-nucleotide repeats) exhibit high mutation rates and greater variability, whereas longer motifs are highly conserved. Hence, perfect SSRs with shorter motif lengths are more informative and the preferred choice for SSR marker development, particularly for genetic diversity analysis and gene mapping studies. These SSR markers produce sharp, distinct banding patterns that facilitate accurate genotyping and reliable allele discrimination.

7. SSR alleles are identified based on the number of PCR amplicons observed by gel electrophoresis. Each unique amplicon/distinct band size represents an allele. The number of distinct bands observed across all genotypes determines the total number of alleles per locus. During SSR genotyping, the diploid organisms show up to two alleles (bands)per locus, while polyploid organisms exhibit more than two bands/alleles due to multiple homologous chromosome sets.

References

1. Karthikeyan A, Shobhana VG, Sudha M, Raveendran M, Senthil N, Pandiyan M, Nagarajan P (2014) Mungbean yellow mosaic virus (MYMV): a threat to green gram (*Vigna radiata*) production in Asia. Int J Pest Manag 60(4):314–324
2. Fujino K, Hirayama Y, Kaji R (2019) Marker-assisted selection in rice breeding programs in Hokkaido. Breed Science 69(3):383–392
3. Powell W, Machray GC, Provan J (1996) Polymorphism revealed by simple sequence repeats. Trends Plant Sci 1:215–222. https://doi.org/10.1016/1360-1385(96)86898-1
4. Wang Z, Weber JL, Zhong G, Tanksley SD (1994) Survey of plant short tandem DNA repeats. Theor Appl Genet 88(1):1–6
5. Botstein D, White RL, Skolnick M, Davis RW (1980) Construction of a genetic linkage map in man using restriction fragment length polymorphisms. Am J Hum Genet 32:314–331
6. Tangphatsornruang S, Somta P, Uthaipaisanwong P, Chanprasert J, Sangsrakru D, Seehalak W, Sommanas W, Tragoonrung S, Srinives P (2009) Characterization of microsatellites and gene contents from genome shotgun sequences of mungbean [*Vigna radiata* (L.) Wilczek]. BMC Plant Biol 9: 137
7. Kitsanachandee R, Somta P, Chatchawankanphanich O, Akhtar KP, Shah TM, Nair RM, Bains TS, Sirari A, Kaur L, Srinives P (2013) Detection of quantitative trait loci for mungbean yellow mosaic India virus (MYMIV) resistance in mungbean [*Vigna radiata* (L.) Wilczek] in India and Pakistan. Breed Sci 63:367–373
8. Kaewwongwal A, Chen J, Somta P, Kongjaimun A, Yimram T, Chen X (2017) Novel alleles of two tightly linked genes encoding polygalacturonase-inhibiting proteins (VrPGIP1 and VrPGIP2) associated with the Br locus that confer bruchid (*Callosobruchus*spp.) resistance to mungbean (*Vigna radiata*) accession. V2709. Front Plant Sci 8:1692
9. Saini P, Sirari A, Gnanesh BN, Mandahal KS, Ludhar NK, Nagpal S, Patel SAH, Akhatar J, Saini P, Pratap A, Bains TS (2023) Assessment of simple sequence repeat (SSR) markers derived from whole-genome sequence (WGS) of mungbean (Vigna radiata L. Wilczek): cross-species transferability and population genetic studies in Vigna species. Horticulturae 10(1):34
10. Afroz M, Teli I, Anakali HK, Rao VT, Hari Y, Sridhar V (2025) Genetic Diversity analysis in mungbean (*Vigna radiata L.*) genotypes using SSR markers. J Adv Biol Biotechnol 28(10):1842–1848
11. Kang YJ, Kim SK, Kim MY, Lestari P, Kim KH, Ha BK, Shim S (2014) Genome sequence of mungbean and insights into evolution within *Vigna species*. Nat Commun 5:5443
12. Jasrotia RS, Yadav PK, Iquebal MA, Bhatt SB, Arora V, Angadi UB, Tomar RS, Jaiswal S, Rai A, Kumar D (2019) VigSatDB: genome-wide microsatellite DNA marker database of three species of Vigna for germplasm characterization and improvement. Database 2019:baz055
13. Kearse M, Moir R, Wilson A, Stones-Havas S, Cheung M, Sturrock S, Buxton S, Cooper A, Markowitz S, Duran C, Thierer T, Ashton B, Meintjes P, Drummond A (2012) Geneious basic: an integrated and extendable desktop software platform for the organization and analysis of sequence data. Bioinformatics 28(12):1647–1649
14. Rozen S, Skaletsky H (2000) Primer 3 on the WWW for general users and for biologist programmers. In: Bioinformatics methods and protocols. Humana Press, Totowa, NJ, pp 365–386
15. Sudha M (2009) DNA isolation protocol for Vigna radiata with free of phenolics. Nat Protoc.167. https://doi.org/10.1038/nprot.2009.167
16. Mayalagu KM, Adhimoolam K, Nallathambi J, Rajagopalan VR, Balasubramanian M, Shihabdeen MNS, Natesan S, Muthurajan R, Manickam S (2023) Development of genome-wide simple sequence repeat markers from whole-genome sequence of mungbean (*Vigna radiata*). Legum Res-An Int J 1:6

Chapter 21

SNP Discovery of 3 K RICE Panel for Specific Traits

Veera Ranjani Rajagopalan, R.C. Sivabharathi, Raveendran Muthurajan and Sudha Manickam

Abstract

Rice is the most important staple food crop cultivated by millions of small-holder farmers. The effective utilization of extensive genetic diversity leads to sustainable rice production. The 3000 Rice Genomes Project is a publicly accessible giga-dataset of genomic sequences with an average coverage depth of 14×, derived from 3000 rice accessions that represent global genetic and functional diversity. Seeds of these accessions are conserved in the International Rice Genebank Collection, and these resources constitute an unprecedented platform for advancing rice science and modern breeding strategies. A key challenge with the available dataset is the need for comprehensive, systematic mining to associate genomic variation with functional traits. SNP discovery from the 3 K rice panel for specific traits provides a robust foundation for dissecting genetic diversity, population structure, and evolutionary relationships in cultivated rice, ultimately enabling the creation of new varieties to support the world population.

Keywords Amino acid, Genome, Rice, Sequence, Variations

1 Introduction

Rice is one of the most important staple food crops globally, feeding half of the world population. Thus, improving grain yield by utilizing the recent advances in genomics and allied disciplines, including high-throughput sequencing, bioinformatics, and integrative OMICS technologies, renders translational research in plants [1, 2]. Over the past decades, various high-throughput OMICS platforms have been established and extensively applied in rice research, facilitating the functional characterization of key genes [3]. Approximately 2296 genes associated with various traits have been cloned and functionally validated. These genes include grain yield (189 genes), fertility and reproductive development (174 genes), growth and development (513 genes), nutrient-use efficiency (207 genes), phytohormone biosynthesis and signaling pathways (472 genes), floral organ development and heading date (276 genes), grain quality traits (63 genes), and stress responsiveness (367 genes) [4].

Nitish Kumar (ed.), *Molecular Markers in Plants*, Methods in Molecular Biology, vol. 3097,
https://doi.org/10.1007/978-1-0716-5663-1_21,

Single-nucleotide polymorphisms (SNPs) are the predominant form of genetic variation in the rice genome and serve as highly informative markers for assessing genetic diversity, trait-associated variation, and population structure. Recent advances in next-generation sequencing technologies (NGS) have enabled genome-wide SNP discovery through whole-genome sequencing, thereby revolutionizing the dissection of complex traits in rice and other crops [5].

The 3 K rice genome Project, coordinated by the International Rice Research Institute (IRRI), represents sequence data from 3024 rice accessions from 89 countries, encompassing five varietal groups, *viz., indica, tropical japonica, temperate japonica, boro/aus,* and *basmati/sadri.* This dataset provides a robust foundation for large-scale identification of novel alleles associated with important rice phenotypes using diverse bioinformatics and genetic approaches. It also enables us to understand the genomic diversity within Oryza sativa in a comprehensive, high-resolution manner.

This chapter presents SNP discovery using the 3 K rice genomes panel to dissect trait-specific genetic variation, describing the genomic resources, experimental materials, analytical pipelines, and key methodological considerations involved. This chapter also emphasizes integrating high-density SNP datasets with high-quality phenotypic information to enable robust candidate gene identification, allele mining, and the effective application of genomics-assisted approaches for rice improvement.

2 Materials

2.1 Fun Rice Genes Database

- The **FunRiceGenes** database is a comprehensively annotated repository of *functionally characterized rice genes*, established to advance functional genomics research and genomic –assisted crop improvement in *Oryza sativa.*
- The database was initially developed by systematically integrating gene function data from major genomic resources, including Oryzabase, PubMed, and the China Rice Data Center [6].
- The initial dataset encompassed approximately 2800 functionally characterized genes and nearly 5000 members of diverse gene families, representing nearly 19.2% of the annotated protein-coding genes in the rice genome [6, 7].
- Over time, the database has included more than 4100 experimentally validated genes and more than 6000 gene family members, providing valuable information for comparative genomic analyses, gene interaction networks, haplotype identification, and translational studies across rice and other cereal crops [7].
- Furthermore, FunRiceGenes provides a user-friendly web interface with advanced search functions, downloadable datasets, and

cross-linking to external databases, substantially enhancing its utility for molecular researchers to focus on agronomic trait improvement.

2.2 Rice SNPseek Database

- A total of 3000 rice germplasm accessions were selected for sequencing, comprising 2466 accessions retrieved from the International Rice Genebank Collection (IRGC) conserved at the International Rice Research Institute (IRRI), and 534 accessions sourced from the China National Crop Gene Bank (CNCGB) at the Institute of Crop Sciences, Chinese Academy of Agricultural Sciences (CAAS).
- These 3000 rice germplasm were from 89 different countries, categorized as 77.1% from centers of rice genetic diversity, 33.9% from Southeast Asia, 25.6% from South Asia and 17.6% from China (Fig. 1a). They have performed the characterization of single nucleotide polymorphism in the 3000 rice genomes by aligning with reference *japonica* Nipponbare genome for the 55,986 genes and have identified 2,043,797 nonsynonymous SNPs in all the 12 chromosomes of the rice genome.
- The project represents a coordinated international effort to analyze and mine the dataset to establish a robust foundation in rice genetics and breeding. The database provides the 3024 rice varieties, high-quality variants, and gene loci [8] (Fig. 1b).

2.3 Sequence Retrieval RFGB Database

- The rapid advancement of high-throughput sequencing technologies and large-scale phenotyping platforms has generated a massive set of genomic, transcriptomic and phenotypic datasets in rice.
- While only a few public repositories provide access to these genome assemblies, this lack of integrated resources linking gene function to agronomically important traits remains a bottleneck.
- The RFGB database was developed to address this critical gap by integrating functional genomics with breeding-oriented datasets. The RFGB database provides the Whole-genome resequencing datasets (3 K Rice Genomes Project), Reference genome assemblies of *Oryza sativa* subspecies, transcriptomic datasets from RNA-seq and microarray, and mutant libraries and gene-edited populations.
- For SNP discovery of 3 K rice genome panel, we retrieve the sequence from the RFGB database (https://rfgbv2.rmbreeding.cn/)

2.4 Sequence Alignment Tools

The sequence of all the genotypes for the gene of interest of the specific trait is retrieved, and then those sequences are aligned with multiple sequence alignment tools (Table 1).

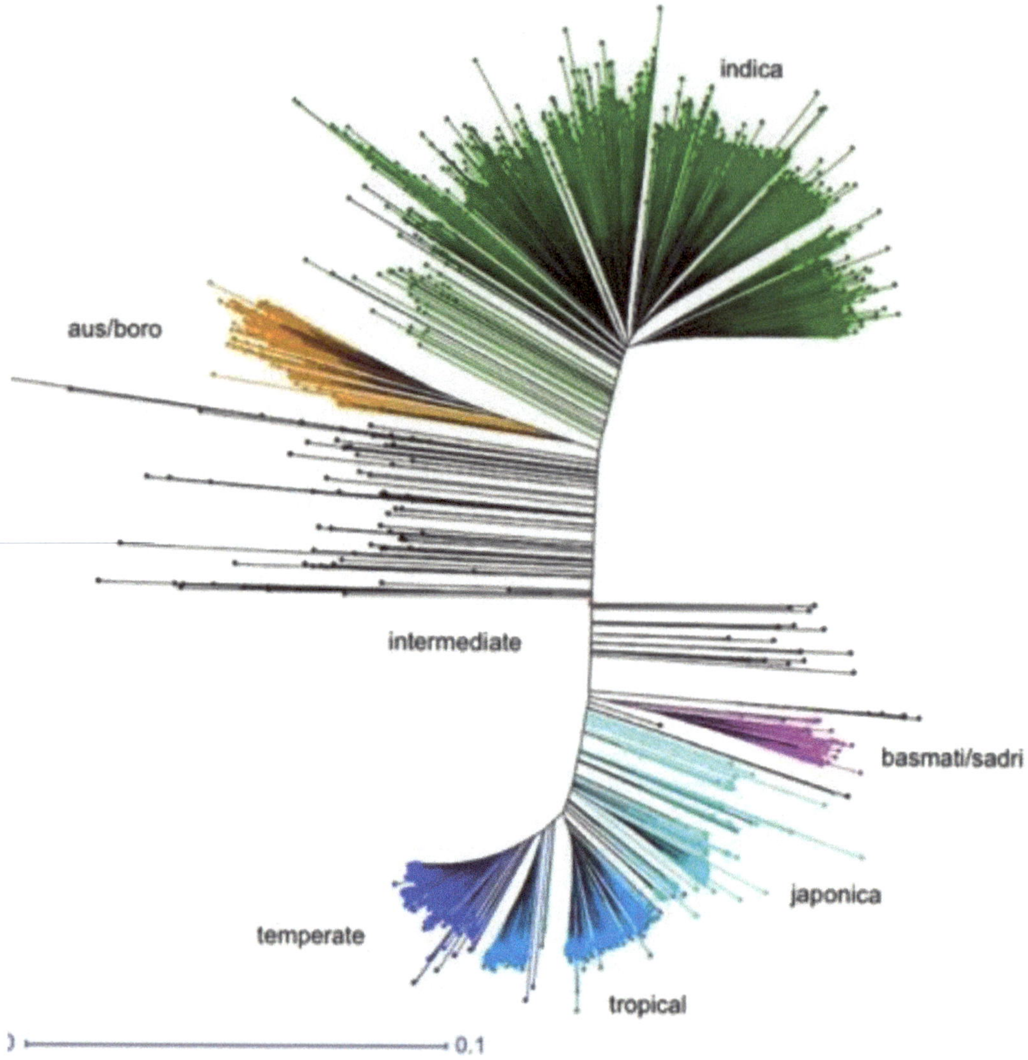

Fig.1 **(a)** represents 3 K rice panel subpopulation; (**b**) Rice SNP database SNP variations publicly available [8]

3 Methods

- The first step in the SNP discovery of 3 K rice panel is choosing the particular gene for the specific trait. It can be done either by a literature survey of metaQTL analysis to get candidate genes or by GWAS and selecting functionally validated genes available in public sources. The Michigan State University Locus ID from the Fun Rice Gene Database was retrieved for the Fun Rice Database (*see* **Note 1**).
- Sequence retrieval of the already selected 3K rice panel for the specific trait of the gene was done from the RFGB

Table 1
List of sequence alignment tools

S. no.	Tools	Aligns	Function	References
1	ClustalW/Clustal Omega	DNA/Protein	Progressive alignments and guide trees	Thompson et al. [9]
2	MUSCLE	DNA/Protein	High-accuracy and fast	Edgar [10]
3	MAFFT	DNA/Protein	Good for large datasets and diverse sequences	Katoh et al. (2005)
4	DIALIGN	DNA/Protein	Useful when sequences have conserved regions separated by unrelated segments	Morgenstern et al. [11]
5	T-COFFEE	DNA/Protein	Combines local and global alignment information	Notredame et al. [12]
6	FSA	DNA/Protein	Large number of sequences with improved accuracy on divergent data.	Schwartz [13]
7	ProbCons	Protein	Probabilistic consistency-scoring well in protein	Mahabhashyam et al. (2005)
8	POA	DNA/Protein	Graph-based alignment approach	Lee et al. [14]

(rice functional genomics and breeding) database, and the Nipponbare sequence can be downloaded from the Ensembl Plants database that describes the details of the gene (*see* **Notes 2 and 3**).

- Once sequences are retrieved, multiple sequence alignment has been done through the BioEdit software. In Bioedit output, we will get a consensus sequence that describes the most common nucleotide or amino acid aligned among all the sequences (*see* **Notes 4 and 5**).
- The variations in exonic regions were only considered since these regions only cause protein change. Amino acid change for the corresponding variations has to be identified through the Expasy Translate tool in order to discriminate between synonymous and nonsynonymous SNPs. The nonsynonynous SNPs were only considered for the further crop breeding program. Figure 2 represents the overall procedure involved in the discovery of SNPs in 3 K rice genome panel with publicly available genomic data (*see* **Notes 6 and 7**).

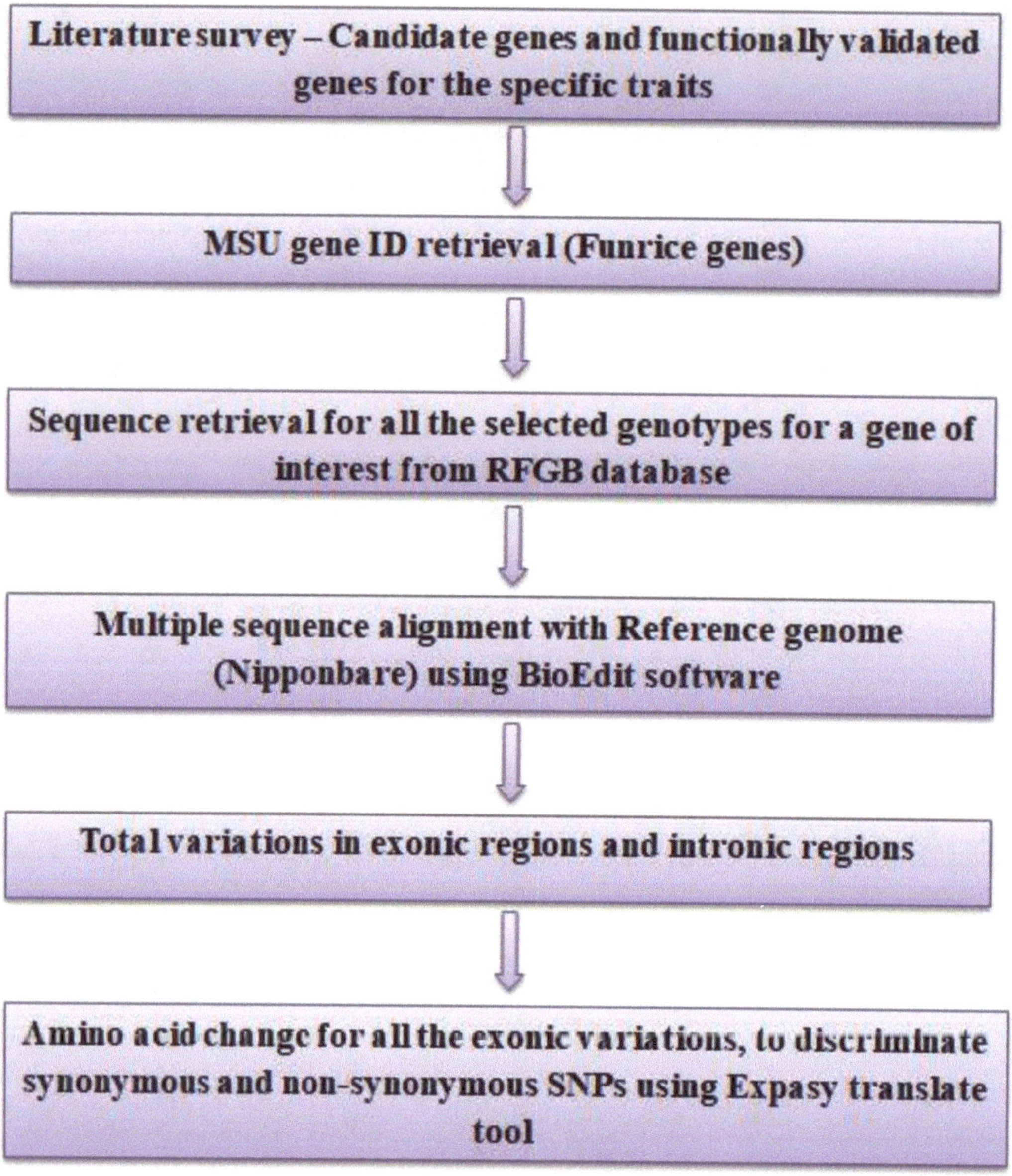

Fig. 2 Overall procedure of SNP discovery of 3 K Rice panel

3.1 Candidate Gene Identification Through Literature Survey for Specific Traits

- Genes with putative function are referred to as candidate genes, regulate the developmental processes of the trait of interest either directly or indirectly, and those were validated with genetic variants through association analysis [15]. The selection of candidate genes governing the trait of interest plays a critical role in the identification and selection of superior individuals for a crop improvement program [16, 17].

- A comprehensive literature survey was conducted to consolidate QTLs for selected traits, followed by meta-QTL analysis [18] to identify consensus genomic regions associated with these traits. Kumari et al. [19] reported 64 meta-QTLs identified through meta-analysis of 330 projected QTLs. Subsequently, identified QTLs with at least two overlapping QTLs grouped on a consensus genetic map were considered robust mQTLs [20]. Table 2 summarizes the diverse genes associated with specific traits in the rice crop, facilitating comparison against the reference genome sequence to identify the SNP variants that were derived from the Rice SNPseek database.

3.2 Sequence Retrieval, Alignment, and Amino Acid Variations

- The sequences of all the selected genotypes of 3 K panel were retrieved from the RFGB database and aligned with the Nipponbare sequence with the help of Bioedit software. We can derive the SNP variants at exonic, intronic, 5′UTR and 3′ UTR regions. Since the variations in exonic regions can only change the protein products, the variations in these regions were prioritized [63].
- The details of a gene, such as the number of exons, introns, 5′ UTR, 3′UTR, and transcript details, are available in Ensembl plants. It is developed in collaboration with international partners in the Gramene and transPLANT projects. Ensembl plants provide genome-scale information for nearly 39 sequenced plant species (Bolser et al. 2017). Figure 3b represents the SNP variation for *OsTB1* gene in chromosome 3 from position 28428504 to 28430462. The amino acid change for all SNP variations was identified using the Expasy Translate tool to differentiate synonymous and nonsynonymous SNPs. Figure 3c represents SNP variation (C/A) at 28429181, which leads to an amino acid change at the 58th position due to a frameshift mutation; as a result, alanine (Nipponbare) changes to asparagine (3K rice genotype).

4 Notes

1. The selection of the gene for the trait of interest could be made either by the candidate gene approach or through a literature survey, from which we will get the functionally validated genes.
2. The locus ID of the 3K panel sequences should be in MSU ID format to retrieve the sequence from the RFGB database, and IRIS ID format should be in the following format: IRIS_ID.
3. The Nipponbare sequence could be downloaded from the Ensembl Plants website. While downloading the Nipponbare sequence, we should download it in accordance with our 3 K

Table 2
List of candidate genes for different traits through a literature survey in rice

Traits	Gene name	Locus ID	SNPs	References
Plant height	*OsWRKY21*	LOC_Os01g60640	3	Zhang et al. [21]
	GA2OX3	LOC_Os01g55240	1	Qin et al. [22]
	SD1	LOC_Os01g66100	2	Cao et al. [23]
	OsBRI1	LOC_Os01g52050	4	Lan et al. [24]
	CCD8	LOC_Os01g54270	5	Alder et al. [25]
	DTH8	LOC_Os08g07740	7	Wei et al. [26]
	Ghd7	LOC_Os07g15770	7	Weng et al. [27]
	SP1	LOC_Os02g44590	1	Li et al. [28]
Number of tillers	*OsCCD8*	LOC_Os01g54270	5	Zou et al. [29]
	OsTB1	LOC_Os03g49880	2	Guo et al. [30]
	OsSPL14	LOC_Os08g39890	3	Luo et al. [31]
	OsMOC1	LOC_Os06g40780	2	Liu et al. [32]
	OsD53	LOC_Os11g01330	3	Patil et al. [33]
	OsMADS57	LOC_Os02g49840	2	Guo et al. [30]
Panicle length	*OsSPL14*	LOC_Os08g39890	3	Miura et al. [34]
	DEP1	LOC_Os09g26999	5	Li et al. [35]
	OsSPL4	LOC_Os02g07780	2	Hu et al. [36]
	LRK1	LOC_Os02g05980	27	Zha et al. [37]
	APO1	LOC_Os06g45460	2	Huang et al. [38]
Grain number per panicle	*APO1*	LOC_OS06G45460	2	Huang et al. [38]
	GHD7	LOC_OS07G15770	7	Weng et al. [27]
	MOC1	LOC_OS06G40780	2	Zhang et al. [21]
	OsSPL14(imp)	LOC_OS08G39890	3	Miura et al. [34]
Grain size	*GW7*	LOC_Os07g41200	4	Ye et al. [39]
	OsSPL13	LOC_Os07g32170	3	Si et al. [40]
	OsSPL16	LOC_Os08g41940	6	Wang et al. [41]
	DEP2	LOC_Os07g42410	3	Abe et al. [42]
	DTH8	LOC_Os08g07740	7	Sun et al. [43]
	Ghd7	LOC_Os07g15770	5	Weng et al. [27]
	LRK1	LOC_Os02g05980	27	Zha et al. [37]
	SP1	LOC_Os11g12740	11	Li et al. [28]

Table 2 (Continued)

Traits	Gene name	Locus ID	SNPs	References
Blast	*Pid3/Pi25*	LOC_Os06g22460	16	Lv et al. [44]
	Pi36	LOC_Os08g05440	119	Liu et al. [45]
	Pi54	LOC_Os11g42010	20	Arora et al. [46]
	Pi2/Pi9	LOC_Os06g17900	25	Su et al. [47]
	Pid2	LOC_OS06G29810	3	Chai et al. [48]
BLB	*Xa10*	LOC_Os11g37620	16	Wang et al. [49]
	Xa21	LOC_Os11g35500	4	Peng et al. [50]
	Xa25	LOC_Os12g29220	10	Zaka et al. [51]
	XB24	LOC_Os01g56470	2	Chen et al. [52]
	Xa1	LOC_Os04g53120	87	Yoshimura et al. [53]
Drought	*OsCIPK12*	LOC_OS01G55450	3	Gao et al. [54]
	OsDREB1C	LOC_Os06g03670	5	Li et al. (2013)
	OsNHK1	LOC_OS07G47100	4	Hyun [55]
	OsSIK1	LOC_OS06G03970	5	Ouyang et al. [56]
Submergence	*SUB1B*	LOC_OS09G11480	7	Singh et al. [57]
	SUB1C	LOC_OS09G11460	14	Singh et al. [57]
Grain quality	*Chalk5*	LOC_Os05g06480	11	Li et al. [58]
	FLO2	LOC_Os04g55230	8	Suzuki et al. [59]
	SSIIIa	LOC_Os08g09230	28	Zhao et al. [60]
	SSIIa	LOC_OS06G12450	7	Fan et al. [61]
	TGW6	LOC_OS06G41850	3	Zhang et al. [62]

panel sequences. Example, If we download UTR, exonic and intronic regions for 3 K panel, same should be followed for Nipponbare.

4. Multiple sequence alignment (MSA) could be done for the files in FASTA format, or it could be in Notepad format to run in the BIOEDIT tool.
5. We should only consider positions that are not showing consensus as SNPs or INDELs for easy identification in the MSA output.
6. The amino acid change in 3 K panel could be performed using CLUSTAL omega website. While feeding the sequences of Nipponbare and 3K sequence, labeling should be given in accordance with the requirements of the website.
7. Care should be taken to find the position of mutation and the type of mutation caused by the SNPs and INDELs.

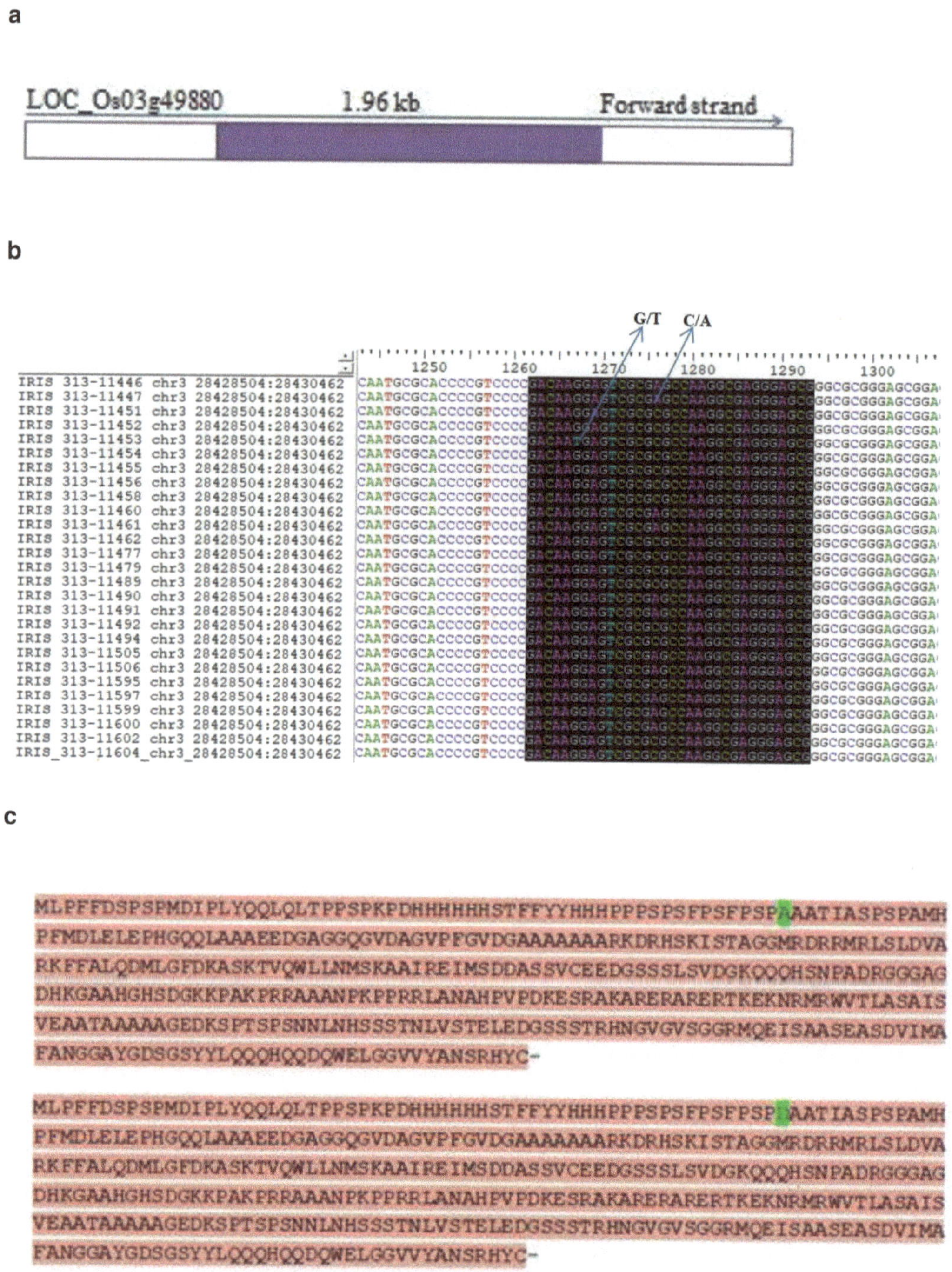

Fig. 3 (**a**) represents *OsTB1* gene with one exon with genome size 1.96 kb and (**b**) represents the SNP variation at 1267 (G/T) and 1276 (C/A) for the gene *OsTB1* in 3 K rice panel (representation) using BioEdit MSA tool. (**c**) Amino acid change of a SNP of *OsTB1* gene

References

1. Varshney RK, Terauchi R, McCouch SR (2014) Harvesting the promising fruits of genomics: applying genome sequencing technologies to crop breeding. PLoS Biol 12(6):e1001 883.3,000
2. Rice Genomes Project (2014) The 3,000 rice genomes project. Gigascience 3(1) https://doi.org/10.1186/2047-217x-3-7
3. Li Y, Xiao J, Chen L, Huang X, Cheng Z, Han B, Wu C (2018) Rice functional genomics research: past decade and future. Mol Plant 11(3):359–380
4. Wing RA, Purugganan MD, Zhang Q (2018) The rice genome revolution: from an ancient grain to Green Super Rice. Nat Rev Genet 19(8):505–517
5. Kwok PY, Chen X (2003) Detection of single nucleotide polymorphisms. Curr Issues Mol Biol 5(2):43–60
6. Yao W, Li G, Yu Y, Ouyang Y (2017) funRiceGenes dataset for comprehensive understanding and application of rice functional genes. Gigascience 7(1):gix119
7. Kumari S et al (2022) New data and new features of the FunRiceGenes (functionally characterized rice genes). Database: 2021 Update. Rice 15:23
8. Mansueto L, Fuentes RR, Borja FN et al (2017) Rice SNP-seek database update: new SNPs, indels, and queries. Nucl Acids Res 45:D1075–D1081
9. Thompson JD, Higgins DG, Gibson TJ (1994) CLUSTAL W: improving the sensitivity of progressive multiple sequence alignment through sequence weighting, position-specific gap penalties and weight matrix choice. Nucleic Acids Res 22(22):4673–4680
10. Edgar RC (2004) MUSCLE: multiple sequence alignment with high accuracy and high throughput. Nucleic Acids Res 32(5):1792–1797
11. Morgenstern B, Koehler S, Stoye J et al (1998) DIALIGN: finding local similarities by multiple sequence alignment. Bioinformatics 14(4):290–294
12. Notredame C, Higgins DG, Heringa J (2000) T-COFFEE: a novel method for fast and accurate multiple sequence alignment. J Mol Biol 302(1):205–217
13. Schwartz AS, Pachter L (2007) Multiple alignment by sequence annealing. Bioinformatics 23:e24–e29
14. Lee C, Grasso C, Sharlow MF (2002) Multiple sequence alignment using partial order graphs. Bioinformatics 18(3):452–464
15. Zhu M, Zhao S (2007) Candidate gene identification approach: progress and challenges. Int J Biol Sci 3(7):420
16. Ikeda M, Miura K, Aya K, Kitano H, Matsuoka M (2013) Genes offering the potential for designing yield-related traits in rice. Curr Opin Plant Biol 16(2):213–220
17. Sivabharathi RC, Rajagopalan VR, Suresh R, Sudha M, Karthikeyan G, Jayakanthan M, Raveendran M (2024) Haplotype-based breeding: a new insight in crop improvement. Plant Sci 346:112129
18. Goffinet B, Gerber S (2000) Quantitative trait loci: a meta-analysis. Genetics 155(1):463–473
19. Kumari A, Sharma D, Sharma P, Sahil, Wang C, Verma V, Agarwal M (2023) Meta-QTL and haplo-pheno analysis reveal superior haplotype combinations associated with low grain chalkiness under high temperature in rice. Front Plant Sci 14:1133115
20. Kumar IS, Nadarajah K (2020) A meta-analysis of quantitative trait loci associated with multiple disease resistance in rice (*Oryza sativa L.*). Plants 9(11):1491
21. Zhang J, Gu M, Liang R, Shi X, Chen L, Hu X, Xu G (2021) OsWRKY21 and OsWRKY108 function redundantly to promote phosphate accumulation through maintaining the constitutive expression of OsPHT1; 1 under phosphate-replete conditions. New Phytol 229(3):1598–1614
22. Qin X, Liu JH, Zhao WS, Chen XJ, Guo ZJ, Peng YL (2013) Gibberellin 20-oxidase gene OsGA20ox3 regulates plant stature and disease development in rice. Mol Plant Microbe Interact 26(2):227–239
23. Cao C, Guo S, Deng P, Yang S, Xu J, Hu T, Jing W (2024) The BEL1-like homeodomain protein OsBLH4 regulates rice plant height, grain number, and heading date by repressing the expression of OsGA2ox1. Plant J 119(3):1369–1385
24. Lan D, Cao L, Liu M, Ma F, Yan P, Zhang X, Luo X (2023) The identification and characterization of a plant height and grain length related gene hfr131 in rice. Front Plant Sci 14:1152196
25. Alder A, Holdermann I, Beyer P, Al-Babili S (2008) Carotenoid oxygenases involved in plant branching catalyse a highly specific conserved apocarotenoid cleavage reaction. Biochem J 416(2):289–296
26. Wei X, Xu J, Guo H, Jiang L, Chen S, Yu C, Wan J (2010) DTH8 suppresses flowering in rice, influencing plant height and yield potential simultaneously. Plant Physiol 153(4):1747–1758
27. Weng X, Wang L, Wang J, Hu Y, Du H, Xu C, Zhang Q (2014) Grain number, plant height, and heading date7 is a central regulator

of growth, development, and stress response. Plant Physiol 164(2):735–747
28. Li S, Qian Q, Fu Z, Zeng D, Meng X, Kyozuka J, Wang Y (2009) Short panicle1 encodes a putative PTR family transporter and determines rice panicle size. Plant J 58(4):592–605
29. Zou J, Zhang S, Zhang W, Li G, Chen Z, Zhai W, Zhu L (2006) The rice HIGH-TILLERING DWARF1 encoding an ortholog of Arabidopsis MAX3 is required for negative regulation of the outgrowth of axillary buds. Plant J 48(5):687–698
30. Guo S, Xu Y, Liu H, Mao Z, Zhang C, Ma Y, Chong K (2013) The interaction between OsMADS57 and OsTB1 modulates rice tillering via DWARF14. Nat Commun 4(1):1566
31. Luo L, Li W, Miura K, Ashikari M, Kyozuka J (2012) Control of tiller growth of rice by OsSPL14 and strigolactones, which work in two independent pathways. Plant Cell Physiol 53(10):1793–1801
32. Liu X, Zhang F, Xun Z, Shao J, Luo W, Jiang X, Wan J (2025) The OsNL1-OsTOPLESS2-OsMOC1/3 pathway regulates high-order tiller outgrowth in rice. Plant Biotech J 23(3): 900–910
33. Patil SB, Barbier FF, Zhao J, Zafar SA, Uzair M, Sun Y, Li X (2022) Sucrose promotes D53 accumulation and tillering in rice. New Phytol 234(1):122–136
34. Miura K, Ikeda M, Matsubara A, Song XJ, Ito M, Asano K, Ashikari M (2010) OsSPL14 promotes panicle branching and higher grain productivity in rice. Nature genetics 42(6):545–549
35. Li J, Qian Q, Liu Y (2024) DEP1 and panicle architecture: influencing rice yield through genetic modulation. Plant Gene Trait 15. https://doi.org/10.5376/pgt.2024.15.0009
36. Hu J, Huang L, Chen G, Liu H, Zhang Y, Zhang R, Ding Y (2021) The elite alleles of OsSPL4 regulate grain size and increase grain yield in rice. Rice 14(1):90
37. Zha X, Luo X, Qian X, He G, Yang M, Li Y, Yang J (2009) Over-expression of the rice LRK1 gene improves quantitative yield components. Plant Biotech J 7(7):611–620
38. Huang L, Hua K, Xu R, Zeng D, Wang R, Dong G, Li Y (2021) The LARGE2-APO1/APO2 regulatory module controls panicle size and grain number in rice. Plant Cell 33(4):1212–1228
39. Ye Y, Wang S, Ren Y, Yang H, Guo J, Jiang H, Liu B (2022) Low grain weight, a new allele of BRITTLE CULM12, affects grain size through regulating GW7 expression in rice. Front Plant Sci 13:997624
40. Si L, Chen J, Huang X, Gong H, Luo J, Hou Q, Han B (2016) OsSPL13 controls grain size in cultivated rice. Nature Genet 48(4):447–456
41. Wang S, Wu K, Yuan Q, Liu X, Liu Z, Lin X, Fu X (2012) Control of grain size, shape and quality by OsSPL16 in rice. Nature Genet 44(8):950–954
42. Abe Y, Mieda K, Ando T, Kono I, Yano M, Kitano H, Iwasaki Y (2010) The SMALL AND ROUND SEED1 (SRS1/DEP2) gene is involved in the regulation of seed size in rice. Genes Genet Syst 85(5):327–339
43. Sun K, Huang M, Zong W, Xiao D, Lei C, Luo Y, Guo J (2022) Hd1, Ghd7, and DTH8 synergistically determine the rice heading date and yield-related agronomic traits. J Genet Genomics 49(5):437–447
44. Lv Q, Xu X, Shang J, Jiang G, Pang Z, Zhou Z, Zhu L (2013) Functional analysis of Pid3-A4, an ortholog of rice blast resistance gene Pid3 revealed by allele mining in common wild rice. Phytopathology 103(6):594–599
45. Liu X, Li Y, Wang L, Liu X, Wang C, Wang L, Pan Q (2010) The effect of the rice blast resistance gene Pi36 on the expression of disease resistance-related genes. Chin Sci Bull 55(18):1881–1888
46. Arora K, Rai AK, Gupta SK, Singh PK, Narula A, Sharma TR (2015) Phenotypic expression of blast resistance gene Pi54 is not affected by its chromosomal position. Plant Cell Rep 34(1):63–70
47. Su J, Wang W, Han J, Chen S, Wang C, Zeng L, Zhu X (2015) Functional divergence of duplicated genes results in a novel blast resistance gene Pi50 at the Pi2/9 locus. Theor Appl Genet 128(11):2213–2225
48. Chai R, Wang J, Wang X, Wen J, Liang Z, Ye X, Pan Q (2022) The Pid family has been diverged into Xian and Geng type resistance genes against rice blast disease. Genes 13(5):891
49. Wang J, Tian D, Gu K, Yang X, Wang L, Zeng X, Yin Z (2017) Induction of Xa10-like genes in rice cultivar Nipponbare confers disease resistance to rice bacterial blight. MPMI 30(6):466–477
50. Peng H, Chen Z, Fang Z, Zhou J, Xia Z, Gao L, Zhang W (2015) Rice Xa21 primed genes and pathways that are critical for combating bacterial blight infection. Sci Rep 5(1):12165
51. Zaka A, Grande G, Coronejo T, Quibod IL, Chen CW, Chang SJ, Oliva R (2018) Natural variations in the promoter of OsSWEET13 and OsSWEET14 expand the range of resistance against *Xanthomonas oryzae* pv. oryzae. PLoS One 13(9):e0203711
52. Chen X, Chern M, Canlas PE, Ruan D, Jiang C, Ronald PC (2010) An ATPase promotes autophosphorylation of the pattern recognition receptor XA21 and inhibits XA21-mediated immunity. Proc Natl Acad Sci USA 107(17):8029–8034

53. Yoshimura S, Yamanouchi U, Katayose Y, Toki S, Wang ZX, Kono I, Sasaki T (1998) Expression of Xa1, a bacterial blight-resistance gene in rice, is induced by bacterial inoculation. Proc Natl Acad Sci USA 95(4):1663–1668
54. Gao C, Lu S, Zhou R, Wang Z, Li Y, Fang H, Cao Y (2022) The OsCBL8–OsCIPK17 module regulates seedling growth and confers resistance to heat and drought in rice. Int J Mol Sci 23(20):12451
55. Hyun T (2020) CRISPR/Cas-based genome editing to improve abiotic stress tolerance in plants. Botanica serbica
56. Ouyang SQ, Liu YF, Liu P, Lei G, He SJ, Ma B, Chen SY (2010) Receptor-like kinase OsSIK1 improves drought and salt stress tolerance in rice (Oryza sativa) plants. Plant J 62(2):316–329
57. Singh A, Singh Y, Mahato AK, Jayaswal PK, Singh S, Singh R, Rai V (2020) Allelic sequence variation in the Sub1A, Sub1B and Sub1C genes among diverse rice cultivars and its association with submergence tolerance. Sci Rep 10(1):8621
58. Li Z, Xi M, Xu Y, Sun X, Tu D, Zhou Y, Yang L (2025) Molecular mechanisms of grain chalkiness variation in rice panicles. Plants 14(2):244
59. Suzuki R, Imamura T, Nonaga Y, Kusano H, Teramura H, Sekine KT, Shimada H (2020) A novel FLOURY ENDOSPERM2 (FLO2)-interacting protein, is involved in maintaining fertility and seed quality in rice. Plant Biotechnol 37(1):47–55
60. Zhao Q, Ye Y, Han Z, Zhou L, Guan X, Pan G, Cheng F (2020) SSIIIa-RNAi suppression associated changes in rice grain quality and starch biosynthesis metabolism in response to high temperature. Plant Sci 294:110443
61. Fan X, Zhu J, Dong W, Sun Y, Lv C, Guo B, Xu R (2017) Comparative mapping and candidate gene analysis of SSIIa associated with grain amylopectin content in barley (*Hordeum vulgare L.*). Front Plant Sci 8:1531
62. Zhang H, Tang S, Wang H, Wang Y, Zhi H, Liu B, Diao X (2024) Genetic diversity of grain yield traits and identification of a grain weight gene SiTGW6 in foxtail millet. Theor Appl Genet 137(4):84
63. Gouda G, Gupta MK, Donde R, Kumar J, Parida M, Mohapatra T, Behera L (2020) Characterization of haplotypes and single nucleotide polymorphisms associated with Gn1a for high grain number formation in rice plant. Genomics 112(3):2647–2657
64. Huang X, Han B (2014) Natural variations and genome-wide association studies in crop plants. Annu Rev Plant Biol 65:531–551
65. Huang X, Wei X, Sang T, Zhao Q, Feng Q, Zhao Y et al (2010) Genome-wide association studies of 14 agronomic traits in rice landraces. Nature Genet 42:961–967
66. Jiang H, Liu S, Kim J et al (2011) PlantQTL-GE: a database for identifying candidate genes in plant quantitative trait loci using gene expression data. Nucleic Acids Res 39: D1114–D1118
67. Kumar A, Sandhu N, Dixit S et al (2019) Marker-assisted selection strategy to pyramid two or more QTLs for quantitative trait improvement in rice. Plant Genome 12:180115
68. Lee SC, Kim SH, Kim SR (2013) Drought inducible OsDhn1 promoter is activated by OsDREB1A and OsDREB1D. J Plant Biol 56(2):115–121
69. Li X, Tao Y et al (2014) Genetic control of rice plant architecture under domestication. Nature Genet 46:1099–1104
70. Bolser DM, Staines DM, Perry E, Kersey PJ (2016) Ensembl plants: integrating tools for visualizing, mining, and analyzing plant genomic data. In: Plant genomics databases: methods and protocols. Springer New York, New York, NY. pp 1–31
71. Wang J, Zhou L, Shi H et al (2020) A single transcription factor promotes both yield and immunity in rice. Science 361:1026–1028
72. Wang W, Mauleon R, Hu Z et al (2018) Genomic variation in 3,010 diverse accessions of Asian cultivated rice. Nature 557:43–49
73. Yano K, Yamamoto E, Aya K et al (2016) Genome-wide association study using whole-genome sequencing rapidly identifies new genes influencing agronomic traits in rice. Nature Genet 48:927–934
74. Zhao K, Tung CW, Eizenga GC et al (2011) Genome-wide association mapping reveals a rich genetic architecture of complex traits in Oryza sativa. Nat Commun 2: 467
75. Do CB, Mahabhashyam MS, Brudno M, Batzoglou S (2005) ProbCons: Probabilistic consistency-based multiple sequence alignment. Genome Res 15(2):330–340
76. Katoh K, Kuma K, Toh H, Miyata T (2002) MAFFT: a novel method for rapid multiple sequence alignment based on fast Fourier transform. Nucleic Acids Res 30(14): 3059–3066
77. Mansueto et al (2017) Rice SNP-seek database update: new SNPs, indels, and queries. Nucl Acids Res 45(D1):D1075–D1081. https://doi.org/10.1093/nar/gkw1135

Check for updates

Chapter 22

Application of Restriction Fragment Length Polymorphism (RFLP)

Muhammad Mumtaz Khan, Rashad Waseem Qadri,
Muhammad Tahir Akram, Abdul Haseeb Ahmed and Abdullah Al-Sadi

Abstract

Restriction fragment length polymorphism (RFLP) was the first reliable genetic marker used in plant molecular studies. It is a DNA-based marker that detects variation in DNA sequence at restriction enzyme sites and produces codominant, locus-specific markers that differentiate homologous and heterozygous genotypes. It provides greater accuracy than trait-based methods. The application of RFLP markers includes developing genetic linkage maps, conducting QTL analysis, characterizing germplasm, and detecting somaclonal variations. Despite being the availability of new molecular markers, RFLP still remained highly valued for its reproducibility and clear Mendelian inheritance patterns. This chapter presents optimized, nonradioactive protocols for DNA isolation, Southern blotting, DIG-labeled probe hybridization, and chemiluminescent detection, offering a practical guideline for integrating RFLP markers into modern plant breeding and genomic research.

Keywords Co-dominant inheritance, Genetic mapping, Molecular markers, QTL analysis, Southern blotting

1 Introduction

DNA sequences located at known positions of chromosomes are recognized as molecular markers. Polymorphism in genetic markers arises from mutations at genomic loci or nucleotide changes, allowing the identification of genetic variation among individuals or species [1]. Molecular markers are widely used due to their abundance and reliability. Genetic markers have made significant contributions to plant breeding and genetics, and they are continually evolving to become more effective and enable programmed genome analysis [2]. Molecular markers enable breeding programs to accelerate the breeding process while achieving goals that are difficult to accomplish through conventional breeding, such as transferring desirable characteristics from exotic germplasm to indigenous cultivars [3].

Nitish Kumar (ed.), *Molecular Markers in Plants*, Methods in Molecular Biology, vol. 3097,
https://doi.org/10.1007/978-1-0716-5663-1_22,

Molecular markers are commonly employed in breeding programs, genome mapping, marker-assisted breeding, and the evaluation of genetic diversity.

RFLP was fundamental to advancing plant molecular genetics because it provided the first reliable method for detecting DNA-level variation in plant genomes. RFLP identifies homozygous and heterozygous genotypes by employing sequence polymorphisms at loci that modify restriction enzyme recognition sites and generate codominant and locus-specific markers [4]. Such qualities overcome the limitations of phenotypic markers, which are influenced by environmental factors and plant developmental phases. RFLP was the cornerstone technique in the early years of plant genetics, as it was widely used for genetic diversity analysis, constructing linkage maps, mapping quantitative trait loci (QTLs), and identifying somaclonal variation [5]. Despite ongoing advances in PCR-based and sequencing-driven marker technologies, RFLP remains highly significant due to its foundational contributions to plant genomics and breeding [6].

2 Materials

2.1 Plant Material and Genomic DNA Isolation

2.1.1 Plant Tissue Preparation for DNA Extraction

1. **Fresh leaves:** Typically, one to two young healthy, disease-free leaves are taken; the recommended sample size varies from 0.5 to 1.0 g (fresh leaves).
2. **Silica-gel dried samples:** The ideal sample size is 100–200 mg of dried leaf tissues. Typically, 1 g of fresh leaf tissues is placed in a polyethylene envelope, which is subsequently dried to yield 100–200 mg dry sample. In general, a fully dried material is preferred for DNA isolation.
3. **CTAB-preserved samples:** This method involves immersing 0.5–1.0 g of fresh leaf tissues in CTAB buffer at a 1:10 tissue-to-buffer volume ratio.
4. **Freeze-dried (lyophilized) samples:** In this method, 50–100 mg of lyophilized tissue are used. Lyophilization retains tissue structure exceptionally well and uses less material than other drying techniques. Lyophilized tissues are easily powdered using a mortar and pestle.
5. **Herbarium/press-dried specimens:** DNA isolation requires 20–50 mg of dried tissues; however, the presence of certain inhibitory substances often compromises DNA purity. To obtain viable DNA, tissue is commonly collected from the midrib or a major vein, as these regions yield higher DNA content compared to the thin lamina.
6. **Liquid nitrogen:** This technique is primarily used with fresh and lyophilized leaf samples to flash-freeze tissues and inhibit

DNase activity. The methods of plant tissue conservation are illustrated in Fig. 1.

2.1.2 Reagents for DNA Extraction

- **CTAB extraction buffer:** The buffer is prepared immediately before use by mixing 100 mM Tris-HCl (pH adjusted to 8.0), thoroughly with 1.4 M NaCl, 20 mM EDTA, 2% CTAB, 0.2% β -mercaptoethanol, and 1% polyvinylpyrrolidone.
- Isopropanol and 70% ethanol, chilled to −20°C (*see* **Note 1**).
- **1× TE buffer:** 10 mM Tris-HCl (pH 8.0), 1 mM EDTA (pH 8.0).
- **RNase A solution:** 10 mg/mL RNase A in 10 mM Tris-HCl (pH 7.5), 15 mM NaCl; heat-inactivated before use.

2.2 Restriction Enzyme Digestion

- High-purity genomic DNA (5–10 μg per reaction).
- Selected restriction endonuclease(s) and matching 10× buffer.
- Nuclease-free water to adjust reaction volume.

2.3 Agarose Gel Electrophoresis and Southern Blotting

- Electrophoresis-grade agarose.
- **10× TBE buffer:** (890 mM Tris base, 890 mM boric acid, 20 mM EDTA, pH 8.0).
- Ethidium bromide solution, gel loading dye, and a DNA Molecular Weight Marker.

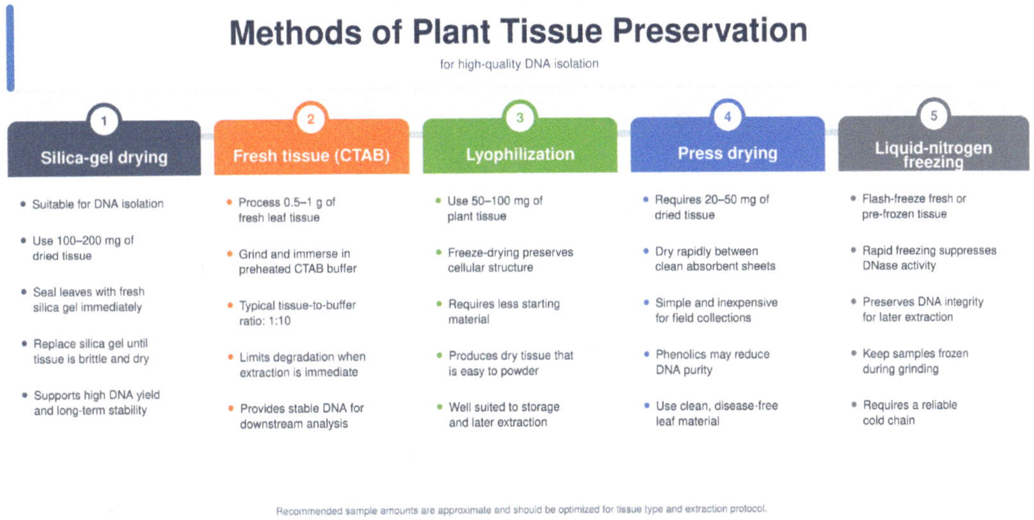

Fig. 1 Different methods are used in plants to preserve tissue for DNA isolation

- **Depurination solution:** 0.25 M hydrochloric acid.
- **Denaturation solution:** 0.5 M sodium hydroxide, 1.5 M sodium chloride.
- **Neutralization solution:** 0.5 M Tris-hydrochloric acid (pH 7.5), 1.5 M sodium chloride.
- **20× SSC stock** (3 M sodium chloride, 0.3 M trisodium citrate dihydrate, pH 7.0); diluted as needed.
- Whatman paper, absorbent paper towels, and a UV crosslinker.

2.4 Probe Labeling, Hybridization, and Detection

- **Probe template:** 25–50 ng of purified amplicon or plasmid DNA.
- **DIG-based random-primed labeling kit** (e.g., DIG High Prime, Roche, or comparable commercial non-radioactive system).
- **Hybridization solution:** 5× SSC, 0.1% N-lauroylsarcosine, 0.02% SDS, 1% blocking reagent.
- **Low-stringency rinse buffer (Buffer I):** 2× SSC and 0.1% SDS.
- **High-stringency rinse buffer (Buffer II):** 0.5× SSC and 0.1% SDS.
- **Maleic acid buffer:** 0.1 M maleic acid, 0.15 M sodium chloride, and buffer pH adjusted to 7.5.
- **Blocking solution:** 1% blocking reagent dissolved in maleic acid buffer.
- **Antibody conjugate:** Antibody conjugate: Anti-digoxigenin–alkaline phosphatase (Fab fragments), used at a 1:10,000 dilutions in blocking solution.
- **Detection buffer:** 0.1 M Tris-HCl, 0.1 M NaCl, pH 9.5.
- **Chemiluminescent substrate:** CDP-Star (ready-to-use) (*see* **Note 2**).

3 Methods

RFLP analysis requires the purification of high-molecular-weight genomic DNA, the hydrolysis of a restriction endonuclease specific to a sequence, the separation of the resulting fragments by electrophoresis, and the transfer of the fragments onto a charged nylon support. The subsequent procedures involve post-hybridization washing, chemiluminescent signal detection, optimized stringency annealing, and the development of a matching probe with nonisotopic tagging [7,8].

3.1 Isolation of Genomic DNA with a Modified CTAB Procedure

A cetyltrimethylammonium bromide (CTAB)-based technique was employed to obtain high-molecular-weight DNA suitable for restriction enzyme digestion, with slight modifications [9, 10]. The steps of the protocol are listed below.

Tissue homogenization: Tissue homogenization involves meticulously grinding a fresh leaf sample weighing 0.5–1.0 g in liquid nitrogen in a pre-cooled mortar and pestle into fine powder.

Lysis: The fine powder is transferred to a 15 mL tube containing 5 mL of CTAB extraction buffer pre-warmed to 65°C and shaken vigorously.

Incubation: The suspension is incubated at 65°C for 45–60 min, with gentle inversions every 10–15 min.

Organic removal: The suspension is removed from the incubator and allowed to cool at ambient temperature, and chloroform: isoamyl alcohol (5 mL; 24:1 v/v) is introduced. The tube is sealed and gradually, repeatedly inverted until a white milky emulsion is evident.

Phase separation: The organic and aqueous layers are separated by centrifugation at 10,000 × *g* for 15 min at room temperature.

Aqueous phase recovery: The proteinaceous interphase is carefully avoided, while the upper, viscous aqueous phase is aspirated using a pipette and transferred to a fresh 15-mL tube.

DNA precipitation: Ice-cold isopropanol (0.7 mL) is added to the tube, and the tube is gradually inverted until thread-like DNA precipitates appear.

Precipitation enhancement: The sample is kept at −20°C for at least 1 h (or overnight) to optimize the recovery of nucleic acids.

Pellet collection: The precipitate is sedimented by centrifugation (12,000 × *g*, 20 min, 4°C), and the supernatant is carefully decanted.

Ethanol wash: The pellet is gently stirred with 1 mL of cooled 70% ethanol before being centrifuged again (10,000 × *g* 5 min 4°C). A micropipette is used to remove any residual ethanol after centrifugation.

Drying: The pellet is air-dried for 10–15 min until no liquid remains in the tube; over-drying might cause the pellet to crack; therefore, it is avoided.

Resuspension: 100 μL of TE buffer is gently pipetted to dissolve the DNA, and is incubated for either 1–2 h at 37°C or overnight at 4°C.

RNase digestion: 2 μL of heat-inactivated RNase A (10 mg/mL) is added, and the mixture is incubated at 37°C for 30 min.

Quantification and integrity check: DNA concentration is assessed spectrophotometrically (A_{260}/A_{280} ratio), and about 200 ng of each sample is electrophoresed on a 0.8% agarose gel. The development of a single, bright, high-molecular-weight band that does not smear is interpreted as evidence of no degradation or of adequate downstream restriction.

3.2 Restriction Endonuclease Digestion

Endonuclease restriction, or restriction endonuclease digestion, utilizes restriction endonuclease enzymes to cleave DNA fragments at specific sites [11, 12]. Endonuclease enzymes, such as EcoRI and HindIII, recognize short, specific sequences of DNA (typically 4–8 base pairs long) and cleave the DNA backbone in or near these particular locations [13]. To produce authentic RFLP patterns, total enzymatic cleavage is crucial. The restriction endonuclease digestion reaction (50 μL) involves 5 μg of pure genome-specific DNA, 5 μL of the manufacturer's 10× restriction buffer, 10–20 U of restriction-selected enzyme, and nuclease-free water, all stored on ice in sterile 0.2 mL tubes. The contents of the tube are gently aspirated using a pipette before being efficiently centrifuged to collect droplets and incubated for 16–18 h at the optimal temperature for enzyme activity at 37°C (*see* **Note 3**). If necessary, the reaction is terminated by heating (e.g., 65°C for 20 min) according to the supplier's instructions. The efficacy of digestion is evaluated by loading 200 ng of digested material and the equivalent quantity of undigested control DNA onto a 0.8% agarose gel. The homogeneous smear signifies complete digestion, whereas the band that persists at the limiting mobility location indicates incomplete cleavage [14].

3.3 Size Fractionation by Agarose Gel Electrophoresis and Southern Transfer

Agarose gel electrophoresis and southern transfer are used to determine the size of the DNA fragments. This method involves the separation of restriction fragments on a 0.8% (w/v) agarose gel stained with 1× TBE. A casting tray with a well-forming comb is set up to hold the concentration, which includes 1.6 g of agarose melted by boiling and allowed to cool to about 60°C, with ethidium bromide (0.5 μg/mL final concentration) for a 200 mL volume. The completely digested DNA sample (in 6× gel loading buffer) is loaded into a single preparative-size well, followed by a DNA size marker in the next column. Electrophoresis is performed at low voltage (1–2 V/cm) overnight or at higher voltage (3–4 V/cm) for 4–5 h, until the bromophenol blue tracking dye has moved at least 10 cm. The gel is examined under UV trans-illumination to assess the uniform distribution of the fragments and photographed with a metric scale for subsequent size estimation.

Gel conditioning: The gel is first immersed in a 0.25 M HCl solution and gently spun for 15 min (depuration). It is then washed down with distilled water and submerged in the denaturation solution for an additional 30 min. Finally, it is rinsed and incubated in neutralizing solution for an additional 30 min to facilitate the transfer of larger fragments.

Capillary blotting: A shallow tray is filled with 20× SSC buffer. A strip of Whatman 3MM paper, moistened with the same buffer, is attached to cover the platform with both ends immersed in it to be used as a wick. The conditioned gel is laid onto the wick, air bubbles are rolled out, and the gel margins are sealed with plastic film to prevent buffer bypass. The gel surface is covered with a positively charged nylon membrane that has been meticulously tailored to its dimensions and soaked in 2× SSC (*see* **Note 4**). Two sheets of 3MM paper (pre-soaked in 2× SSC) are layered above, after a 5–8 cm layer of absorbent paper towels, a glass plate, and a 500-g weight. The transfer could operate between 16 and 24 h.

DNA immobilization: Following disassembly, the membrane is marked with a soft pencil to record well positions, briefly rinsed in 2× SSC to dislodge agarose debris, and air-dried. DNA is covalently cross-linked to the membrane by UV irradiation (e.g., 1200 × 100 μJ/cm^2). Dried membranes can be stored between sheets of filter paper at ambient temperature until required.

3.4 Non-Radioactive Probe Labeling

A digoxigenin (DIG)-labeled probe is developed from purified template DNA (25–50 ng of a PCR product or linearized plasmid). The template is denatured at 95°C for 5 min and immediately chilled on ice (*see* **Note 5**). Denatured DNA is transferred to a vial containing DIG-High Prime (Roche) reaction mix, and the volume is adjusted to 20 μL with the kit's components. The labeling process is incubated for one to 3 h at 37°C following a quick centrifugation and vortexing. Heating the probe to 65°C for 10 min terminates the reaction; it can then be stored at −20°C for subsequent use (*see* **Note 6**).

3.5 Hybridization and Post-Hybridization Washes

The membrane having immobilized DNA is moistened in 2× SSC, rolled, and inserted into a hybridization tube with the DNA-facing side inward. Pre-hybridization is carried out by adding 10–20 mL of pre-warmed (65°C) pre-hybridization Buffer. After pre-hybridization, rotate the tube in a hybridization oven at 65°C for 2 h (*see* **Note 7**).

The DIG-labeled probe is denatured by boiling for 5 min, followed by instantaneous cooling on ice. Hybridization involves diluting the denatured probe into 5–10 mL of fresh, pre-warmed pre-hybridization buffer (which serves as the hybridization buffer). The pre-hybridization solution is discarded, and the probe-containing buffer is readily added. Hybridization unfolds overnight (16–18 h) at 65°C under continuous rotation.

After hybridization, the probe solution is recovered and may be reused after storage at −20°C. The membrane is subjected to two consecutive washes at room temperature in 50 mL of Low stringency wash buffer (5 min each, with rotation). After that, 50 mL of pre-heated (65°C) high stringency wash buffer is used for two intense washes, extending 15 min each.

3.6 Chemiluminescent Detection

The membrane is washed with Maleic Acid Buffer for 1–2 min. The membrane is gently orbitally agitated for 30 min at room temperature in 50 mL of blocking solution to saturate nonspecific binding sites. The membrane is immersed in a new blocking solution containing anti-digoxigenin antibodies conjugated to alkaline phosphatase

(Anti-DIG-AP, Fab fragments) at a 1:10,000 dilution for 30 min while being gently shaken continuously. The unbound antibody conjugate is removed by washing twice with 50 mL of maleic acid buffer for 15 min.

The membrane is set DNA-side up on a clear, translucent plastic sheet after being equilibrated for 2–5 min with 20 mL of detection buffer. The surface is evenly pipetted with ten, 2 mL CDP-star chemiluminescent substrate. After 5 min of incubation at room temperature, the additional substrate is drained and replaced with new plastic wrap, taking care to prevent air bubbles. After wrapping, the membrane is put in an autoradiography cassette. It is commonly advised that a sheet of X-ray film be exposed to the membrane under safe lighting conditions; the initial exposure time should be between 1 and 5 min, with subsequent changes made based on signal intensity. The film is developed in accordance with typical darkroom procedures. Polymorphic fragments are visually evaluated, and binary data (1 = band present, 0 = band absent) is recorded for further linkage or diversity analysis (*see* **Note 8**).

4 Notes

1. **DNA quality:** Complete restriction digestion necessitates high-molecular-weight pure DNA. Incomplete digestion will result in smear bands or erroneous banding patterns.
2. **Radioactive safety:** If ^{32}P-labeled probes are used instead of DIG, institutional radiation safety standards must be followed for handling, shielding, and disposing. One should be able to operate behind appropriate shields while keeping an eye on contamination.
3. **Enzyme selection:** Select enzymes with recognition sites of 4–6 bp to detect the most significant polymorphisms. Several enzymes will most likely be investigated to discover polymorphism loci in specific samples.
4. **Membrane handling:** Always wear gloves to avoid contamination and nuclease insertion. Do not touch the surface of the membrane to which DNA is bound.
5. **Probe specificity:** The probe's specificity must be confirmed using BLAST analysis or sequencing to rule out cross-hybridization with nontarget sequences, which might result in excessive background noise.
6. **Storage:** Labeled probes can be kept at –20°C for several months. Stored hybrid membranes can be stripped and re-probed.

7. **Hybridization temperature:** The hybridization temperature is determined by the probe's GC concentration and the required stringency. An increase in temperature results in higher stringency; for background, reduce the wash temperature or SSC concentration in the wash buffers.
8. **Signal strength:** If the signal is feeble, increase the probe concentration, hybridization time, or film exposure time. If the background is high, additional wash stringency and blocking are required.

Acknowledgments

The authors would like to thank Sultan Qaboos University, Oman, the University of Agriculture, Faisalabad, Pakistan, and PMAS Arid Agriculture University, Rawalpindi, Pakistan, for providing the facilities required for this study.

References

1. Singh VP, Slathia PS, Sharma P, Singh P (2026) Mechanisms for genetic variations. In: Genome analysis. Academic Press. pp 517–542
2. Kumar R, Das SP, Choudhury BU, Kumar A, Prakash NR, Verma R, Mishra VK (2024) Advances in genomic tools for plant breeding: harnessing DNA molecular markers, genomic selection, and genome editing. Biol Res 57: 80
3. Mondal R, Kumar A, Gnanesh BN (2023) Crop germplasm: current challenges, physiological-molecular perspective, and advance strategies towards development of climate-resilient crops. Heliyon 9:e12973
4. Jajoriya R, Sharma R, Meena VK, Vadithya AS, Singh K, Kumar P, Nathawat TS (2025) Emergence of molecular marker from RFLP to SNP: a review. Plant Arch 25:2062–2074
5. Biswas P, Kumar N (2023) Application of molecular markers for the assessment of genetic fidelity of in vitro raised plants: current status and future prospects. In: Kumar N (ed) Molecular marker techniques. Springer, Singapore, pp 233–256
6. Singh R, Das S, Dash L, Kumar S (2025) Role of DNA markers (RFLP, RAPD, SSR) in marker-assisted selection for crop improvement and biotechnology regulation. In: Elements of plant biotechnology. Alphine Publications, pp 109–122
7. Babu CV (2023) Molecular biology techniques in agricultural biotechnology. In: Uike AR, Singh NP, Yankanchi S, Anand R, Kontheti M (eds) Advances in biotechnology, vol 1. Stella International TM Publication, p 181
8. Rowan AK, Davenport RJ, Snape JR, Fearnside D, Barer MR, Curtis TP, Head IM (2005) Development of a rapid assay for determining the relative abundance of bacteria. Appl Environ Microbiol 71:8481–8490
9. Doyle JJ, Doyle JL (1987) A rapid DNA isolation procedure for small quantities of fresh leaf tissue. Phytochem Bull 19:11–15
10. Akram MT, Khan QRW, Jaskani MJ, Awan FS (2019) Ampelographic and genetic characterization of grape genotypes collected from Potohar region of Pakistan. Pak. J Agric Sci 56:595–605
11. Lone MA (2025) Restriction enzymes and vectors. In: Molecular genetics: structures, mechanisms, and functions. Apple Academic Press, pp 101–118
12. Hazarika G, Dutta R, Saikia DP, Malakar D, Hazorika M, Bora M (2024) A comprehensive exploration of restriction enzymes and their applications in molecular biology: a review. World J Adv Res Rev 21:2399
13. Roopashree MS, Venkatesh KH (2025) Genetic engineering tools: restriction enzymes and vectors. In: Molecular genetics: structures, mechanisms, and functions. Apple Academic Press, pp 75–99
14. Miura K, Iwashita T (2025) Selective deletion or preservation of tissue components via enzymatic digestion monitored by scanning acoustic microscopy. Sci Rep 15: 29300

Chapter 23

Application of Molecular Markers for Trait-Specific Breeding of Aroma Accumulation in Rice

Suvarna Rani Chimmili, K. Varadhan, Kanakachari Mogilicherla, P. Ranjith, K. Sruthi, V. Manasa, K. Basavaraj, Satyaswara Jasudasu Gompa, K. Suneetha, S.K. Mangrauthia, S.V. Sai Prasad, R.M. Sundaram and M. Srinivas Prasad

Abstract

The translation of genetic variation into breeding gain relies on an integrated and sequential pipeline comprising high-quality DNA extraction, precise locus amplification and sequencing, accurate variant discovery, conversion of polymorphisms into diagnostic markers, high-throughput genotyping, and systematic integration into crossing and selection programs. This chapter presents a comprehensive framework for the development and application of molecular markers for aroma improvement in rice, with broader relevance to yield and stress-resilience traits in Oryza sativa. Rice aroma, a premium grain quality trait influencing consumer preference and market value, is primarily associated with the accumulation of 2-acetyl-1-pyrroline (2-AP), which results from loss-of-function mutations in the BADH2 gene. Advances in molecular genetics have enabled the identification of functional polymorphisms within BADH2, facilitating the development of robust functional and linked markers that ensure high selection accuracy, codominant discrimination, and environmental stability for marker-assisted selection (MAS) and marker-assisted backcross breeding (MABB). The deployment of high-throughput SNP and KASP genotyping platforms further accelerates early-generation selection, rapid aroma introgression into elite cultivars, and gene pyramiding with other agronomic traits.

Keywords Aromatic rice, 2-AP, BADH2, Molecular markers, Trait-specific breeding

1 Introduction

Oryza sativa L. (rice) is the staple food for more than half of the global population, playing a central role in food security, cultural heritage, and agricultural economies, particularly in Asia [1]. Beyond yield and adaptability, grain quality traits have become decisive factors in varietal acceptance and market value [2, 3]. Among these traits, aroma is especially important, as it defines premium rice types such as Basmati and Jasmine, which command

Nitish Kumar (ed.), *Molecular Markers in Plants*, Methods in Molecular Biology, vol. 3097,
https://doi.org/10.1007/978-1-0716-5663-1_23,

higher prices in both domestic and international markets [4]. The characteristic fragrance of aromatic rice is primarily attributed to the volatile compound 2-acetyl-1-pyrroline (2-AP), which imparts a popcorn- or pandan-like scent even at very low concentrations due to its extremely low odor threshold [5]. Consequently, aroma has become a major breeding objective aligned with evolving consumer preferences [6, 7].

Despite their premium status, most traditional aromatic cultivars are agronomically inferior, often exhibiting low yield potential, tall plant stature, lodging susceptibility, and sensitivity to biotic and abiotic stresses [8]. These constraints limit large-scale cultivation and reduce farmer adoption [9]. Conventional breeding for aroma has relied on sensory-based cooked-rice evaluation and biochemical assays such as potassium hydroxide (KOH) leaf tests; however, these methods are subjective, environmentally influenced, destructive, and unsuitable for early-generation screening [10]. Although gas chromatography-based quantification of 2-AP provides greater precision, it is costly, labor-intensive, and impractical for high-throughput breeding programs [11].

Advances in molecular genetics have clarified that rice fragrance is largely controlled by a single recessive gene, with loss-of-function mutations in the BADH2 (betaine aldehyde dehydrogenase 2) gene leading to 2-AP accumulation [12]. This discovery revolutionized aroma breeding by enabling the development of gene-specific DNA markers and facilitating marker-assisted selection (MAS) and marker-assisted backcross breeding (MABB) for precise introgression into elite, high-yielding backgrounds while minimizing linkage drag [13]. Diverse marker systems, including SSRs, InDels, SNPs, functional markers, and Kompetitive Allele-Specific PCR (KASP) platforms, have since enhanced selection accuracy and efficiency [14, 15]. Overall, integrating molecular markers with conventional breeding has accelerated the development of high-quality aromatic rice varieties adapted to modern agricultural demands.

2 Materials

Aromatic rice germplasm, micropipettes, sterile microtubes, eppendorf tubes, centrifuge, water bath, ice box, CTAB (Table 1), TE buffer (Table 2), 50× TAE buffer (Table 3), Chloroform:isoamyl alcohol (24:1), Databases (NCBI GenBank, Gramene database, rice genome annotation databases), Software tools (BLAST, Primer design tools like IDT, Primer3, NCBI Primer-BLAST), Genomic DNA, Selected molecular markers linked to aroma trait, PCR reagents (Taq DNA Polymerase, dNTP Mix, MgCl2), primers, Agarose gel electrophoresis system, Tris-HCl (pH 8.3–8.8), KCl, NaCl,

Table 1
Composition of CTAB extraction buffer

Component	Concentration
CTAB	2%
Tris-HCl (pH 8.0)	100 mM
EDTA (pH 8.0)	20 mM
NaCl	1.4 M
PVP	1%
1%	0.2% (added fresh before use)

The CTAB extraction buffer was prepared by dissolving CTAB in distilled water with heating at 60–65 °C, followed by the addition of Tris-HCl, EDTA, and NaCl to obtain the desired final concentrations. The solution volume was adjusted with distilled water, and β-mercaptoethanol was added freshly just before use to ensure optimal activity

Table 2
Composition of TE buffer

Component	Concentration
Tris-HCl	10 mM
EDTA	1 mM

Tris-HCl and EDTA were mixed, and the final volume was adjusted to the required level using sterile distilled water

Table 3
50× TAE buffer composition

Component	Volume/concentration
Tris base	242 g
Glacial acetic acid	57.2 ml
EDTA (0.5 M, pH 8.0)	100 ml
Autoclaved distilled water	Adjusted to 1 l

Tris base was dissolved in distilled water, followed by the addition of glacial acetic acid and EDTA solution, and the volume was adjusted to 1 l to prepare a concentrated stock solution. The working solution was subsequently obtained by diluting the stock to a 1× concentration

DTT, Acrylamide solution, Bis-acrylamide, Ammonium persulfate (APS), TEMED, SDS, Field or greenhouse facilities, Pollination tools for hybridization, Phenotyping tools for agronomic and quality traits, Seed storage facilities, Physical mutagens (e.g., gamma rays) or chemical mutagens (e.g., EMS), Pollination control materials (bags, tags).

3 Methods

3.1 Genomic DNA Isolation (CTAB Method)

1. Young, healthy leaf tissues from selected aromatic and non-aromatic rice genotypes were collected for genomic DNA isolation [16].
2. Approximately 100 mg of leaf tissue was homogenized in liquid nitrogen or directly in preheated CTAB extraction buffer to obtain a fine slurry (*see* **Notes 1 and 2**).
3. The homogenate was incubated in CTAB buffer at 65 °C for 30–45 min to ensure efficient cell lysis and release of genomic DNA.
4. An equal volume of chloroform:isoamyl alcohol (24:1) was added, mixed gently, and centrifuged to separate phases, and the aqueous phase was carefully transferred to a fresh tube (extraction repeated if required).
5. Genomic DNA was precipitated from the aqueous phase using chilled isopropanol, followed by centrifugation to obtain a DNA pellet.
6. The DNA pellet was washed with 70% ethanol, air-dried, and dissolved in TE buffer (*see* **Note 2**); RNase treatment was performed when necessary to remove RNA contamination.
7. The quality and integrity of the extracted DNA were assessed by agarose gel electrophoresis, and high-quality DNA samples were used for downstream molecular marker analysis (*see* **Note 3**).

3.2 Primer Designing

1. Nucleotide sequences corresponding to genomic regions associated with the aroma trait were retrieved from public databases using reported accession numbers and chromosomal coordinates, with special emphasis on the BADH2 gene located on chromosome 8, along with its flanking regions (*see* **Note 4**).
2. Target regions containing polymorphic sites such as SNPs and insertion–deletion mutations linked to aroma expression were identified, and microsatellite repeat motifs were detected for SSR marker development (*see* **Note 5**).
3. Retrieved sequences were aligned and analyzed to confirm polymorphisms between aromatic and non-aromatic genotypes and to select suitable regions for marker development (*see* **Note 6**).
4. Primer design tools such as Primer3 and Primer-BLAST were used to generate candidate primer pairs by specifying target regions and amplification parameters.
5. Key primer design parameters, including primer length, GC content, melting temperature, and expected amplicon size,

were optimized to ensure specific and efficient amplification (*see* **Note 7**).

6. Multiple primer pairs were generated for each target region and subjected to in silico specificity analysis to avoid non-specific binding and off-target amplification.
7. Selected primer pairs were synthesized commercially and initially validated using parental genomic DNA samples through PCR, including gradient PCR where necessary to determine optimal annealing temperatures.
8. Amplification products were analyzed by agarose gel electrophoresis to verify expected fragment size and specificity, and primers showing clear, reproducible amplification were finalized for marker validation and breeding applications.

3.3 Polymerase Chain Reaction (PCR)

1. PCR amplification was performed using genomic DNA extracted from aromatic and non-aromatic rice parental lines and validation populations.
2. Reaction mixtures were prepared under sterile and chilled conditions (Table 4), containing template DNA, forward and reverse primers, dNTPs, PCR buffer, $MgCl_2$, Taq DNA polymerase, and nuclease-free water in the required final volume (*see* **Note 8**).
3. Components were mixed gently to ensure uniform reagent distribution, and negative controls without template DNA were included to monitor contamination.
4. Amplification was conducted in a thermal cycler with an initial denaturation step followed by cyclic denaturation, primer annealing, and extension to amplify aroma-associated genomic regions (Table 5).

Table 4
Composition of PCR reaction mixture

Component	Volume (µl)
Genomic DNA (50 ng)	1 µl
10× PCR Buffer	2.5 µl
$MgCl_2$ (25 mM)	1.5 µl
dNTP mix (10 mM each)	0.5 µl
Forward primer (10 µM)	0.5 µl
Reverse primer (10 µM)	0.5 µl
Taq DNA polymerase (5 U/µl)	0.2 µl
Nuclease-free water	Up to 25 µl

Table 5
PCR conditions

Steps	Temperature	Time	Cycles
Initial denaturation	95 °C	5 min	1
Denaturation	95 °C	30 s	35
Annealing	55–65 °C[a]	30 s	
Extension	72 °C	45 s	
Final extension	72 °C	7 min	1
Hold	4 °C	∞	–

[a]Annealing temperature optimized based on primer design

5. Annealing temperature was selected based on primer melting temperature (Tm), and conditions were optimized when necessary to enhance specificity.
6. Extension was carried out at the optimal temperature for DNA polymerase activity, followed by a final extension step to ensure complete synthesis of PCR products.
7. Optimization of PCR conditions was performed using gradient PCR by adjusting annealing temperature, $MgCl_2$ concentration, and template DNA quantity to obtain clear and specific amplification (*see* **Note 9**).
8. Amplified products were analyzed by agarose gel electrophoresis, and optimized conditions yielding clear, reproducible bands were standardized for large-scale genotyping and aroma marker validation (*see* **Note 10**).

3.4 Agarose Gel Electrophoresis

1. Agarose gel was prepared by dissolving agarose powder in 1× TAE buffer through heating until a clear solution was obtained, then cooled to 50–60 °C before adding nucleic acid stain (if pre-staining was followed).
2. The molten gel was poured into a casting tray with a comb to form wells and allowed to solidify at room temperature.
3. After solidification, the comb was removed, and the gel was placed in an electrophoresis chamber containing 1× TAE running buffer.
4. PCR products were mixed with loading dye and loaded into the wells along with an appropriate DNA ladder.
5. Electrophoresis was carried out at a constant voltage to separate DNA fragments based on size.

6. Following electrophoresis, the gel was visualized under UV light using a gel documentation system to detect amplified DNA fragments.
7. Banding patterns corresponding to expected product sizes were recorded, documented, and preserved for marker scoring and aroma breeding analysis.

3.5 Molecular Marker Identification (SSR, SNP, QTL for Aroma Trait)

1. The target trait, aroma accumulation in rice, was defined based on breeding objectives, and its genetic basis was reviewed through published literature to identify major genes and associated loci.
2. The BADH2 gene on chromosome 8 was identified as the primary candidate responsible for fragrance and selected as the main focus for marker development (Table 6).

Table 6
Summary of representative molecular markers used for aroma detection: marker type, target genomic region, primary applications, and references

Marker name	Marker type	Target region	Application	References
RM223	SSR	Chromosome 8 (linked to aroma locus)	Genetic diversity analysis, preliminary screening	[17]
RM42	SSR	Chromosome 8 (linked to aroma locus)	Germplasm characterization	[17]
RM515	SSR	Chromosome 8 (linked to aroma locus)	Linkage analysis, diversity studies	[17]
BAD2	Functional InDel	BADH2 exon 7 (8-bp deletion)	Foreground selection for aroma	[18,19]
BADH2-Indel	Functional InDel	BADH2 exon 7	MAS in segregating populations	[19,20]
BADEX7	Functional marker	BADH2 exon 7	Aroma allele discrimination	[20]
BADH2-SNP	SNP	BADH2 coding region	Allele-specific selection	[20]
KASP-BADH2	KASP SNP	BADH2 gene	High-throughput breeding applications	[20]

3. Previously reported QTLs associated with aroma and grain quality traits were compiled to identify additional genomic regions relevant for marker targeting.
4. Molecular markers such as SSRs, SNPs, InDels, and functional gene-based markers linked to the BADH2 locus or aroma-related QTLs were identified through literature surveys and database mining (*see* **Note 11**).
5. Genomic and coding sequences of BADH2 and its flanking regions were retrieved and analyzed to detect polymorphic sites suitable for marker development.
6. Sequence comparisons between aromatic and non-aromatic rice genotypes were performed to identify SNPs and insertion–deletion variations associated with aroma expression (*see* **Note 12**).
7. Candidate markers were shortlisted based on linkage strength to aroma, proximity to BADH2, reproducibility, and amplification efficiency, with priority given to functional markers targeting known BADH2 mutations.
8. Selected markers were validated using aromatic and non-aromatic parental lines, and those showing clear polymorphism and strong association with the aroma phenotype were finalized and documented for breeding applications.

3.6 Marker Validation

1. Plant materials representing aromatic and non-aromatic phenotypes, including parental lines, segregating populations, and breeding lines, were selected for marker validation.
2. Polymorphic parental lines identified during initial marker screening were included as reference controls (*see* **Notes 12 and 13**).
3. Genomic DNA was extracted using the CTAB method, and DNA quality was assessed before PCR amplification.
4. Candidate molecular markers were amplified through PCR, and allelic variations were detected using agarose gel electrophoresis.
5. Polymorphic markers distinguishing aromatic and non-aromatic parental lines were identified based on clear banding differences.
6. Segregating populations (F_2 and backcross generations) were genotyped using selected polymorphic markers, and individual plants were phenotypically classified for aroma through sensory evaluation.
7. Genotypic data were compared with phenotypic observations to assess marker–trait association, and co-segregation analysis was performed using chi-square tests to confirm Mendelian inheritance patterns.

8. Markers, including functional markers targeting mutations in the BADH2 gene, were validated across multiple generations and diverse genetic backgrounds, and reliable markers were finalized for marker-assisted breeding applications.

3.7 Molecular Marker Analysis

1. Molecular marker analysis was conducted using PCR-amplified DNA fragments targeting aroma-associated loci and other genomic regions of interest.
2. PCR products from SSR and SNP markers were separated using agarose gel electrophoresis or high-resolution systems to detect fragment size variations.
3. Banding patterns were visualized under UV light, and fragment sizes were determined using a DNA ladder to assess allelic variation among genotypes.
4. Marker alleles were scored based on the presence/absence of bands or fragment size differences between aromatic and non-aromatic genotypes.
5. RAPD analysis was performed using single arbitrary primers, and polymorphisms were identified based on differential banding patterns.
6. RFLP analysis involved restriction enzyme digestion of genomic DNA, followed by electrophoresis and, when necessary, Southern blot hybridization to detect polymorphic loci.
7. AFLP analysis included restriction digestion, adapter ligation, selective amplification, and high-resolution fragment separation to identify polymorphisms.
8. Polymorphic markers associated with the aroma trait were identified through comparative analysis of parental lines and validation populations, and reliable markers were selected for statistical analysis and breeding applications (Fig. 1).

4 Notes

1. Young rice leaves ground in liquid nitrogen yield high-quality, high-molecular-weight DNA suitable for downstream molecular applications.
2. The use of fresh β-mercaptoethanol, careful chloroform extraction, and complete ethanol removal is essential for obtaining pure DNA and ensuring successful PCR amplification.
3. Extracted DNA should be stored at −20 °C for short-term use or −80 °C for long-term stability and integrity.

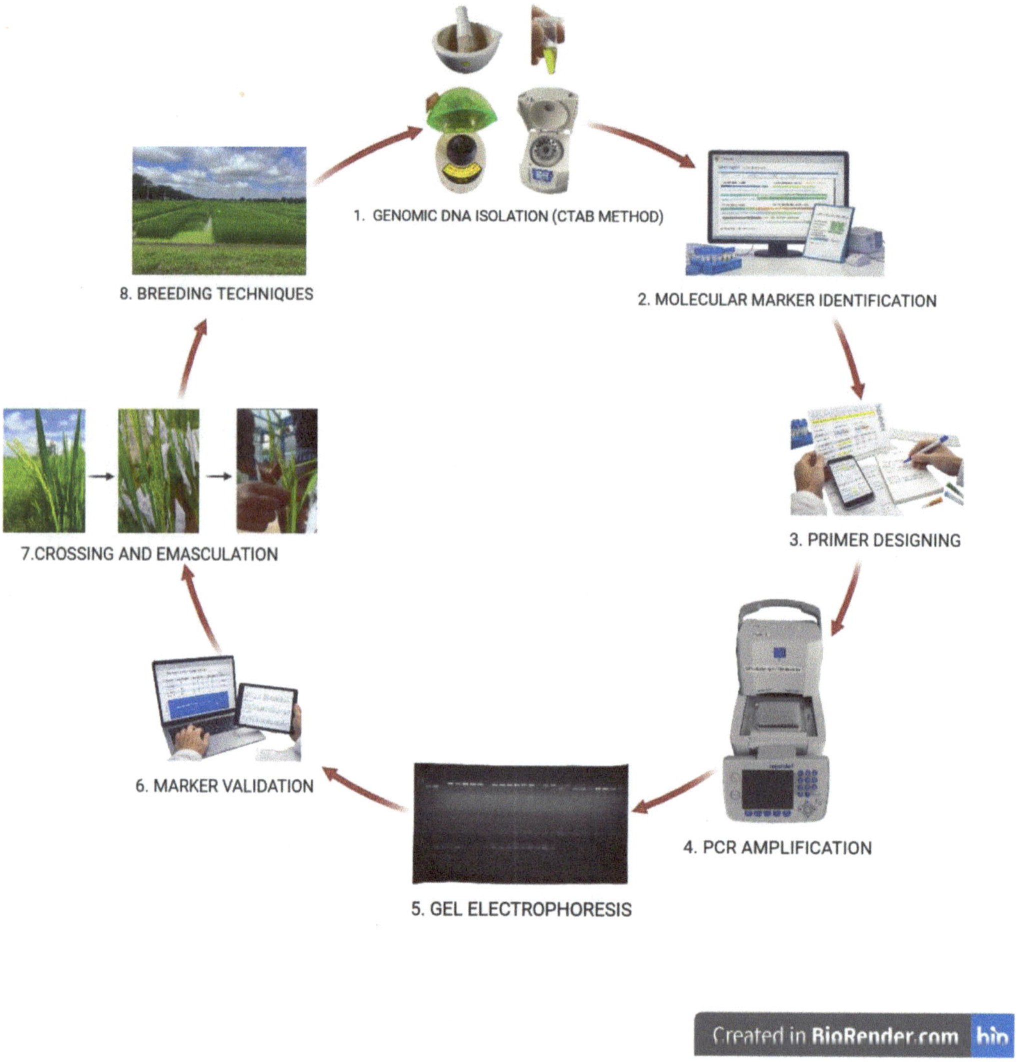

Fig. 1 Overview of molecular breeding methods for the development of aromatic rice

4. The BADH2 gene is the primary determinant of fragrance in rice, and the aroma trait is generally recessively inherited.
5. Functional markers targeting polymorphisms within BADH2 provide the highest accuracy for aroma detection compared to linked SSR markers.
6. Markers located closer to the target gene minimize recombination risk, improving reliability in marker-assisted selection.
7. Proper primer design, including balanced GC content, similar melting temperatures, and targeting functional polymorphic regions, ensures specificity and amplification efficiency.

8. PCR optimization through gradient PCR, appropriate Mg^{2+} concentration, correct annealing temperature, and controlled template quantity is critical for consistent results.
9. Inclusion of negative controls, replication, and uniform PCR conditions ensures reproducibility and prevents contamination.
10. Agarose gel concentration, appropriate DNA ladder selection, controlled voltage, fresh running buffer, and proper staining ensure clear band resolution and accurate genotyping.
11. SSR and SNP markers provide higher reproducibility than RAPD markers, while functional gene-based markers remain the most reliable for validation.
12. Parental polymorphism screening, accurate phenotyping, precise genotype documentation, and proper sample handling are essential before large-scale population analysis.

Acknowledgments

KM acknowledges the Ramalingaswami re-entry grant (BT/RLF/Re-entry/11/2023) from the Department of Biotechnology, Govt. of India.

References

1. Chu Z (2024) The role of rice in global food security: strategies for sustainable production. J Rice Res 12(3):411. https://doi.org/10.4172/2375-4338.1000411
2. Abdelsalam KM, Shaalan AM, AbouEl-Soud GM, El-Dalil MA, Marei AM, El-Moneim DA, El-Banna AA, Lamlom SF, Abdelghany AM (2025) Comprehensive quality profiling and multivariate analysis of rice (Oryza sativa L.) cultivars: integrating physical, cooking, nutritional, and micronutrient characteristics for enhanced varietal selection. BMC Plant Biol 25(1):492. https://doi.org/10.1186/s12870-025-06438-5
3. Jukanti AK, Rathod S, Neeraja CN, Krishnan SG, Aleena D, Mahesh G, Prasanna G, Sundaram RM (2026) Grain quality characterization of indigenous rice (Oryza sativa L.) landraces using mixed clustering and selection index approaches. Discov Food 6(1):26. https://doi.org/10.1007/s44187-025-00715-w
4. Islam M, Akter NA, Islam N, Hasan N, Patwary A, Khatun S, Maniruzzaman S, Ali L (2025) The journey of aromatic rice: genetic diversity, global trade, and prospects for improvement: a review: the journey of aromatic rice. Ann Bangladesh Agric 29(2):265–290. https://doi.org/10.3329/aba.v29i2.84511
5. Chen Y, Hua X, Li S, Zhao J, Yu H, Wang D, Yang J, Liu L (2024) Aromatic compound 2-acetyl-1-pyrroline coordinates nitrogen assimilation and methane mitigation in fragrant rice. Curr Biol 34(15):3429–3438. https://doi.org/10.1016/j.cub.2024.06.063
6. Concepcion JCT, Garson MJ, Fitzgerald MA (2025) Metabolite quantitative trait loci mapping for fragrance and fatty acid composition in rice (Oryza sativa L.). Rice 18(1):1–20. https://doi.org/10.1186/s12284-025-00790-8
7. Zhu D, Shao Y, Chen M, Yu Y, Zeng Y, Shao L, Chen H, Qin Y (2026) Integrated HS-GC-IMS and E-nose analysis reveals aroma differentiation in representative Chinese high-quality rice varieties. Food Chem 507:148251. https://doi.org/10.1016/j.foodchem.2026.148251
8. Shahidullah SM, Shirazy BJ, Quais MK, Tahjib-Ul-Arif M, Rhaman MS (2025) Unveiling global research dynamics in aromatic rice: bibliometric

insights and future directions. Discov Plant 2(1):261. https://doi.org/10.1007/s44372-025-00348-4
9. Bhambri MC, Kumar S (2025) Studies on growth and yield behavior of short grain aromatic traditional, improved scented and fortified quality rice varieties under organic production system. J Adv Biol Biotechnol 28(8):1–11. https://doi.org/10.9734/jabb/2025/v28i82677
10. Zhou J, Zhang A, Xu W, Cheng C, Niu F, Sun B, Cao L, Zhang J, Chu H (2024) Rapid detection of rice fragrance allele badh2-E7 by recombinant polymerase amplification (RPA). Biosci Methods 15(1):1–8. https://doi.org/10.5376/bm.2024.15.0001
11. Cai Y, Pan X, Zhang D, Yuan L, Lao F, Wu J (2024) The kinetic study of 2-acetyl-1-pyrroline accumulation in the model system: an insight into enhancing rice flavor through the Maillard reaction. Food Res Int 191:114591. https://doi.org/10.1016/j.foodres.2024.114591
12. Chandanshive S, Mathure S, Nadaf A (2024) Multiple mutations in BADH2 gene reveal the novel fragrance allele in indica rice (Oryza sativa L.). J Genet 103(1):13. https://doi.org/10.1007/s12041-024-01464-y
13. Ye J, Wang K, Wang Y, Zhao Z, Yan Y, Yang H, Zhang L, Hu Z, Shi Z, Sun D, Bai J (2025) Improving panicle blast resistance and fragrance in a high-quality japonica rice variety through breeding. Front Plant Sci 15:1507827. https://doi.org/10.3389/fpls.2024.1507827
14. Asante MD, Ofosu KA, Frimpong F, Alphonso DK, Nartey E, Obeng AE, Bam RK, Gamenyah DD, Ribeiro PF, Manilal W (2025) Effectiveness of KASP-SNP markers in selecting for grain quality traits in rice. Plant Gene 43:100503. https://doi.org/10.1016/j.plgene.2025.100503
15. Fang H, Huang H, Yu L, Wang L, Lou J, Qi Y (2025) Development and application of an SNP marker for high-throughput detection and utilization of the badh2 gene in rice breeding. Genes 16(10):1132. https://doi.org/10.3390/genes16101132
16. Pérez-Pazos JV, Romero-Ferrer JL, Berdugo-Cely JA (2023) Selection of a CTAB protocol for high-quality DNA extraction in Oryza sativa L. validated for application in genotyping process based on Illumina sequencing. J Crop Sci Biotechnol 26(4):433–446. https://doi.org/10.1007/s12892-022-00188-8
17. Burman M, Nair SK, Sao A, Gauraha D, Sharma D (2023) SSR marker-based genetic diversity and marker-trait association analysis in aromatic rice (Oryza sativa) landraces. Indian J Agric Sci 93(4):365–370. https://doi.org/10.56093/ijas.v93i4.133177
18. Bradbury LM, Fitzgerald TL, Henry RJ, Jin Q, Waters DL (2005) The gene for fragrance in rice. Plant Biotechnol J 3(3):363–370. https://doi.org/10.1111/j.1467-7652.2005.00131.x
19. Sakthivel K, Sundaram RM, Rani NS, Balachandran SM, Neeraja CN (2009) Genetic and molecular basis of fragrance in rice. Biotechnol Adv 27(4):468–473. https://doi.org/10.1016/j.biotechadv.2009.04.001
20. Singh G, Gopala Krishnan S, Kumar A, Vinod KK, Bollinedi H, Ellur RK, Nagarajan M, Bhowmick PK, Madhav SM, Singh K, Singh AK (2022) Molecular profiling of BADH2 locus reveals distinct functional allelic polymorphism associated with fragrance variation in Indian aromatic rice germplasm. Physiol Mol Biol Plants 28(5):1013–1027. https://doi.org/10.1007/s12298-022-01181-9

Chapter 24

Development and Application of Single Nucleotide Polymorphism (SNP) Markers for the Identification of Pest and Disease Resistance in Sugarcane

Sowparnika Binu, Janani Sree Sivakumar, Lanka Naga Krishna Ushasree, Vasanth Krishnan, Kanakachari Mogilicherla and Dhandapani Gurusamy

Abstract

Sugarcane (*Saccharum* spp.) is an economically important crop cultivated worldwide for sugar and bioenergy production. However, its productivity is significantly affected by various pests and diseases, which lead to substantial yield and quality losses. Traditional breeding approaches for developing resistant varieties are often time-consuming and complicated due to the complex polyploid genome of sugarcane. In recent years, molecular marker technologies have provided powerful tools for the rapid identification and selection of desirable traits in crop improvement programs. Among the available molecular markers, single-nucleotide (SNP) markers have emerged as highly informative, abundant, and reliable markers for genetic analysis and marker-assisted breeding. Advances in next-generation sequencing (NGS) technologies have greatly facilitated the discovery and high-throughput genotyping of SNPs across the sugarcane genome. These markers can be effectively used to identify genomic regions associated with resistance to major sugarcane pests and diseases. The development and application of SNP markers provide valuable tools for understanding genetic variation and identifying genomic regions associated with pest and disease resistance in sugarcane. These markers enable accurate and high-throughput genotyping, facilitating the detection of resistance-related loci and improving the efficiency of molecular breeding programs. The use of SNP markers, therefore, plays a crucial role in accelerating the development of improved sugarcane varieties with enhanced resistance to major biotic stresses, ultimately contributing to sustainable sugarcane production.

Keywords Sugarcane, Pests and diseases, DNA library, Illumina sequence, Axiom 100K SNP array, SNP markers, Marker-assisted selection, Sustainable sugarcane production

1 Introduction

Sugarcane (*Saccharum* spp.) is a tropical perennial Hatch-Slack cycle (C_4 cycle) grass and one of the world's most important sugar and bioenergy crops, thriving in warm climates with abundant moisture and well-drained loamy soils [1]. Globally, India and Brazil together account for more than half of total sugarcane production, underscoring the crop's economic and strategic importance in both tropical and

Nitish Kumar (ed.), *Molecular Markers in Plants*, Methods in Molecular Biology, vol. 3097,
https://doi.org/10.1007/978-1-0716-5663-1_24,

subtropical regions [2]. Commercial cultivation is primarily carried out through vegetative propagation using stem cuttings planted in trench, ridge and furrow, or ring pit systems with healthy nursery cane as seed material, while the spaced transplanting technique (STP) improves planting efficiency by chipping single buds with nodal tissue, treating them with fungicide, and raising seedlings in polybags for subsequent field transplantation [3]. In parallel, micropropagation through apical meristem culture has enabled large-scale production of disease-free seed cane, thereby improving purity of the variety and crop health [4]. SNP markers are increasingly used to identify genomic regions associated with resistance to major sugarcane diseases such as rust, red rot, leaf scald, and yellow leaf disease. In addition to diseases, several insect pests, such as the Sugarcane Top Borer and Sugarcane Early Shoot Borer, also cause significant damage to the crop [5].

Genetic improvement of sugarcane is complicated by its exceptionally large (~10 Gb), highly polyploid, and aneuploid genome, derived from interspecific hybridization between *Saccharum officinarum* and *S. spontaneum* [6]. SNPs can also be identified using other molecular techniques, such as high-throughput techniques like GBS, as well as techniques like cleaved amplified polymorphic sequences (CAPS), derived CAPS (dCAPS), allele-specific PCR (ASP), ligase chain reaction, single-nucleotide amplified polymorphism (SNAP), single-strand conformation polymorphism (SSCP), and dideoxy fingerprinting [7]. The genome's complexity long constrained marker development and genetic analysis; however, recent advances in target enrichment sequencing and high-throughput genotyping platforms, particularly the Axiom 100K SNP array, have substantially expanded genome-wide SNP discovery and genotyping capacity [10].

Comparative genomics approaches leveraging the sorghum reference genome have facilitated large-scale SNP identification and mapping of quantitative trait loci (QTLs) associated with major diseases such as rust, leaf scald, and yellow leaf virus [8]. Concurrently, genome-wide association studies (GWAS) and marker-assisted selection (MAS) have demonstrated increasing feasibility in sugarcane breeding, enabling more precise dissection of genomic regions governing disease resistance and other complex agronomic traits [9]. Among molecular marker systems, single-nucleotide polymorphisms (SNPs) have emerged as the most versatile due to their abundance, genome-wide distribution, and compatibility with high-throughput platforms, surpassing earlier systems such as RFLP, RAPD, AFLP, SSR, and InDel markers in efficiency and scalability [10].

2 Materials

Parental lines of Sugarcane genotypes (*see* **Note 1**) and fresh young leaf samples should be used as plant material. The materials required include micropipettes, sterile microtips, microcentrifuge tubes, water bath, liquid nitrogen, and a refrigerated centrifuge.

Table 1
Composition of CTAB extraction buffer

Component	Final concentrationFor (1 l)
Cetyltrimethylammonium bromide (CTAB)	2%
Polyvinylpyrrolidone (PVP-40)	2%
NaCl	1.4 M
EDTA (pH 8.0)	20 mM
Tris-HCl (pH 8.0)	100 mM
β-mercaptoethanol	0.2%
Distilled water	Up to 1 l

2.1 Chemicals and Reagents

For genomic DNA extraction and analysis, include CTAB extraction buffer (Table 1), β-mercaptoethanol, phenol: chloroform: isoamyl alcohol (25: 24: 1), isopropanol, absolute ethanol (100%), 70% ethanol, RNase A, TE buffer (pH 8.0), agarose, and nuclease-free water. Molecular biology reagents such as *Taq* DNA polymerase, along with PCR components, should be used for DNA amplification. The equipment required for DNA quality assessment and downstream analysis includes an agarose gel electrophoresis unit, gel documentation system, Nanodrop spectrophotometer, Qubit fluorometer, PCR thermal cycler, and Bioanalyzer.

2.2 Preparation of Buffers and Reagents

1 M Tris (pH 8.0): 100 ml of 1 M Tris should be prepared by dissolving 12 g of Tris base in 60 ml of distilled water. After dissolving, the pH should be adjusted to 8.0 using concentrated HCl. Then, the volume made up to 100 ml with distilled water and autoclaved at 15 psi pressure for 20 min.

0.5 M EDTA (pH 8.0): 100 ml of 0.5 M EDTA should be prepared by dissolving 18.6 g of EDTA in 75 ml of distilled water. NaOH pellets can be added to dissolve the EDTA as it takes time to dissolve completely. The pH should be adjusted to 8.0 using 1 N NaOH. Finally, the volume made up to 100 ml with distilled water and autoclaved at 15 psi pressure for 20 min.

5 M NaCl (pH 8.0): 5.84 g of NaCl should be added to 80 ml of distilled water and dissolved completely finally made up to 100 ml and autoclaved at 15 psi pressure for 20 min.

5% CTAB: 100 ml of 5% CTAB should be prepared by dissolving 5 g of CTAB in 80 ml distilled water and thereby 2 g of NaCl should be added to it. Then the final volume made up to 100 ml and autoclaved at 15 psi pressure for 20 min.

Chloroform:isoamyl alcohol (24:1): 100 ml of chloroform (24): isoamyl alcohol (1) should be prepared by adding 4 ml of isoamyl alcohol in 96 ml of chloroform and mix well. This mixture should be stored at 4 °C.

Phenol:chloroform:isoamyl alcohol (25:24:1): 100 ml of phenol (25): chloroform (24): isoamyl alcohol (1) should be prepared by adding 2 ml of isoamyl alcohol in 48 ml of chloroform with 50 ml of phenol and mix well. This mixture should be stored at 4 °C.

0.1× TE (pH 8.0): 20 ml of 0.1× TE should be prepared by adding 0.02 ml of 1 M Tris (pH 8.0) and 0.004 ml of 0.5 M EDTA (pH 8.0) to 18 ml. After dissolving properly, the volume made up to 20 ml and autoclaved at 15 psi pressure for 20 min.

Preparation of 50× TAE buffer: Dissolve 242.0 g of Tris base in 700 ml Milli-Q water with gentle heating. Then, 57.1 ml of glacial acetic acid, followed by 100 ml of 0.5 M EDTA (pH 8.0) should be added. Mix thoroughly and make the final volume to 1000 ml with Milli-Q water. Adjust the pH to 7.6–7.8 then autoclave the buffer and store at room temperature.

Ethidium bromide (EtBr) preparation: Ethidium bromide (EtBr) is a potent mutagen used to visualize nucleic acids, typically prepared as a 10 mg/ml stock solution in water (stored at in the dark) and used at a 0.5–1.0 μg/ml working concentration for agarose gels. It is added directly to molten agarose or used for post-electrophoresis staining to detect DNA/RNA via fluorescence under UV light (*see* **Note 3**).

- **Stock solution preparation (10 mg/ml)**
  - Dissolve 1.0 g of EtBr powder in 100 ml of distilled water. Store in a dark, light-proof container or wrapped in aluminum foil at 4 °C and handle with gloves or use fume hood to avoid inhaling dust.
- **Working solution preparation (0.5–1.0 μg/ml):**
  - **In-gel staining**: add 5–10 μl of the 10 mg/ml stock to 100 ml of molten agarose (<60 °C) for a final concentration of 0.5–1.0 μg/ml.
  - **For post-staining**: Dilute the stock solution in electrophoresis buffer (1× TAE or TBE) to a final concentration of 0.5–1.0 μg/ml and immerse the gel for 20–30 min.

CTAB extraction buffer: To prepare 1 litre of 2% CTAB extraction buffer (Table 1), dissolve 12.1 g Tris base in 700–800 ml distilled water and adjust the pH to 8.0 with concentrated HCl. Add 7.44 g EDTA and 81.76 g NaCl, stirring until fully dissolved. Slowly incorporate 20 g CTAB while gently heating to 60–65 °C to ensure complete dissolution, and optionally add 2 g PVP-40 (0.2%) if required. After all components are dissolved, make up the volume to 1 l with distilled water and mix thoroughly. Sterilization is usually unnecessary, but filtration may be performed if needed. Just before use, add freshly prepared β-mercaptoethanol to a final concentration of 0.2–0.5% (2–5 ml per liter), and store the buffer at 4°C.

3 Methods

3.1 Genomic DNA Extraction

3.1.1 PROTOCOL I: Genomic DNA Extraction by CTAB [11–13]

1. **Sample preparation and grinding:** Approximately 100–200 mg of fresh sugarcane leaf tissue should be taken to ground a fine powder using liquid nitrogen in the chilled mortar and pestle.
2. **CTAB buffer addition:** The powdered tissue should be transferred and mixed with pre-warmed CTAB extraction buffer (65°C) supplemented with β-mercaptoethanol. It was subsequent washed with 10 μl of proteinase K.
3. **Incubation for cell lysis:** The mixture should be incubated and gently mixed at 65 °C for 30–60 min to ensure proper cell lysis.
4. **Organic extraction:** An equal volume of phenol:chloroform:isoamyl alcohol (25:24:1) should be added and mixed thoroughly to remove proteins and contaminants.
5. **Centrifugation and phase separation:** The mixture should be centrifuged at 12,000 rpm for 10 min, and the clear aqueous phase should be carefully collected and transferred into a new microcentrifuge tube.
6. **DNA precipitation:** Genomic DNA should be precipitated and allowed to settle by adding ice-cold absolute ethanol or isopropanol and incubating at −20°C for 1 h.
7. **Ethanol washing:** The DNA pellet should be recovered and washed with 70% ethanol to remove residual salts.
8. **DNA resuspension:** The pellet should be air-dried and dissolved in TE buffer (pH 8.0).
9. **RNase treatment:** RNase A treatment should be performed at 37 °C for 30 min to remove RNA contamination (*see* **Notes 4 and 5**).

3.1.2 PROTOCOL II: Extraction of Genomic DNA by CTAB Method [11–13]

1. **Sample preparation and grinding:** Approximately 100 mg of young sugarcane leaf tissue should be taken and ground a fine powder in liquid.
2. **CTAB buffer addition:** The powdered tissue should be transferred and mixed with preheated CTAB extraction buffer (65°C).

Table 2
50× TAE buffer preparation for gel electrophoresis (1 l) (pH 7.6–7.8)

Requirements	Volume for (1 l)
Tris base (2 M)	242.2 g
Glacial acetic acid (1 M)	57.1 ml
EDTA (0.5 M)	100 ml

3. **Incubation for cell lysis:** The mixture should be incubated at 65°C for 30 min to facilitate cell lysis and release genomic DNA.
4. **Organic extraction:** An equal volume of chloroform:isoamyl alcohol (24:1) should be added to remove proteins and secondary metabolites.
5. **Centrifugation and phase separation:** The mixture should be centrifuged at 12,000 rpm for 10 min, and the aqueous phase should be transferred to a new tube.
6. **DNA precipitation:** DNA should be precipitated and incubated using cold isopropanol and incubating at −20°C for 30–60 min.
7. **Ethanol washing:** The DNA pellet should be recovered by centrifugation and washed with 70% ethanol.
8. **DNA resuspension:** After air drying, the pellet should be dissolved in TE buffer.

3.1.3 Analysis of Extracted Genomic DNA by Agarose Gel Electrophoresis [12]

1. **Gel preparation:** Agarose (1%) should be weighed and dissolved thoroughly in an appropriate volume of 1× TAE buffer until solution becomes fully transparent and uniform.
2. **Staining:** Once the melted agarose cools slightly (Still in liquid form), 5 μl of ethidium bromide should be added to the solution. This stain intercalates with DNA fragments and enables direct visualization of DNA bands after electrophoresis.
3. **Gel casting:** The agarose TAE mixture should be poured gently into a clean gel casting tray fitted with a comb. The gel should be left undisturbed to solidify. Casting the gel in a horizontal tray is essential for proper separation of bands.
4. **Gel setup:** After solidification, adhesive tapes or side seals from the tray are removed. The gel is placed carefully into the electrophoresis chamber and fully submerged in 1× TAE running buffer (Table 2) ensuring uniform electrical contact.
5. **Sample preparation:** DNA samples should be mixed with 6× gel loading dye (Table 3), followed by a brief pulse centrifugation to ensure proper mixing. The tracking dye will help to visualize sample loading and migration during electrophoresis.

Table 3
Preparation of DNA sample for gel electrophoresis analysis

Sample	Volume (10 µl)
DNA	2.0 µl
6× DNA gel loading dye (Thermo Fisher Scientific)	5.0 µl
Nuclease-free water	3.0 µl

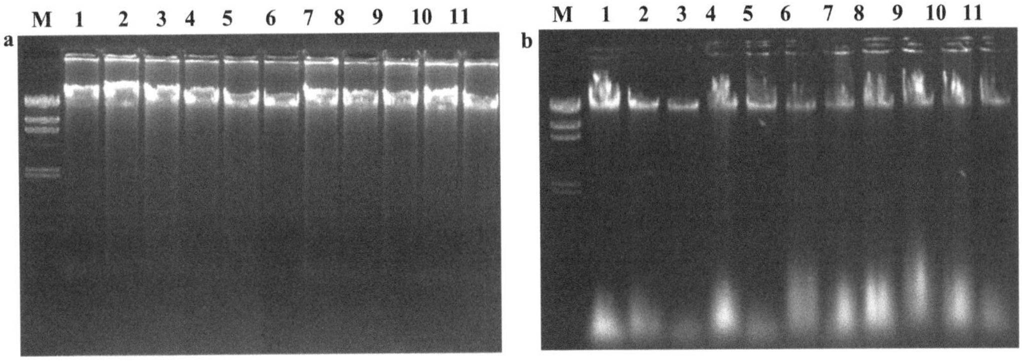

Fig. 1 Representative agarose gel electrophoresis image of genomic DNA extracted from leaf tissue. **a** Genomic DNA extracted using liquid nitrogen, followed by treatment with β-mercaptoethanol, Proteinase K, and RNase A showing high quality, intact genomic DNA without detectable RNA contamination. **b** Genomic DNA extracted using liquid nitrogen and β-mercaptoethanol only, showing intact genomic DNA with visible RNA contamination due to the omission of Proteinase K and *RNase* A treatments. Lane M: λ DNA/*Hind*III marker; Lanes 1–11: Genomic DNA extracted from leaf tissue [11]

6. **Sample loading:** Using a micropipette, DNA samples along with a 100 bp DNA ladder should be slowly dispensed into the wells to avoid well damage or sample leakage. Careful loading maintained accurate band patterns.
7. **Electrophoresis run:** The system is connected to a power supply and electrophoresis should be carried out at a constant voltage, ensuring even migration of amplified DNA fragments. Proper voltage maintains clear separation of DNA bands.
8. **Visualization:** After completion of the run, the gel should be transferred and visualized under a UV transilluminator (Fig. 1) (*see* **Notes 6 and 7**).

3.1.4 Assessment of DNA Purity by Nanodrop Spectrophotometer

1. **Instrument initialization:** The Nanodrop Spectrophotometer (ThermoFischer Scientific, USA) should be switched on and initialized according to the manufacturer's instructions. Select the dsDNA measurement mode in the instrument software.

2. **Pedestal cleaning:** The measurement pedestal should be cleaned and wipe using a lint free tissue and wipe gently with distilled water or 70% ethanol to remove any residues from the previous samples.
3. **Blank preparation:** Pipette 1–2 µl of nuclease-free water or TE buffer (pH 8.0) should be added onto the lower measurement pedestal and lower the measurement arm to perform blank calibration.
4. **Blank calibration:** The blank function should be selected and calibrated in the software to set the baseline absorbance for the measurement.
5. **Pedestal cleaning after blank:** After blanking, the pedestal should be wiped carefully with a clean lint free tissue to remove the blank solution.
6. **Sample loading:** Pipette 1–2 µl of the extracted genomic DNA sample onto the lower measurement pedestal.
7. **Measurement setup:** Lower the measurement arm gently so that the DNA sample spreads evenly between the optical surfaces.
8. **Absorbance measurement:** The sample should be measured and recorded at 260 nm and 280 nm.
9. **Data recording:** The instrument will automatically calculate the DNA concentration and A260/A280 purity ratio, and will be displayed on the screen.
10. **Cleaning between samples:** After measurement, the pedestal should be cleaned with lint-free tissue before analyzing the next sample.
11. **Interpretation of results:** DNA samples showing an A260/A280 ratio of approximately 1.8–2.0 should be considered to be good purity and suitable for downstream molecular analyses such as Polymerase Chain Reaction and SNP marker analysis [13].

3.2 Method 1: SNP Identification by NGS-Based Sequencing

3.2.1 Library Preparation

Library preparation is a critical step in next-generation sequencing (NGS) in which genomic DNA is processed into fragments with sequencing adapters. This allows the DNA fragments to be amplified and sequenced efficiently for downstream SNP identification.

3.2.2 Restriction Digestion

1. Genomic DNA should be quantified and normalized using fluorometric method, and each sample is normalized to 150 ng.

Table 4
Restriction enzymes used for double digestion

Enzyme name	Recognition sequence (5′-3′)	Type
Sphl-HF	GCATGC	Type II
Mbol	GATC	Type II

2. The restriction digestion reaction should be prepared by adding 150 ng genomic DNA, 10× compatible restriction buffer, restriction buffer, restriction enzymes Sphl-HF and Mbol restriction enzyme (high-fidelity engineered enzyme, New England Biolabs, USA), and nuclease-free water to obtain the required final reaction volume (Table 4).
3. The reaction mixture should be gently mixed and briefly centrifuged to collect the contents at the bottom of the tube. The reaction should be incubated at 37 °C for 9 min to achieve complete digestion.
4. Restriction enzymes should be inactivated and heated at 65 °C for 20 min.
5. 1.5× AMPure XP beads (Beckman Coulter, Inc, United States) should be added and mixed with the digested product for purification.
6. The mixture should be thoroughly mixed and incubated at room temperature to allow DNA fragments to bind to the magnetic beads.
7. The tubes should be placed and separated on a magnetic stand, and the supernatant should be carefully discarded.
8. The beads should be washed with freshly prepared 70% ethanol without disturbing the pellet, and then briefly air-dried to remove residual ethanol.
9. The purified digested DNA should be eluted and collected in an appropriate volume of nuclease-free water or elution buffer.
10. The fragmentation efficiency should be assessed and confirmed using a Fragment Analyzer System, confirming the presence of a uniform DNA fragment distribution with sizes below approximately 3 kb, indicating successful digestion.
11. The digested and purified genomic DNA should be used for downstream library preparation [14].

A) Fragmentation

- Purified genomic DNA fragmentation should be carried out and sheared into fragments suitable for sequencing (150–200 bp) by mechanical or enzymatic fragmentation.

Table 5
Composition of end repair reaction mixture

Components	Volume
Fragmented DNA	40 μl
10× end repair buffer	5 μl
End repair enzyme mix	5 μl
Nuclease-free water	50 μl

- Later, the fragment size should be analyzed and confirmed by the Bioanalyzer.

B) End repair

- This step should be performed to convert overhangs into blunt ends, where fragments of high-quality DNA, reaction mixture (50 μl) should be added (Tables 5 and 6) and incubated at 20 °C for 30 min.

C) A-tailing: This step should be carried out and used to add adenine overhangs to the 3′ ends, where the reaction mixture should be added (Tables 7 and 8) to the vial containing the DNA and incubated at 37 °C for 30 min, 70 °C for 5 min for enzyme inactivation.

D) Adapter ligation: A compatible adaptor obtained from illumina platform of concentration 15 μM should be added and ligated to the A-tailed DNA along with other reaction components (60 μl), and the mixture should be incubated and maintained at 20 °C for 15 min (Tables 9 and 10) and incubated at 20 °C for 15 min.

E) Cleanup

- Purification of DNA should be performed and carried out using AMPure XP Beads at a ratio of 60 μl beads for 60 μl reaction, where beads should be added and mixed with the DNA.
- The mixture should be incubated and allowed for 5 min at room temperature and place on magnetic stand. The supernatant should be removed and discarded and it should be washed twice with 200 μl 80% ethanol, later air dry for 5–7 min and elute in 30 μl nuclease-free water (Table 11).

3.2.3 Library PCR

The purified DNA should be suspended and mixed in a vial containing the PCR master mix (Table 12) and load it in a thermocycler to assess the amplification of adapter-ligated ends [15].

1. PCR reagents should be prepared and maintained on ice and the reaction should be set up and assembled in a sterile PCR tube.
2. 2× High-Fidelity PCR Master Mix should be added and mixed to the tube.

Table 6
Incubation for end repair reaction

Step	Temperature (°C)	Time
End repair reaction	20 °C	30 min
Enzyme inactivation	75 °C	5 min

Table 7
Composition of A-tailing reaction mixture

Components	Volume
End-repaired DNA	45 μl
10× A-tailing buffer	5 μl
Klenow fragment (exo-)	3 μl
Total reaction volume	53 μl

Table 8
Incubation of A-tailing reaction

Ste dap	Temperature (°C)	Time
A-tailing reaction	37 °C	30 min
Hold	4 °C	∞

Table 9
Composition of illumina adaptor reaction mixture

Components	Volume
A-tailed DNA	50 μl
10× ligation buffer	5 μl
T4 DNA ligase	3 μl
Illumina adapter	2 μl
Total reaction volume	60 μl

3. Adapter-ligated genomic DNA should be added and used as template.
4. 1 μl of each forward (10 μM; universal adapter primer) and reverse (10 μM; indexed/barcoded primer, if required) primer should be added to the reaction mixture.

Table 10
Incubation conditions for adapter ligation reaction

Step	Temperature	Time
Adapter ligation	30 °C	10 min
Hold	4 °C	∞

Table 11
Composition of DNA cleanup reaction using AMPure XP beads

Component	Volume
DNA reaction mixture	60 μl
AMPure XP beads (Beckman coulter, USA)	60 μl
80% ethanol (wash)	200 μl (×2 washes)
Nuclease-free water	30 μl

Table 12
Composition of PCR reaction mixture

Component	Final concentration	Volume (50 μl)
2× high-fidelity PCR master mix	1×	25 μl
Forward primer (10 μM)	0.2 μM	1 μl
Reverse primer (10 μM)	0.2 μM	1 μl
Adapter-ligated DNA	5–20 ng	10–15 μl

5. Nuclease-free water should be added and adjusted to make up the final reaction volume to 50 μl.
6. Mix the reaction gently by pipetting and briefly spin down and collect the contents at the bottom.
7. The tubes should be placed in a thermal cycler and perform initial denaturation at 95 °C for 3 min.
8. Denaturation at 95 °C for 30 s, followed by annealing at 55–65 °C for 30 s (depending on primer Tm) and extension at 72 °C for 30 s (adjusted according to fragment length), steps should be carried out and repeated for 30 cycles (Table 13).
9. Final extension should be performed at 72 °C for 5 min, and then hold the reaction at 4 °C until further processing.

Table 13
PCR reaction condition

Step	Temp-time	Cycles
Initial denaturation	98 °C/3 min	1
Denaturation	95 °C/15 s	
Annealing	60–65 °C/30 s	40
Extension	72 °C/1 min	
Final extension	72 °C/5 min	1
Hold	4 °C/∞	1

10. PCR products should be purified and eluted using magnetic beads or gel purification and elute the purified library in 20–30 μl of nuclease-free water or elution buffer (*see* **Notes 19 and 20**).
11. DNA should be quantified and analyzed using fluorometric methods such as Qubit, analyze the fragment size distribution by agarose gel electrophoresis or a Bioanalyzer, and confirm the absence of primer dimers while verifying the expected library size [16].

3.2.4 Library Quality Assessment

1. The final sequencing libraries should be subjected and analyzed using a Fragment Analyzer System.
2. The sequencing libraries should show and contain DNA fragment sizes ranging from 400 to 600 bp.
3. The fragment distribution should consist and include an insert size ranging from 260 to 460 bp, along with an adapter region of approximately 140 bp (Table 14).

3.2.5 Illumina Sequencing

A) Sequencing on illumina NextSeq500 (70 bp paired end reads)

1. Libraries should be pooled and loaded onto the illumina NextSeq500 platform. Sequencing should be performed and generated to produce 75 bp single-end reads.
2. Raw reads should be quality trimmed and retained to 70 bp and an average of 4 million reads per sample should be obtained and recorded.

B) Sequencing on illumina NovaSeq6000 (150 bp paired end reads)

1. Libraries should be normalized and pooled, and it should be loaded and sequenced on the Illumina NovaSeq6000 Platform. Sequencing should be carried out and generated using a 150 bp paired-end chemistry.

Table 14
Library fragment size distribution

Parameter	Size range (bp)
Total library fragment size	400–600 bp
Insert size	260–460 bp
Adaptor/barcode size	140 bp
Method used	Fragment analyzer

2. The run should generate and produce an average of 10 million reads per sample. High sequencing depth and coverage should be achieved for all samples [17].

3.2.6 Sequencing Data Analysis

1. Raw sequencing reads should be first subjected to stringent quality control, including trimming of adapter sequences and removal of low-quality bases and reads.
2. Reads that fail to meet predefined quality thresholds should be discarded to minimize errors in downstream analyses.
3. High-quality reads should be aligned and mapped to a reference sugarcane genome or transcriptome assembly.
4. In some cases, a complete or high-quality reference genome is unavailable, reads are aligned to a closely related species or to a de novo assembled reference.
5. Read alignment facilitates the identification of homologous regions, repetitive elements, and structurally complex genomic regions.
6. Aligned reads should be processed for SNP discovery using variant-calling pipelines optimized for polyploid genomes (*see* **Notes 22 and 23**).
7. Identified SNPs should be filtered and selected based on read depth, base quality, allele frequency, and missing data thresholds to reduce false-positive variants.
8. The resulting high-confidence SNP dataset will provide a dense and informative marker resource for downstream genetic analyses, including association mapping, genomic selection (GS), and marker-assisted breeding (Table 15) (*see* **Note 21**) [18].

3.2.7 Data Output and Coverage

The sequencing output from both platforms should be evaluated and analyzed for reading the number and quality. The NovaSeq6000 dataset (150 bp paired ends) should be provided for higher read depth and coverage compared to the NextSeq500 dataset (Ne-70) (*see* **Note 10**). Sequencing efficiency should be evaluated by comparing

Table 15
Bioinformatics tools used for discovering SNP

Tool/software	Analysis purpose	Link	Reference
FastQC	Raw read quality assessment	https://shorturl.at/qcAJO	[19]
SAMtools	Alignment file processing	http://www.htslib.org/	[14]
Genome Analysis Toolkit	Variant calling (SNP) detection	https://gatk.broadinstitute.org	[20]
BCF tools	Variant file processing	http://samtools.github.io/bcftools/	[20]
VCF tools	Variant filtering and population-level filtering	https://vcftools.github.io/index.html	[21]
PLINK	Population genetics	https://www.cog-genomics.org/plink/	[22]

read length, sequencing depth, and the effect of single-end and paired-end data.

1. Reads generated from the dataset should be trimmed to 70 bp while retaining paired-end information and high coverage, this dataset should be designated as 70 bp paired ends and compared with the original 150 bp paired-end data.
2. To examine the influence of read number per sample, 4 million reads per individual should be randomly subsampled and generated from the 70 bp paired-end (PE) dataset, producing the 70 bp paired-end with medium coverage (MC) dataset (PE/MC).
3. The 70 bp PE/MC dataset should be compared with the original 70 bp single-end dataset to assess the performance of paired-end reads relative to single-end reads.
4. In addition, the 150 bp paired-end dataset should be down sampled and generated from high to medium coverage to generate 150 bp PE/MC, allowing assessment of sequencing depth independent of read length (*see* **Note 11**) [1].

3.2.8 Protocol Robustness Evaluation

- To evaluate the robustness of the protocol, a new library of samples NA 78–724 and LCP 85–384 should be prepared from a new DNA extraction using the protocol described above and should be sequenced using the Illumina NovaSeq6000 platform [150 bp paired-end (PE) reads, ~4 million reads, referred to as medium sequencing coverage (MC)].
- In addition, a random subsampling of 150 bp reads should be performed to obtain a medium sequencing coverage dataset for the sugarcane genotypes NA 78–724 and LCP 85–384. In this

Table 16
Technical replicate strategy for robustness assessment [23]

Genotype	Dataset 1	Dataset 2	Sequencing platform and type
NA 78–724	New MC library	Subsampled No-150 MC	Illumina NovaSeq6000, 150 bp PE
LCP 85–384	New MC library	Subsampled No-150 MC	Illumina NovaSeq6000, 150 bp PE
NA 78–724	Independent library preparation	In silico subsampled dataset	Illumina NovaSeq6000, 150 bp PE
LCP 85–384	Independent library preparation	In silico subsampled dataset	Illumina NovaSeq6000, 150 bp PE

way, two datasets should be generated in independent experiments, which should be used as replicates for the assessment of the protocol robustness (Table 16).

3.2.9 SNP Discovery: High-Quality SNP Filtering

1. Pre-processed BAM files should be used and analyzed for variant calling for all samples.
2. Initial variant calling should be performed and generated using GATK (v3.8).
3. SAMtools (v1.9.1) should be used and applied for BAM file handling and variant processing support.
4. Raw VCF file should be generated and obtained containing all detected variants.
5. SNP_S should be extracted from the raw variant dataset if both SNP_S and InDels are present.
6. Compress the raw VCF file using bgzip, Index the compressed VCF file using BCF tools index to enable rapid access (*see* **Notes 8 and 9**).
7. Assess the total number of raw SNP_S (2,420,550 SNP_S). Remove SNP_S with low overall quality score, filter out SNP_S with low read depth and retain only biallelic SNP_S by removing multi-allelic loci.
8. Apply genotype quality filtering to remove unreliable genotype calls and remove SNP_S with a high percentage of missing data across samples.
9. Samples should be removed for showing excessive missing genotypes, if necessary and apply minor allele frequency filtering to eliminate rare variants.
10. Filter SNP_S based on maximum mean depth to avoid paralogous regions. Perform population-level missing data filtering step to monitor SNP retention.

11. Convert the final filtered VCF into the required downstream analysis formats. Export the high-confidence SNP dataset for population The filtered dataset should be converted [23].

3.2.10 Sample Level and Allele Frequency Filtering

SNPs with a missing rate which is greater than 0.3 and a minor allele frequency (MAF) below 0.05 should be removed using BCFtools, ensuring adequate representation and informativeness across the population. This step reduced the dataset to 197,994 SNPs.

3.2.11 Variant Quality Filtering

A second round of filtering should be performed with the thresholds QUAL < 30.0, QD < 2.0, FS > 60.0, MQ < 40.0, and SOR > 4.0 to eliminate low-confidence variants and loci with poor sequencing or mapping quality. After this step, 91,054 SNPs should be retained (*see* **Notes 12** and **13**).

3.2.12 Hardy-Weinberg Equilibrium Filtering [23]

1. The filtered dataset should be converted to PLINK format using VCF tools, and SNPs showing significant deviation from Hardy-Weinberg equilibrium should be excluded using PLINK (v1.9), thereby removing loci potentially affected by genotyping errors or strong selection.
2. This resulted in a final set of 55,750 high-quality SNPs, which should be used for subsequent population genetic analyses and marker development.
3. To maintain compatibility across different analytical platforms, chromosome identifiers should be standardized during file conversion using BCF tools annotate. All bioinformatics analyses should be conducted in a Linux environment, with the necessary software paths configured as environment variables to ensure smooth execution of the pipeline (*see* **Note 14**).

3.2.13 SNP Filtering and "Diploidization" Processing for the Polyploid Genome

Selection of SNPs [22]: Candidate SNPs loci obtained from the initial SNP discovery dataset should be selected for single copy filtering to eliminate interference from homologous regions present in the polyploid sugarcane genome.

1. **Extraction of flanking sequences**
   For each SNP locus, 100 bp upstream and 100 bp downstream flanking sequences should be retrieved from the sequence data for locus-specific assessment.
2. **Alignment with reference genome**
   The extracted flanking sequences should be aligned against the reference genome of *Saccharum officinarum* LA-Purple genome to determine their genomic distribution and copy number status.
3. **Identification of single copy loci**
   SNP loci showing a unique best match position in the genome or exhibiting low overall sequence homology should be re-

tained, while loci mapping to multiple genomic locations or falling within highly repetitive or homologous regions should be discarded (*see* **Note 2**).

4. **Physical position prioritization**
The physical locations of the retained SNPs should be recorded and prioritized to ensure selection of genomically unique and well distributed loci for downstream marker development.
5. **Primer design for locus specific amplification**
PCR primers should be designed from the filtered flanking regions, ensuring that primer binding sites avoided highly homologous sequences and repetitive elements to achieve specific amplification of the target locus (*see* **Note 25**).
6. **Validation of biallelic behavior**
The selected SNP loci should be validated across the test genotypes, and only those loci that produced clear, stable, and reproducible biallelic clustering patterns without complex multi-allelic or dosage-based signals should be retained.
7. **Diploidization of SNP markers**
The final set of SNP_S, originating from single-copy genomic regions and exhibiting simple biallelic segregation, should be considered as genetically "diploidized" markers within the complex polyploid background.
8. **Utilization for marker panel development**
These filtered and diploidized SNP loci should be subsequently used for PCR amplification and LDR probe design to enable highly accurate and reproducible genotyping in sugarcane (*see* **Notes 17 and 18**).

3.2.14 Linkage Map Construction [22]

- The filtered SNP markers should be used to construct a high-density genetic map using the software JoinMap growing and ordering should be performed based in recombination frequency and logarithm of odds (LOD) scores.
- A LOD threshold value of 10 should be applied to assign markers into linkage groups.
- The resulting linkage groups correspond to the number of chromosomes of the species under study.
- Due to the complex polyploid nature of sugarcane, the number of linkage groups obtained may vary depending on the mapping population and marker density.
- Unlike diploid species, modern sugarcane cultivars exhibit high Ploidy levels (Often 8–12×), large and complex genome structure.

- As a result, the linkage analysis generally produces a large number of linkage groups, often exceeding the basic chromosome number, reflecting homologous and homoeologous chromosome sets.

3.2.15 Marker Development and Genome Anchoring [22]

The identified SNP markers served as highly informative genetic markers for sugarcane genome analysis.

These SNP markers should be utilized to:

1. Anchor genome scaffolds to pseudo chromosomes enabling the integration of fragmented genome assemblies into chromosome-scale structures.
2. Order genomic sequences based on recombination frequencies derived from linkage analysis, thereby improving genome assembly accuracy.
3. Orient genomic regions ensuring the correct alignment and directional placement of scaffolds within the genome.

3.3 Method 2: High-Throughput SNP Genotyping Using Axiom Sugarcane 100K SNP Array

3.3.1 SNP Genotyping Using Axiom 100K Array (ThermoFischer Scientific, USA) [7, 10]

1. High-quality genomic DNA samples should be extracted using the CTAB method and quantified using Qubit or Nanodrop spectrophotometer. DNA samples should be normalized to a concentration of 50–100 ng/μl, ensuring an A260/A280 ratio of 1.8–2.0.
2. Approximately 200–250 ng of genomic DNA should be used for whole-genome amplification. The amplification reaction should be prepared by adding DNA to the amplification master mix and incubated in a thermal cycler for 20–24 h to ensure sufficient amplification of genomic fragments.
3. The amplified DNA should be subjected to enzymatic fragmentation by adding a fragmentation enzyme mix and incubating under recommended conditions (typically at 37 °C). Fragmentation should generate DNA fragments in the size range of approximately 25–125 bp.
4. Fragmented DNA should be end-labeled using appropriate labeling reagents provided in the Axiom 2.0 reagent kit. The labeling reaction should be incubated to allow incorporation of biotin-labeled nucleotides.
5. The labeled DNA samples should be transferred onto the Axiom Sugarcane 100K SNP array plate. Hybridization should be performed by incubating the array at controlled temperature

Table 17
Software pipeline for axiom SNP genotyping analysis

Software/tool	Output
Axiom analysis suite	Genotyping calls, QC metrics
Affymetrix power tools	Processed genotype files
TASSEL	GWAS results
R software	Graphs, plots, and summaries

conditions for 23–24 h to allow specific binding of DNA fragments to complementary SNP probes.

6. Following hybridization, the array should be subjected to automated washing and staining using the GeneTitan Multi-Channel Instrument (ThermoFisher Scientific, USA) to remove non-specifically bound DNA and enhance signal detection.
7. The processed array should be scanned using the GeneTitan instrument to capture fluorescence intensities corresponding to each SNP locus. Raw image data should be generated and stored as Cell intensity (CEL) files.
8. The CEL files should be imported into Axiom Analysis Suite or Affymetrix Power Tools software for genotype calling. SNP genotypes should be assigned using the Axiom GT1 clustering algorithm, classifying alleles into homozygous and heterozygous categories (Table 17).

3.3.2 Linkage Mapping and QTL Identification Using SNP Markers: Construction of Linkage Map

1. High-quality SNP markers should be selected from the filtered SNP dataset for linkage analysis.
2. A suitable mapping population such as biparental or F1 population should be used for genotyping.
3. Genotyping data for all individuals in the population should be generated using SNP markers.
4. Markers showing high missing data or segregation distortion should be filtered and removed prior to analysis.
5. Markers showing high missing data or segregation distortion should be filtered and removed prior to analysis.
6. Pairwise recombination frequencies, SNP markers should be grouped into linkage groups representing chromosomes.
7. Based on recombination frequencies, SNP markers should be grouped into linkage groups representing chromosomes.
8. Maker ordering within each linkage group should be performed using suitable algorithms.
9. Genetic distances between markers should be estimated using mapping functions.

3.3.3 Utilization of QTL Regions for Marker Development [22]

1. SNP markers located within or near the identified QTL regions should be selected for further analysis.
2. Flanking sequences or selected SNP markers should be extracted for primer and probe design.
3. The SNP markers associated with resistance traits should be prioritized.
4. These SNP markers should be used for downstream marker development and validation.

3.3.4 Marker Development: Multiple Marker Typing System [23]

1. Start with the high-confidence SNP dataset and reference genome sequence. Retrieve chromosome position for each SNP locus.
2. Extract 100 bp upstream and 100 bp downstream sequences for every SNP. Use this sequence for specificity assessment and primer/probe design.
3. Screen the ±100 bp flanking regions against the complete variant dataset. Retain only those loci containing the target SNP alone, and then discard loci having additional SNP_S or InDels within the flanking region (*see* **Note 24**).
4. Align the flanking sequences to the reference genome using BLAST. Select loci showing a single unique genomic hit and remove loci with multiple high-similarity matches.

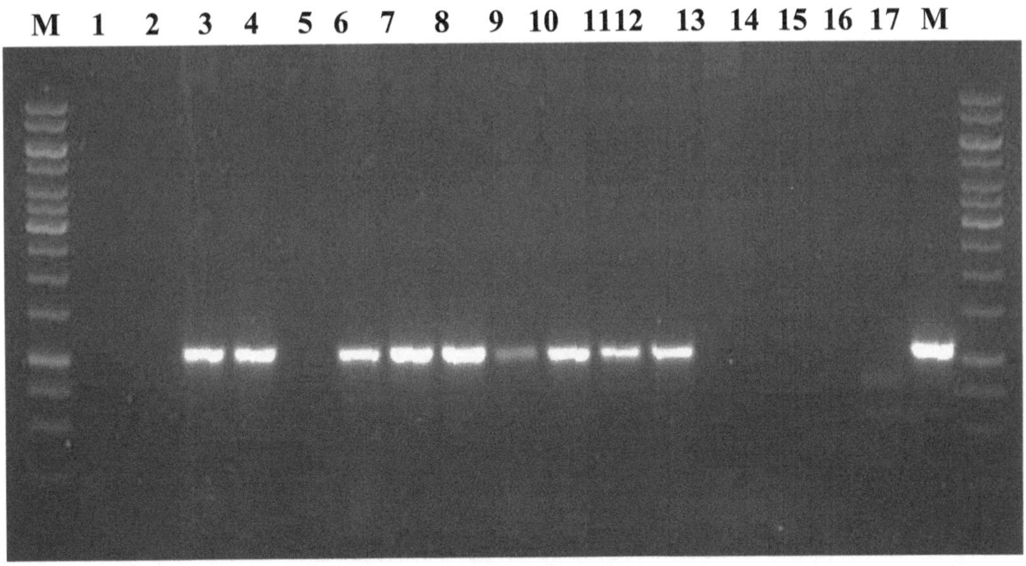

Fig. 2 Representative agarose gel electrophoresis image showing PCR-based validation of a selected InDel (insertion/deletion) marker in a representative set of 17 genotypes. PCR amplification of the selected InDel marker revealed polymorphic banding patterns among the genotypes. Lane M: 1 kb DNA ladder; Lanes 1, 2, 5, 13, 14, 15, and 16: absence of the target marker allele; Lanes 3, 4, 6–12, and 17: presence of the target marker allele [24]

5. Compile SNP_S passing both filters and finalize the list of candidates SNP_S for assay conversion.
6. Design locus-specific primers to amplify short fragments containing the SNP and perform ligation detection reaction using PCR products as template, and detect allele-specific ligation products by gel or capillary electrophoresis.
7. Screen all converted markers across representative genotypes and retain markers showing clear, reproducible and unambiguous genotyping results. Record the successfully developed markers.
8. Group validated markers based on amplicon size and primer compatibility, Optimize multiplex PCR and LDR conditions. Organize markers into four multiplex panels for high-throughput genotyping (Fig. 2) (*see* **Note 15**).

3.3.5 Multiple SNP Typing Kit

- Amplification primers, splice primers (Tables 18 and 19) for primer information) and Ligase Detection Reaction (LDR) probes are designed based on the sequence information of the target SNPs.
- The amplification primers are used to amplify the target region containing the SNP locus, while the LDR probe specifically recognizes both alleles of the SNP (the reference allele and the variant allele).
- The design ensures that the end of the probe matches the allele exactly.
- The fluorescent signal intensity of the PCR-LDR is analyzed in conjunction with the genotype data to determine the SNP genotypes in different samples (*see* **Note 16**).

3.3.6 SNP Genotyping and Detection [25]

1. After confirming successful PCR amplification, purify the PCR products of required to remove excess primers, nucleotides, and enzymes.
2. Transfer an appropriate volume of the amplified PCR products onto a clean microcentrifuge tube for SNP analysis.
3. Perform SNP detection using suitable genotyping methods such as Sanger sequencing, KASP genotyping, or other allele-specific genotyping techniques.
4. For sequencing-based SNP detection, prepare the purified PCR products according to the sequencing facility guidelines and submit them for sequencing.
5. Once sequencing results are obtained, download the nucleotide sequence data in FAST format.

Table 18
Multiplex PCR reaction ligation primer sequences [26]

Primer name	Primer sequence	Length	PCR length(bp)
100F	GCAGGACGCAATTCTGTTATCAT	23	100
100R	TAGTGACTTGTTGGCATGTTTGTC	24	100
101F	GCCACAACATTCAAATACCCACA	23	91
101R	AGAGCCGCTTGAGCTAGCAAT	21	91
102F	GAACTATGGAAATTTAGGTATTTGGTGCT	29	108
102R	GAGTATTGATCAATAAAAGACGGTGTGT	28	108
103F	CCTGGCCATATGTTCTAAGGATAGC	25	98
103R	ATTGACAGAGCAGTAGTGCCACCAG	25	98
104F	GTCTGTCCTTAACACTCCTGCTCCTT	26	181
104R	AAGTGCACATGTAGTGCAACAACC	24	181
107F	TTCCATGCGTAACTGTCATAGCTG	24	107
107R	TCGTCCATGTGGTGTGACCAGTA	23	107
10F	GTTTGTTCACACCAAAATTGAAGGA	25	124
10R	CAAATGAATGATCAAAGTGTAATGTAGG	28	124
110F	GGATGATTTCAAACTTTGGACGTGTAG	27	90
110R	AGATTTTACTGTTTGGTTGGATGCTTG	27	90
117F	GATGCCAAAATGATAAAGCACTGCTAC	27	115
117R	AGGAACTTTGGGCTTCTTATTCCAG	25	115
118F	GAAGATAAGTTTCAAGGCACTGGTTTAG	28	123
118R	CGAGTACCCTGTGGCAAGAGCTA	23	123

Primer names indicate the locus number followed by primer orientation. F denotes the Forward primer, and R denotes the Reverse primer used for PCR amplification.

6. Import the sequence files into sequence analysis software such as MEGA (Molecular Evolutionary Genetics Analysis) or BioEdit (Biological Sequence Alignment Editor) (Table 20).
7. Align the obtained sequences with the reference sequence to identify single-nucleotide variations at specific loci.
8. Examine the aligned sequences carefully to determine the presence of SNP_S by observing base substitutions at the target positions.
9. Record the SNP genotype for each sample and compile the data for further analysis
10. Use the SNP genotype data to differentiate plant genotypes and identify resistance-associated alleles in crops.

Table 19

Primer sequences for multiple PCR reaction [26, 27]

Primer name	Sequence (5′- 3′)	PCR length(bp)
100F3	AAAGGCAAGAGTCGACAAACATGTTTTTTTTTTTTTTTTTT TTTTTTTTTTTT[NED]	53
100FC	TGTACGTGATACTGACAAGTTGTACATGATATACACATATT AAGCTTAAAAGTAATTTGGTGACATCTTTGAC	73
100FG	TGTACGTGATACTGACAAGTTGTACATGATATACACATATT TTAAGCTTAAAAGTAATTTGGTGACATCTTTGAG	75
101F3	GCAGTTGCAGCTTAAGCATCTTCTCTTTTTTTTTTTTTTT TTTTTTTTTTTTTTTTTT[FAM]	59
101FC	TGTACGTGATACTGACAAGTTGTACATGATATACACATAT TATTCAAATACCCACACTAACCTCTAAATGAC	72
101FT	TGTACGTGATACTGACAAGTTGTACATGATATACACATAT TTTATTCAAATACCCACACTAACCTCTAAATGAT	74
102R3	AAGATCGAAAGCACCAAATACCTAAATTTTTTTTTTTTTTT TTTTTTTTTTTTTTTTTT [VIC]	59
102RC	TGTACGTGATACTGACAAGTTGTACATGATATACACATAT TTACTAAATGAGCAACAATAAATGCTCAAATGACTGG	77
102RT	TGTACGTGATACTGACAAGTTGTACATGATATACACATAT TTTACTAATATGAGCAACAATAAATGCTCAAATGACTGA	79
103F3	ATCCTGGTGGCACTACTGCTCTTTTTTTTTTTTTTTTT [NED]	39
103FA	TGTACGTGATACTGACAAGTTGTACATGATATACACATAT TGATAAATTAGCTGGATGCCTGGAGAA	67
103FC	TGTACGTGATACTGACAAGTTGTACATGATATACACATATT TTGATAAATTAGCTGGATGCCTGGAGAC	69
104F3	ATGATTGGGCATTTTTAATACGTGTGTTTTTTTTTTTTTT TTTTTTTTTTTTTTTTTT[NED]	59
104FA	TGTACGTGATACTGACAAGTTGTACATGATATACACATATT TTTATCGTGTTTTGGAAACTACCTTTAATATGCGTA	77
104FT	TGTACGTGATACTGACAAGTTGTACATGATATACACATAT TTAATTATCGTGTTTTGGAAACTACCTTTAATATGCGTT	79
107R3	AAATGCTGTAGTCCTGTAGCTATCAGTAACTTTTTTTTTT TTTTTTTTTTTTTTT[PET]	56
107RC	TGTACGTGATACTGACAAGTTGTACATGATATACACATATT GGGAACATGCACGATGCCATG	62
107RG	TGTACGTGATACTGACAAGTTGTACATGATATACACATATT TTGGGAACATGCACGATGCCATC	64

Table 19 (continued)

Primer name	Sequence (5′- 3′)	PCR length(bp)
10R3	ATAAAATGGACCGAAGGGAGTACTTTACTTTTTTTTTTTT TTTTTTTTTTTTTTTTTT[VIC]	57
10RC	TGTACGTGATACTGACAAGTTGTACATGATATACACATAT TCGTGTCATTCTAACTTGCGCAAG	64

• Primer sequences are shown in the 5′- 3′ direction. Primer names consist of a locus number followed by primer orientation and primer type. • The numbers such as 100, 101, 102 indicate the specific SNP locus or marker, while F and R denote the Forward and Reverse Primers, respectively. • The additional letters, such as C, T, G, or A, represent common or allele-specific primers used for SNP discrimination. • The fluorescent labels FAM, VIC, NED, and PET indicated in brackets represent different reporter dyes used for detection of amplified fragments during capillary electrophoresis. • These dyes emit fluorescence at different wavelengths, enabling simultaneous detection of multiple PCR products in multiplex assays. • The Poly-T tails (TTTT) present in certain primers are included to modify fragment length and improve separation of PCR products during electrophoresis analysis. • The universal sequence present at the 5′ end of several primers represents a common tail used to enhance amplification efficiency. The length column indicated the total number of nucleotides present in each primer.

Table 20
Software pipeline for sequence alignment and SNP identification

Software	Link
MEGA	https://www.megasoftware.net/
BioEdit	https://bioedit.software.informer.com/

4 Notes

1. Sugarcane possesses a highly polyploid and heterozygous genome, which complicates SNP discovery, accurate allele dosage estimation, and reliable genotype calling.
2. The abundance of repetitive DNA sequences and homoeologous loci further creates ambiguity in read alignment and marker interpretation.
3. If bands are faint check the working concentration and ensure proper mixing in agarose and, if background fluorescence is high, reduce EtBr concentration or wash the gel briefly with water before visualization.
4. Efficient grinding of tissue in liquid nitrogen is critical (see DNA extraction CTAB method). Incomplete grinding leads to poor cell lysis and low DNA recovery.
5. A modified CTAB method containing PVP and β-mercaptoethanol is strongly recommended for sugarcane tissues. These components neutralize phenolic compounds that otherwise bind and degrade DNA.

6. If the agarose gel appears uneven and contains bubbles or wells are damages, it may affect DNA migration. Ensure agarose is completely dissolved, avoid bubble formation while pouring and insert/remove the comb carefully.
7. Faint or smeared bands may occur due to improper loading, incorrect voltage, or uneven mixing of ethidium bromide and loading dye. Ensure proper DNA dye mixing, correct buffer concentration and maintain a stable voltage during electrophoresis.
8. Compress VCF files using bgzip and create an index using BCF tools to enable faster access and efficient downstream analysis.
9. Adjust variant calling parameters in Genome Analysis Toolkit (GATK) to apply stricter thresholds for base quality and mapping quality. Increasing the minimum quality score can help eliminate sequencing artefacts.
10. Poor library quality, improper pooling, or incorrect fragment size can lead to low read counts or poor coverage. Ensure accurate library quantification, proper fragment size and correct normalization before loading onto the sequencing.
11. Errors may occur if read trimming, subsampling, or dataset labeling is not performed correctly. Ensure proper quality trimming, accurate random subsampling, and correct dataset identification to obtain reliable sequencing efficiency comparisons.
12. Poor alignment quality or incorrect parameter settings in GATK may lead to false SNP detection. Ensure properly aligned BAM files, adequate sequencing depth, and correct variant calling parameters before generating the VCF file.
13. Excessive missing genotypes or very rare variants may reduce the reliability of downstream analyses. Apply appropriate filtering thresholds for missing rate and minor allele frequency.
14. Errors may occur while compressing, indexing, or converting VCF files using tools such as BCF tools, SAMtools, or PLINK. Ensure correct file formats, proper indexing, and compatible software versions in the Linux environment before running the pipeline.
15. If primers amplify multiple regions, SNP typing may produce ambiguous results. Ensure proper BLAST verification of flanking sequences and design highly specific primers targeting unique genomic regions.
16. Poor PCR amplification or inefficient ligation detection reaction can lead to weak fluorescent signals. Optimize PCR conditions, primer concentration, and multiplex reaction parameters to obtain clear and reproducible SNP detection.

17. Some SNPs may align to multiple regions due to the highly polyploid and repetitive nature of the sugarcane genome.
18. Primers designed from homologous or repetitive regions may produce multiple PCR products. To avoid this, design primers from unique flanking regions and optimize PCR conditions to obtain clear and reproducible amplification.
19. This may occur due to poor template quality, incorrect annealing temperature, or degraded primers. Ensure high-quality adapter ligated DNA optimize annealing temperature according to primer and to verify the primer integrity.
20. Excess primer concentration or improper PCR conditions may cause unwanted products. Optimize primer concentration, adjust annealing temperature, and perform proper purification to obtain clean libraries.
21. Errors may occur due to low-quality SNP data, missing genotypes, or incorrect parameter settings in software. Ensure high-quality filtered SNP datasets and verify software parameter before performing phylogenetic, PCA, and population structure analyses.
22. Low-quality reads or untrimmed adapter sequences can lead to incorrect alignment and false SNP detection.
23. Improper reference genome selection or insufficient read depth may cause inaccurate variant detection. Use a suitable reference genome and apply strict filtering criteria for reliable SNP identification.
24. This may occur due to inaccurate phenotypic data or environmental variation. Ensure precise phenotyping, proper validation across populations and testing under multiple environmental conditions before applying SNP markers in marker-assisted selection.
25. Avoid SNPs/indels, repetitive elements, and highly homologous regions during primer design. Perform in-silico specificity checks, redesign primers if non-specific amplification occurs, and confirm SNPs by PCR and sequencing or genotyping assays.

References

1. Msomba BH, Ndaki PM, Joseph CO (2024) Sugarcane sustainability in a changing climate a systematic review on pests, diseases, and adaptive strategies. Front Agron 6:1423233
2. Perez Contreras S, Hernandez Rosas F, Lizardi Jimenez MA, Herrera Corredor JA, Baltazar Bernal O, Avalos de la Cruz DA, Hernandez Martínez R (2025) Sugarcane industry by products: a decade of research using biotechnological approaches. Recycling 10(4):154
3. OECD/FAO (2025) OECD-FAO Agricultural Outlook 2025-2034, OECD Publishing, Paris/FAO, Rome. https://doi.org/10.1787/601276cd-en.

4. Solomon S, Rao GP, Malik D (2026) Energizing economies the global impact of Indian sugar. J Sugar Tech 28:354–359
5. Raja Babu C, Rajanbabu V, Balaji T, Masilamani P, Jasmine RS, Nageswari R, Anitha R, Thirumalaiselvi K (2025) Evaluation of different planting methods for sustainable sugarcane production. A comprehensive review. Plant Sci Today 12(2):1–12.
6. Nagargade M, Tyagi V, Singh A, Upadhyay P, Rathore SS (2025) Advanced seed multiplication and planting techniques for sustainable sugarcane. Indian Farming 75(12):65–68
7. Gupta V, Sharma S (2025) Advancements in sugarcane tissue culture. A review. Agric Rev 46(4):681–685
8. Sekhar VC, Varma PK, Saritha R, Adilakshmi D, Vasavi R, Rao CM (2026) Evaluation of meristem and tissue culture techniques for the management of yellow leaf disease in sugarcane. J Adv Biol Biotechnol 29(1):55–62
9. Sandhu KS, Shiv A, Kaur G, Meena MR, Raja AK, Vengavasi K, Mall AK, Kumar S, Singh PK, Singh J, Hemaprabha G (2022) Integrated approach in genomic selection to accelerate genetic gain in sugarcane. Plants 11(16):2139
10. You Q, Yang X, Peng Z, Islam MS, Sood S, Luo Z, Comstock J, Xu L, Wang J (2019) Development of an Axiom Sugarcane100K SNP array for genetic map construction and QTL identification. Theor Appl Genet 132(10):2829–2845
11. Suganya S, Lakshmi S (2018) Optimizing a protocol for DNA isolation of leaf *Saccharum officinarum*. J Pharmacog Phytochem 7(1): 1165–1170
12. Bermudez Guzman MJ, Guzman Gonzalez S, Orozco Santos M, Velazquez Monreal JJ, Buenrostro Nava MT, Michel Lopez CY (2016) Optimizing a protocol for DNA isolation of leaf *Saccharum officinarum*. Rev Mex Cienc Agric 7(4):767–778
13. Chansri S, Burns P, Nilsang S, Kositratana W, Watcharachaiyakup J (2024) Comparison of three genomic DNA extraction methods from sugarcane for detection of sugarcane white leaf phytoplasma. Nat Life Sci Commun 23(4):e2024052
14. Li H, Handsaker B, Wysoker A (2009) The sequence alignment/map format and SAMtools. Bioinformatics 25(16):2078–2079
15. Purcell S, Neale B, Todd-Brown K (2007) PLINK A tool set for whole-genome association and population-based linkage analyses. Am J Hum Genet 81(3):559–575
16. Chaudhary D, Jeena AS, Pal N, Rohit R, Kumar S (2024) DNA fingerprinting and genetic diversity analysis in advanced clones of sugarcane (*Saccharum sp.* complex) using STMS markers. Braz Arch Biol Technol 67:e24231022
17. Head SR, Komori HK, LaMere SA, Whisenant T, Van Nieuwerburgh F, Salomon D, Ordoukhanian P (2014) Library construction for next-generation sequencing overviews and challenges. BioTechn 56(2):61–77
18. Healey AL, Garsmeur O, Lovell JT, Shengquiang S, Sreedasyam A, Jenkins J, Plott CB, Piperidis N, Pompidor N, Llaca V, Metcalfe CJ (2024) The complex polyploid genome architecture of sugarcane. Nature 628(8009):804–810.
19. Andrews S (2010) FastQC: A Quality Control Tool for High Throughput Sequence Data. https://www.bioinformatics.babraham.ac.uk/projects/fastqc/.
20. McKenna A, Hanna M, Banks E (2010) The Genome Analysis Toolkit A MapReduce framework for analyzing next-generation DNA sequencing data. Genome Res 20(9):1297–1303
21. Danecek P, Bonfield JK, Liddle J (2021) Twelve years of SAMtools and BCFtools. GigaScience 10(2):giab008
22. Clevenger J, Chavarro C, Pearl SA, Ozias-Akins P, Jackson SA (2015) Single nucleotide polymorphism identification in polyploids a review example and recommendations. Mol plant 8(6):831–846
23. McGowan SE, Albertini F, Thomas SJ, Lomax AJ (2015) Defining robustness protocols a method to include and evaluate robustness in clinical plants. Phy Med Bio 60(7):2671–2684
24. Parida SK, Kalia S, Pandit A, Nayak P, Singh RK, Gaikwad K, Mohapatra T (2016) Single nucleotide polymorphism in sugar pathway and disease resistance genes in sugarcane. Plant Cell Rep 35(8):1629–1653
25. Zhao W, Wang Y, Yang Z, Zhao J, Huang C, Huang G, Xu L, Liu J, Zhao Y, Zhang Y, Deng Z, Zhao X (2026) Development and application of an LDR-based SNP panel for high-resol genotyping and varity identifi in sugarcane 16(3):343
26. Pearman WS, Urban L, Alexander A (2022) Commonly used Hardy Weinberg equilibrium filtering schemes impact population structure inferences using RADseq data. Mol Ecol Resour 22(7):2599–2613
27. Manimekalai R, Suresh G, Govinda Kurup H, Athiappan S, Kandalam M (2020) Role of NGS and SNP genotyping methods in sugarcane improvement programs. Crit Rev Biotech 40(6):865–880

Check for updates

Chapter 25

Development and Applications of Single Nucleotide Polymorphisms (SNPs) for the Improvement of Fiber Traits in Cotton Crops

Tanu Priya Natarajan, Gokul Murugesan, Janani Sree Sivakumar, Dhandapani Gurusamy, Kanakachari Mogilicherla and Vasanth Krishnan

Abstract

Single-nucleotide polymorphisms (SNPs) are among the most stable and widely used molecular markers in plant genomics and breeding. Especially in cotton (*Gossypium hirsutum* L.), SNP markers play a critical role in identifying genetic variations associated with different development traits such as fiber length, strength, micronaire, and elongation. This chapter describes the workflow of SNP marker development in a comprehensive manner targeting fiber development in cotton. The process is initiated by high-quality genomic DNA isolation followed by sequencing carried out through high-throughput sequencing approaches such as Whole Genome Sequencing (WGS), Genotyping-by-Sequencing (GBS), and Restriction Site-Associated DNA Sequencing (RAD-seq). Using the reference genome *Gossypium hirsutum* L., the sequencing reads are subjected to various bioinformatics analyses such as quality control, alignment, and SNP calling to identify true SNPs. Functional SNPs associated with genes are prioritized for further analysis. Selected SNPs are validated through PCR amplification and sequencing, followed by conversion into markers for accurate genotyping of germplasm and breeding populations. Overall, this protocol provides the discovery and validation of SNP markers for fiber development in cotton. Consequently, the validated markers are effectively applied in genetic studies, QTL mapping, and marker-assisted breeding strategies aimed at improving fiber quality traits in cotton.

Keywords Cotton, Fiber quality, Genomic DNA, Single nucleotide polymorphism, Next generation sequencing, CASP marker

1 Introduction

Molecular markers have revolutionized the field of plant genetics and breeding by facilitating the identification of genetic variations present in DNA sequences associated with important agronomic traits [1]. Among different types of molecular markers, single-nucleotide polymorphism (SNP) is the most abundant and stable type of genetic variation present in the genome of plants compared to other types of molecular markers like RFLP, RAPD, AFLP, SSR,

Nitish Kumar (ed.), *Molecular Markers in Plants*, Methods in Molecular Biology, vol. 3097,
https://doi.org/10.1007/978-1-0716-5663-1_25,

and ISSR. SNPs are nucleotide variations at specific positions in the genome among individuals of a particular species. Because of their abundance and stability, SNPs are considered appropriate for large-scale genotyping applications [2]. The discovery of SNP markers has been significantly accelerated with the help of Next Generation Sequencing technologies. Next Generation Sequencing technologies include whole-genome resequencing, transcriptome sequencing or RNA sequencing, genotyping by sequencing, restriction site-associated DNA sequencing, and SNP arrays. These technologies can efficiently identify thousands to millions of SNPs in the genome of plants. After their identification and validation, SNPs can be used with different genotyping technologies like KASP, TaqMan, HRM, and SNP chips [3]. SNPs can also be identified using other molecular techniques, such as high-throughput techniques like GBS, as well as techniques like cleaved amplified polymorphic sequences (CAPS), derived CAPS (dCAPS), allele-specific PCR (ASP), ligase chain reaction, single-nucleotide amplified polymorphism (SNAP), single-strand conformation polymorphism (SSCP), and dideoxy fingerprinting [4]. It acts as a guide in identifying functional SNPs linked to different traits in a reference system, thus enabling the identification of differences in different members of the same species [1]. Reference genomes in association with NGS techniques are the backbone of cotton genomics, as they enable SNP identification in different germplasm and genotypes of cotton. Large sets of SNPs were employed in association mapping techniques to identify quantitative trait loci (QTLs) linked to fiber quality traits in *Gossypium hirsutum* L., thus enabling insights into candidate genes related to biological processes such as fiber elongation, cell wall formation, and hormone regulation in cotton [5]. For example, a genome-wide association study employing GBS revealed over 200 SNPs linked to fiber yield traits in upland cotton, which equates to more than 130 QTLs and thousands of candidate genes expressed during different developmental stages in the fibers, ovule, and *seeds* [6]. SNP markers have also proven to be very useful in the dissection of complex traits like fiber quality. SNP variation has the potential to be grouped and linked to traits expressed during critical developmental stages, demonstrating the potential of SNPs to reveal trait-linked loci. This information could be very useful in marker-assisted selection and genomic breeding approaches, especially to improve fiber quality in cotton [7]. Upland cotton productivity faces numerous environmental stresses during its growth stage, and SNP variation among genotypes could be a reflection of the differential response to environmental stresses [1]. Globally, cotton is an important crop for the textile industry as a source of natural fibers, with *G. hirsutum* being allotetraploid and accounting for almost 95% of global cotton production due to its high productivity and adaptability. Increasing population trends along with the rising need for textiles emphasize the importance of increasing the productivity of cotton to improve the profitability of the farmers [7]. There are six allotetraploid species and 46 diploid species of the genus *Gossypium*. Though upland cotton is widely cultivated, the quality of the fiber must be improved to meet the standards of modern spinning technology [8]. SNP markers have now been found to play a vital role in the understanding of genetic diversity in the plant genome. Moreover, the application of NGS along with advanced bioinformatics tools has now sped up the use of SNP markers. Accordingly, the present chapter is focused on providing a protocol for the development of SNP markers.

2 Materials

Cotton sample (boll/fiber/leaf), Genomic DNA extraction buffer (see Table 1), liquid nitrogen, mortar and pestle, microfuge tubes, micropipettes, microtips, ice bucket, water bath, 70% ethanol, centrifuge, β-mercaptoethanol, chloroform: isoamyl alcohol (24:1), isopropanol, TE buffer, TAE buffer, agarose powder, DNA gel loading dye, DNA ladder (100 bp), ethidium bromide, gel documentation system, Nanodrop spectrophotometer (ThermoFischer Scientific, USA). Illumina HiSeq 2000 platform, MinElute columns (QIAGEN, Valencia, California, USA), Qubit dsDNA HS Assay Kit (Thermo Fisher Scientific, USA), AccuBlue High Sensitivity Quantitation Kit (Biotium Fremont, California, USA), KAPA3G Plant PCR Kit Kapa (Biosystems, Roche, Basel, Switzerland).

1 M Tris (pH 8.0): 100 ml of 1 M Tris should be prepared by dissolving 12 g of Tris base in 60 ml of distilled water. After that, the pH should be adjusted to 8.0 using concentrated HCl. Then, the volume should be made up to 100 ml with distilled water and autoclave at 15 psi pressure for 20 min.

0.5 M EDTA (pH 8.0): 100 ml of 0.5 M EDTA should be prepared by dissolving 18.6 g of EDTA in 75 ml of distilled water. NaOH pellets should be added to dissolve the EDTA as it taken time for dissolving completely. The pH should be adjusted to 8.0 using

Table 1
Genomic DNA extraction buffer (for 100 ml)

Requirements	Volume (100 ml)
Milli-Q water	38 ml
2% CTAB	20 ml (5% stock)
1.4 M NaCl	28 ml (5 M stock)
100 mM Tris. HCl	10 ml (1 M stock)
20 mM EDTA	4 ml (0.5 M stock)

1 N NaOH. Finally, the volume made up to 100 ml with distilled water and autoclave at 15 psi pressure for 20 min.

5 M NaCl: 5.84 g of NaCl should be added in 80 ml of distilled water and finally made up to 100 ml and autoclaved at 15 psi pressure for 20 min.

5% CTAB: 100 ml of 5% CTAB should be prepared by dissolving, 5 g of CTAB in 80 ml distilled water and thereby add 2 g of NaCl to it. Then the volume should be made up to 100 ml and autoclave at 15 psi pressure for 20 min.

Chloroform:isoamyl alcohol (24:1): 100 ml of chloroform (24): isoamyl alcohol (1) should be prepared by adding 4 ml of isoamyl alcohol in 96 ml of chloroform, mix and store at 4 °C.

0.1× TE (pH 8.0): 20 ml of 0.1 X TE should be prepared by adding 0.02 ml of 1 M Tris (pH-8.0) and 0.004 ml of 0.5 M EDTA (pH 8.0) to 18 ml. Dissolve properly and the volume made up to 20 ml and autoclave at 15 psi pressure for 20 min.

Preparation of 50× TAE buffer: Dissolve 242.0 g of Tris base in 700 ml Milli-Q water with gentle heating. Add 57.1 ml glacial acetic acid, followed by 100 ml of 0.5 M EDTA (pH 8.0). Mix thoroughly and make the final volume to 1000 ml with Milli-Q water. Adjust pH to 7.6–7.8 and autoclave the buffer and store at room temperature.

Ethidium bromide (EtBr) preparation: Ethidium bromide (EtBr) is a potent mutagen used to visualize nucleic acids, typically prepared as a 10 mg/ml stock solution in water (stored at in the dark) and used at a 0.5–1.0 μg/ml working concentration for agarose gels. It is added directly to molten agarose or used for post-electrophoresis staining to detect DNA/RNA via fluorescence under UV light.

Stock solution preparation (10 mg/ml) for ethidium bromide

- Dissolve 1.0 g of EtBr powder in 100 ml of distilled water. Store in a dark, light-proof container or wrapped in aluminum foil at 4 °C and handle with gloves or use fume hood to avoid inhaling dust.
- Working solution preparation (0.5–1.0 μg/ml).
- In-gel staining: add 5–10 μl of the 10mg/ml stock to 100 ml of molten agarose (<60 °C) for a final concentration of 0.5–1.0 μg/ml.
- For post-staining: dilute the stock solution in electrophoresis buffer (1× TAE or TBE) to a final concentration of 0.5–1.0 μ g/ml and immerse the gel for 20–30 min.

3 Methods

3.1 Sample Preparation and Conditioning [9]

1. A total of 15–20 phenotypically contrasting cotton genotypes should be selected based on fiber quality traits. The selected genotypes included lines exhibiting high fiber length and

strength as well as lines with comparatively lower fiber quality parameters (*see* **Note 1**).

2. Seeds should be sown under uniform agronomic conditions in a randomized block design (RBD) with three biological replicates per genotype. Standard crop management practices should be followed to minimize environmental variation.
3. Harvest fully opened and mature bolls from the middle fruiting branches to minimize positional variation, while avoiding pest-damaged or immature samples.
4. Separate the lint from seeds using a laboratory gin. Care should be taken to ensure minimal mechanical damage and to prevent fiber breakage during the ginning process.
5. Before testing, fiber samples should be conditioned under standard textile laboratory conditions:
   - Temperature: 21 ± 1 °C.
   - Relative humidity: 65 ± 2%.
   - Duration: 24–48 h.
6. This ensures moisture equilibrium, which directly affects fiber properties.

3.2 Phenotypic Evaluation [7]

1. Manually harvest a total of 25 naturally opened bolls from each plot. After ginning, collect 10–15 g of lint fiber from each sample and subject to fiber quality evaluation.
2. Measure fiber quality traits using the HVI 1000 automatic fiber testing system (Uster Technologies, Switzerland), which is widely used in cotton fiber research and industry for rapid and standardized fiber assessment.
3. The instrument determines fiber length by analyzing the optical length distribution of a prepared fiber beard sample mounted in the HVI clamp. Fiber strength is measured by clamping a fiber bundle and stretching it until breakage, with the maximum force recorded and expressed as centiNewton per tex (cN/tex). Fiber micronaire, an indicator of fiber fineness and maturity, is measured by compressing the fiber sample in a chamber and assessing air permeability resistance.
4. Record the following fiber quality parameters such as Fiber Elongation (FE, %), Fiber Micronaire (FM), Fiber Strength (FS, cN/tex), Fiber Length (FL; Upper Half Mean Length, mm), Fiber Uniformity (FU, %).
5. To minimize environmental effects, Best Linear Unbiased Predictors (BLUPs) for fiber quality traits should be estimated using the lme4 package in R (https://github.com/lme4/lme4/), with genotype and genotype × environment interaction

modeled as random effects and environment treated as a fixed effect.

6. The BLUP values should be subsequently used for association mapping analysis. Pearson's correlation coefficients among fiber traits and analysis of variance (ANOVA) should be performed in R (version 4.5.2; R Core Team, Vienna, Austria).
7. Broad-sense heritability (H^2) for each trait should be calculated using variance components estimated from the mixed model, following standard procedures estimated using the lmer function in the lme4 package of R.

3.3 Extraction of Genomic DNA

3.3.1 Extraction of Genomic DNA by CTAB Method [10–14]

1. **Sample freezing**: Weigh 1 g of cotton tissue and freeze it in liquid nitrogen.
2. **Tissue pulverization**: Grind the frozen tissue into fine powder using pre-chilled mortar and pestle in the presence of liquid nitrogen (*see* **Notes 2–4**).
3. **CTAB lysis**: Approximately 350 g of the powder should be distributed in each of three micro fuse tubes (micro fuse tubes must be chilled in liquid nitrogen before and after weighing (*see* **Note 5**). Spatula should be pre-chilled and used to transfer the ground powder). Equal volume (w/v) of pre-hot (65 °C) 2× CTAB buffer should be added (see Table 1).
4. **Organic extraction and phase separation**: Mix well by inversion in vigorously and equal volume of (700 μl v/v of ice-cold chloroform: Isoamyl alcohol (24: 1) should be added.
5. **Centrifugation**: Mix thoroughly by inversion to form an emulsion, and centrifuge for 5 min at 14,000 rpm at room temperature.
6. **Aqueous layer transfer**: Then, using the cut tips upper aqueous content should be transferred to new microfuge tubes.
7. **CTAB enrichment**: Measure the volume, 1/5 volume of 5% CTAB solution should be added, and mix well by gentle inversion.
8. **Secondary organic extraction**: Equal volume of chloroform: Isoamyl alcohol (24: 1) should be added and centrifuge for 5 min at 14,000 rpm at room temperature.
9. **Centrifugation**: Using cut tips, upper aqueous content should be transferred to new centrifuge tubes, then add equal volume of CTAB Precipitation buffer, and mix gently by inversion.

10. **Centrifugation**: Centrifuge for 1 min at 14,000 rpm, discard the supernatant, and keep the tube by inverted position in tissue paper and dry it for 30 min.
11. **Pellet dissolution**: Dissolve the pellet in 200 μl of high-salt TE buffer and gently tap the tubes and it should be incubated at 4 °C fridge.
12. **DNA precipitation**: Next day, tap the tube gently and make sure the pellet dissolved completely, and then 2.5 volumes of ice-cold 95% ethanol should be added, and mix gently by inversion.
13. **Centrifugation**: Centrifuge for 10 min at 14,000 rpm at 4 °C and discard the supernatant.
14. **Ethanol washing**: Add 500 μl of 70% ethanol to the pellet, and centrifuge for 5 min at 14,000 rpm at 4 °C.
15. **Pellet drying**: Air-dry or vacuum-dry the pellet and dissolve in 50 μl of 0.1× TE.
16. **DNA analysis**: Leave it for some time at room temperature or 4 °C then analyze the isolated genomic DNA by 0.8% agarose gel electrophoresis and Nanodrop spectrophotometer (*see* **Note 6**).

3.3.2 Extraction of Genomic DNA Through Modified CTAB Method (for Plants Containing High Polysaccharide and Polyphenol Component)

1. **Tissue homogenization**: 1.0 g of cotton tissue should be ground using with chilled mortar and pestle, and transfer the powder to polypropylene tube (30 ml) using chilled spatula (50 mg of PVP for 0.5 g tissue should be added when grinding the tissue) (*see* **Notes 2–4**).
2. **Cell lysis**: 10 ml of DNA extraction buffer (preheated at 60 °C) should be added (see Table 1), mix the tubes by inversion, keep at 60 °C for 45 min, and mix once in 15 min. Take the tubes and bring to room temperature (RT) (*see* **Note 5**).
3. **Organic extraction**: Add 10 ml of chloroform: Isoamyl alcohol (24: 1) and mix by inversion until it forms emulsion.
4. **Phase separation**: Centrifuge the tubes for 10 min at 6000 rpm (25 °C), and repeat the steps one more time to remove the cloudiness from the samples.
5. **DNA release**: Carefully transfer the supernatant to fresh tubes, add and mix ½ volume at 5 M NaCl, add 2.5 volumes of 95% ethanol (pre-chilled at 4 °C), mix the tubes, and then keep at −20 °C for an hour.
6. **Pellet recovery**: Spin at 6000 rpm for 10 min (25 °C), discard the aqueous solution, wash the pellet with 70% ethanol, and spin for 5 min. Air dry the pellet for 1 h, add 300 μl of 0.1× TE to dissolve the pellet, and keep for 4 °C overnight.

7. **RNase treatment**: Dissolved DNA should be transferred to 1.5 ml Eppendorf tube, and 5 µl of RNase A (10 mg/ml) should be added and incubated for 1 h at 37 °C water bath.
8. **Phenol purification**: Extract equal volume of neutral phenol and two times with ether solution to remove phenol solution completely.
9. **DNA precipitation**: Add 1/10 volume of 3 M sodium acetate pH 5.2 and 2.5 volume of 90% ethanol to the solution for precipitation of genomic DNA. Keep the tubes overnight at −70 °C.
10. **Ethanol washing**: Next day, spin the tubes for 12,000 rpm, 10 min at 4 °C, discard the supernatant, and wash the pellet twice with 70% ethanol (pre-chilled). Finally dry the pellet and dissolve with 100 µl of 0.1× TE buffer.
11. **DNA quality assessment**: The quality and quantity of the extracted genomic DNA should be analyzed by 0.8% agarose gel electrophoresis and Nanodrop spectrophotometer (*see* **Note 6**).

3.4 Analysis of Extracted Genomic DNA by Agarose Gel Electrophoresis [10]

1. **Agarose preparation**: Weigh 0.5 g of Agarose and melt thoroughly in 50 ml of 1× TAE (see Table 2) buffer until solution becomes fully transparent and uniform.
2. **DNA staining**: Once the melted agarose cools slightly (still in liquid form), 5 µl of ethidium bromide should be added and mix gently to enable visualization of DNA fragments.
3. **Gel casting**: The agarose TAE mixture should be poured gently into a clean gel casting tray fitted with a comb. The gel should be left undisturbed to solidify at room temperature. Casting the gel in a horizontal tray is essential for proper separation of bands.
4. **Electrophoresis setup**: After solidification, remove adhesive tapes or side seals from the tray. The gel should be placed carefully into the electrophoresis chamber and fully submerge in 1× TAE running buffer (see Table 2), ensuring uniform electrical conduction.

Table 2
50× TAE buffer preparation for gel electrophoresis (1 l) (pH 7.6–7.8)

Requirements	Volume for (1 l)
Tris base (2 M)	242.2 g
Glacial acetic acid (1 M)	57.1 ml
EDTA (0.5 M)	100 ml

Table 3
Preparation of DNA sample for gel electrophoresis analysis

Sample	Volume (10 μl)
DNA	2.0 μl
6× DNA gel loading dye (Thermo Fisher Scientific)	5.0 μl
Nuclease-free water	3.0 μl

5. **Sample preparation**: DNA samples should be mixed with 6× gel loading dye (see Table 3), followed by a brief pulse centrifugation to ensure proper mixing. The tracking dye helps visualize sample loading and migration during electrophoresis.
6. **Sample loading**: Using a micropipette, DNA samples along with a 100 bp DNA ladder should be slowly dispensed into the wells to avoid well damage or sample leakage. Carefully load the band patterns.
7. **Electrophoretic separation**: The system should be connected to a power supply and electrophoresis is carried out at a constant voltage, ensuring even migration of amplified DNA fragments. Proper voltage maintains clear separation of DNA bands.
8. **Band visualization**: After completion of the run, the gel should be transferred to a UV transilluminator where bands stained with ethidium bromide are observed (Fig. 1A).

3.5 Quantification of Genomic DNA by Using Nanodrop Spectrophotometer 1000 (ThermoFischer Scientific, USA)

1. **Cleaning the pedestals**: Open the sampling arm and clean the upper and lower pedestals with a dry, wipe preferably Kimwipe.
2. **Blanking**: Pipette 2 μl of thc blanking buffer (the same buffer used to resuspend your DNA, such as TE or water) onto the lower pedestal. Close the arm and click blank in the equipment.
3. **Sample loading**: Wipe the pedestals clean. Pipette 2 μl of your DNA sample onto the pedestal, lower the arm and enter the Sample ID.
4. **Measurement**: Click measure. The software displays the concentration (in ng/μl) and purity ratios (260/280 and 260/230).
5. **Cleanup**: Wipe the pedestal with a clean wipe between samples and perform a final clean with water after all measurements are finished.
6. **Interpretation**: Optimum range of A260/A280 and A260/A230 are 1.8–2.0 and 2.0–2.2, respectively (Fig. 1B).

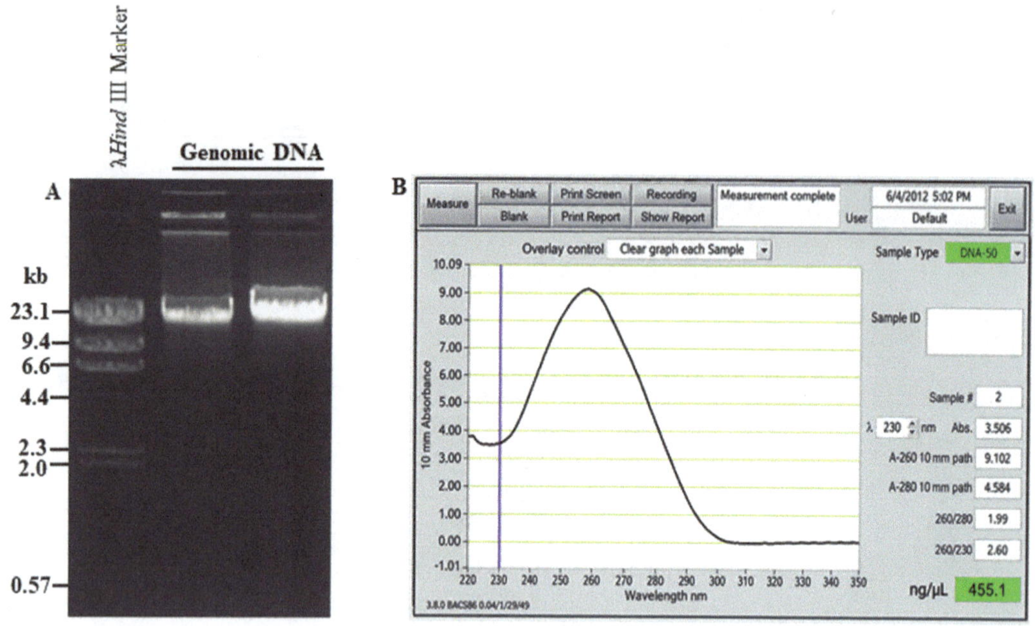

Fig. 1 Analysis of genomic DNA extracted from the cotton tissues by Agarose gel electrophoresis and Nanodrop Spectrophotometer. **A** 0.8% Agarose gel image shows the quantity and integrity of genomic DNA extracted from the cotton tissues. **B** Image shows the quantity and purify analysis of extracted cotton genomic DNA by Nanodrop1000 Spectrophotometer (ThermoFisher Scientific, USA) [12]

7. **Contamination**: Range below 1.8 and 2.0 in A260/A280 and A260/A230 indicates protein, phenol contamination, and salt contamination present in the sample, respectively.

3.6 Method 1: SNP Identification by WGS-Based Sequencing

3.6.1 Whole Genome Sequencing [15]

1. **Genomic library preparation**

   To prepare a good-quality genomic library, high-quality genomic DNA should be used.

2. **DNA fragmentation**

   Purified genomic DNA from all samples should be sheared into ~200–800 bp fragments by passing compressed nitrogen gas (30 psi) through a nebulizer for 1–2 min, then samples should be collected, purified, and size-selected for library preparation.

3. **End repair (*see* Note 16)**

   Fragmented genomic DNA should be subjected to end-repair to convert overhangs into blunt- ended DNA fragments suitable for adapter ligation (see Tables 4 and 5).

Table 4
Requirements of end repair reaction

Requirements	Volume (50 μl)
Fragmented DNA	30 μl
End repair mix (ERM)	10 μl
End repair buffer (ERB)	5 μl
Nuclease-free water to the final volume of	50 μl

Table 5
Incubation for end repair reaction

Step	Temperature (°C)	Time
End repair reaction	20 °C	30 min
Enzyme inactivation	75 °C	5 min

Table 6
Requirements of A-tailing reaction

Requirements	Volume (42.5 μl)
End-repaired DNA	30 μl from previous step
A-tailing buffer	12.5 μl
Total reaction volume	42.5 μl

Table 7
Incubation for A-tailing reaction

Step	Temperature	Time
Adapter ligation	30 °C	10 min
Hold	4 °C	∞

4. **A-tailing (3′ adenylation)**
   After end repair, A-tailing is also done in the same tube; the preparation and components are mentioned (see Tables 6 and 7).
5. **Adapter ligation**
   After end repair and A-tailing adapter ligation should be done and for 50 μl reaction 500 ng–1 μg DNA is used (see Tables 8 and 9) (*see* **Note 17**). Purify Ligation products on 0.5%, 1%, or 2% agarose gels, each of which target a specific range of insert sizes.

Table 8
Requirements of adapter ligation reaction

Step	Temperature °C	Time
A-tailing reaction	37 °C	30 min
Hold	4 °C	∞

Table 9
Incubation conditions for adapter ligation reaction

Requirements	Volume (50 μl)
A-tailed DNA	42.5 μl
Ligation mix (LM)	2.5 μl
Indexed TruSeq adapter	2.5 μl
Ligation buffer (LB)	2.5 μl
Total volume	50 μl

Table 10
PCR reaction conditions for library preparation

Steps	Conditions
Initial denaturation	98 °C for 30 s
Denaturation	98 °C for 10 s
Annealing	60 °C for 30 s
Extension	72 °C for 30 s
Final extension	72 °C for 5 min
Hold	4 °C
No. of cycles	8–12

3.6.2 Construction of Libraries with Different Insert Sizes (see Notes 18 and 19)

- **Short-insert libraries (250–800 bp)**

1. DNA yield after ligation may be low, so PCR amplification (see Tables 10 and 11) should be performed to increase the amount of library.
2. Diversity of fragments should be preserved to ensure enough material for sequencing.

- **Long-insert libraries (>5 kb)**
  PCR is often avoided or minimized because

Table 11
Library construction parameters for short-insert and long-insert sequencing libraries

Library type	Fragment size (bp)	Adapter-ligatedsize (bp)	Concentration (ng/μl)
Short-insert	250–800	350–900	10–50
Short-insert	250–800	350–900	10–50
Long-insert	2000–40,000	+120 bp	5–20

1. Long fragments, prone to PCR bias and preferential amplification of shorter fragments.
2. To scaffold and connect the distant genomic regions, original fragment length should be maintained.

3.6.3 Post-PCR Cleanup and Quality Check

Remove primers, nucleotides, and enzymes using magnetic beads Agencourt AMPure XP beads (Beckman Coulter, Brea, California, USA) or column purification. Size selection should be carried out to remove small PCR artifacts or adapter dimers.

- Check library size distribution using Bioanalyzer (Agilent Technologies, Santa Clara, CA, USA).
- Measure concentration using Qubit dsDNA HS Assay Kit (Thermo Fisher Scientific, USA).
- Ensure library has enough material for sequencing without over-representation of short fragments.

3.6.4 Sequencing Using Illumina Platform

All prepared libraries should be sequenced using the Illumina sequencing platform NextSeq 2000 (https://shorturl.at/qDkaP) and paired-End reads from each library are obtained (*see* **Note** 7).

3.6.5 Raw Read Quality Filtering

Filter the raw reads to remove unusable reads. Discard the reads if they contained the following conditions.

1. Reads containing ≥10% "N" Bases.
2. Reads with low-quality data (the ASCII value of base-64 ≤7) for 65% of the bases for short insert (<2 kb) or 80% of the bases for long insert (≥2 kb).
3. Reads containing >10-bp adapter sequences.
4. Reads with >10 bp that overlapped between two ends of reads of short inserts.
5. Reads with identical sequences at the two ends.

3.6.6 Genome Assembly

The filtered reads should be assembled into the cotton genome using the software SOAP de novo (https://bio.tools/soapdenovo).

Table 12
Genome assembly statistics of the assembled genome

Parameter	Value
Genome size assembled	~2.17 Gb
Contig N50	~80 kb
Scaffold N50	~764 kb

- **Assembly steps involve**

1. Short reads should be assembled into contigs.
2. Contigs should be connected into scaffolds using paired-end information.
3. Long-insert libraries helped link distant contigs and assembly parameter are shown (see Table 12).

3.6.7 SNP Discovery

1. To identify genetic variation, sequencing reads should be compared with the assembled genome.
2. Mapping population used a recombinant inbred line (RIL) population should be generated from a cross between different parental genome and the RIL individuals should be analyzed.

3.6.8 SNP Identification

To detect single-nucleotide polymorphisms, use Genome Analysis Toolkit (GATK) (https://shorturl.at/Kn5OG) (*see* **Note 8**).

Steps include:

1. Reads should be aligned to the reference genome.
2. Identify positions where nucleotides differ.
3. Record these positions as SNPs. The output should be generated in VCF format.

3.6.9 SNP Filtering and Bin Mapping

Not all detected SNPs are reliable. To remove false SNPs, use bin mapping (*see* **Notes 11 and 12**).

1. Adjacent SNPs showing identical inheritance patterns should be grouped into bins.
2. Statistical models should be applied to correct errors.
3. To remove parental genotype inference, use Maximum Parsimony of Recombination (MPR) (https://www.ub.edu/dnasp/).
4. To identify parental genotype validation, use Bayesian inference (BEAST X v10.X version v10.5.0).
5. Hidden Markov Models (HMM) (http://hmmer.org/) should be used for correction of progeny genotype and recombination detection.

6. SNPs grouped into bins should be adjusted. Bins used to anchor scaffolds to the chromosomes.

3.6.10 Genetic Linkage Map Construction

1. The filtered SNP markers should be used to construct a high-density genetic map using the software JoinMap (https://shorturl.at/eDLxy). Marker grouping and ordering must be performed based on recombination frequency and logarithm of odds (LOD) scores.
2. A LOD threshold value of 10 should be applied to assign markers into linkage groups.
3. The resulting linkage groups should be corresponded to the number of chromosomes of the species under study.
4. In the case of cotton, the linkage analysis typically produces 26 linkage groups, which represent the 26 chromosomes of the cotton genome. The final genetic map consists of a large set of SNP markers distributed across these linkage groups, providing a high-density framework for downstream analyses such as SNP marker development.

3.6.11 Marker Development and Genome Anchoring

The SNP markers served as genetic markers (*see* **Note 10**).
They can be used to:

1. Anchor genome scaffolds to chromosomes.
2. Order genome sequences.
3. Orient genomic regions.
4. Together, these SNP markers create a high-density marker framework for genetic analysis.

3.7 Method 2: SNP Identification Based on Genotyping-By-Sequencing [16]

3.7.1 Plant Material and DNA Extraction

To prepare a good-quality genomic library, high-quality genomic DNA should be used.

3.7.2 Restriction Digestion

For this 250 ng of genomic DNA from all the samples should be digested with 10 units of restriction enzymes *Hin*P1I or *Bsr*GI at 37 °C for 2 h. *Hin*P1I is a CpG methylation-sensitive four-base cutter (G|CGC), while *Bsr* GI is a methylation-insensitive six-base cutter (T|GTACA).

3.7.3 Adapter Ligation

1. Adapter ligations should be carried out in 50-μl reactions in the presence of 40 pmol of the restriction-enzyme appropriate combination of annealed, 6-bp bar-coded, Illumina-compatible P5 and P7 adapters and 1600 cohesive-end ligation units of T4 DNA ligase (New England Biolabs, Ipswich, Massachusetts, USA) for 1 h at 37 °C.

2. Active restriction enzymes still present in these reactions, a final incubation for 30 min at 37 °C should be performed to cleave any chimeric ligation products Purified, digested (*see* **Notes 24 and 25**), or sheared DNA fragments from previous steps (typically end-repaired and A-tailed, between genomic DNA fragments (the adapters should be designed to abolish the *Hin* P1I or *Bsr* GI recognition site upon ligation). For performing the adapter ligation, the following components are added in fresh microfuge tubes mentioned (see Table 13).

3.7.4 Size Selection

The above adapter ligated samples should be subjected size verification on 2.5% agarose gel electrophoresis to obtain libraries that ranged from *Hin*P1I 190–210 bp and *Bsr*GI 160–290 bp.

3.7.5 Library Pooling

All samples should be pooled and purified using MinElute columns (QIAGEN, Valencia, California, USA) (*see* **Note 21**).

- **Pooling libraries**

1. Combine adapter-ligated DNA libraries in desired proportions (e.g., equal ng amounts per sample) into a single tube.
2. Mix gently by pipetting or brief vortex.

- **Library purification**
  Prepare binding buffer: add 5 volumes of PB buffer MinElute columns (QIAGEN, Valencia, California, USA) to 1 volume of pooled DNA. For instance, 50 μl pooled DNA should be added to 250 μl PB buffer and mix thoroughly by pipetting.
- **Apply to spin column**

1. Place a MinElute spin column in a collection tube. Then, transfer the PB mixture containing DNA to the column.
2. Centrifuge at ≥13,000 × *g* for 1 min and discard flow-through.

- **Washing**

1. Add 750 μl of PE buffer to the MinElute columns (QIAGEN, Valencia, California, USA).

Table 13
Requirements of adapter ligation reaction

Requirements	Volume (50 μl)
DNA fragments	20 μl
P5 & P7 adapter mix	1 μl
T4 DNA ligase buffer	4 μl
Nuclease-free water to the final volume of	50 μl

2. Centrifuge at ≥13,000 × *g* for 1 min and discard flow-through.
3. Repeat wash if the sample has high salt or contaminants.

- **Dry column**

1. Centrifuge the empty column at ≥13,000 × *g* for 1 min to remove residual wash buffer.

- **Elution**

1. Place the column in a clean 1.5 ml tube and add 10–30 μl of nuclease-free water directly to the center of the silica membrane. Incubate at room temperature for 1 min.
2. Centrifuge at ≥13,000 × *g* for 1 min to elute purified DNA.

- **Post-purification**

1. Measure DNA concentration.
2. Assess fragment size distribution with Agilent 2100 Bioanalyzer (Agilent Technologies, Santa Clara, CA, USA).
3. The purified, pooled library is now ready for sequencing on Illumina platforms.

- **Removal of plastid DNA contamination**

1. Download a complete chloroplast genome of the species or a closely related cotton species (*see* **Note 26**).
2. For instance, *Gossypium hirsutum* L., chloroplast genome from NCBI GenBank (https://www.ncbi.nlm.nih.gov/genbank/) in FASTA file format.

3.7.6 Identify Restriction Sites (Optional)

Use in silico restriction digestion software (e.g., NEBcutter (https://nc3.neb.com/NEBcutter/), EMBOSS (https://emboss.sourceforge.net/) to identify restriction enzyme cut sites in the plastid genome (*see* **Note 27**).

3.7.7 Map Reads to Plastid Genome

Align raw sequencing reads (FASTQ) to the plastid genome reference using a short-read aligner, such as Burrows-Wheeler Aligner (https://bio-bwa.sourceforge.net/), Bowtie2 (https://bowtie-bio.sourceforge.net/bowtie2/), or minimap2 (https://minimap2.com/).

3.7.8 Single-Stranded DNA Preparation

Size-selected fragments should be treated with NEBNext Fill-In and ssDNA Isolation module (New England Biolabs, Ipswich, MA, USA), as a template for PCR amplification (*see* **Note 9**) with Illumina-compatible primers PE_PCR_Primer_1.01 and PE_PCR_Primer_2.01 mentioned (see Table 14).

Table 14
Adapter PCR primers used for illumina sequencing library amplification [16]

Primer name	Primer sequence (5′-3′)	Length
PE_PCR_Primer_1.01	AATGATACGGGACCACCGAGATCTACACTCTTTC CCTACACGACGACGCTCTTCCGATCT	60
PE_PCR_Primer_2.01	CAAGCAGAAGACGGCATACGAGATCGGTCTCGG CATTCCTGCTGAACCGCTCTTCCGATCT	61

Table 15
PCR reaction conditions

Steps	Conditions
Initial denaturation	98 °C for 30 s
Denaturation	98 °C for 12 s
Annealing	65 °C for 30 s
Extension	72 °C for 30 s
Final extension	72 °C for 1 min
No. of cycles	
*Bsr*GI	20 cycles
*Hin*P1I	30 cycles

3.7.9 PCR Amplification

Once after the size selection the prepared libraries should be amplified using illumina-specific compatible primers under the following temperature conditions mentioned (see Table 15).

3.7.10 Library Cleanup

The library should be cleaned followed by PCR amplification to increase library concentration using Agencourt AMPure XP beads (Beckman Coulter, Brea, California, USA).

- **Library quantification**

1. The all-prepared libraries should be measured using AccuBlue High Sensitivity Quantitation Kit (Biotium Fremont, California, USA).
2. VICTOR X3 Multilabel Plate Reader should be measured to determine the exact concentration for sequencing.

- **Library normalization**
  Finally, the libraries should be diluted to 10 nM concentration, which is the optimum concentration for the sequencing platform.

3.7.11 Sequencing

Samples must be diluted to 10 nM and used for sequencing on the Illumina GAII and/or HiSeq 2000 instrument (Illumina, San Diego, California, USA).

3.7.12 Raw Sequence Data Import and Quality Trimming

Raw paired-end reads should be imported into the Geneious bioinformatic package for initial data processing (https://www.geneious.com/). Perform quality trimming using the following parameters

1. Quality threshold: $P=0.05$.
2. Low-quality bases should be removed from the sequence ends.
3. Reads should be filtered to ensure high-confidence base calls.
4. After trimming, all reads should be standardized to a length of 66 bp to ensure uniformity during downstream analysis using Stacks ver. 0.998.

3.7.13 RAD-Tag Processing and Barcode Removal

1. Sequences should be pre-processed using the process _ rad-tags script, in which the 6-bp barcodes should be sorted and removed, yielding 60-bp fragments.
2. The process radtag options included: -c (clean data) and -q (discard low-quality reads).

3.7.14 De-Novo Assembly of RAD Loci

Processed FASTQ files should be assembled using the "denovo_map.pl" scripts with the help of following parameters mentioned (see Table 16). This process generates RAD loci ("stacks") representing unique genomic regions.

3.7.15 SNP Detection and Data Storage

The output data files should be loaded to MySQL tables (Oracle Corporation, Redwood Shores, California, USA), and SNPs between cotton taxa should be annotated in pairwise manner and we obtained a raw sequences and stacks (loci) for each sample.

3.7.16 Identification of Gene-Rich Regions

1. To determine whether GBS markers located in gene-rich regions, sequence similarity searches should be conducted using BLASTN.
2. To search the unique *Bsr*GI unique stacks and unique *Hin*P1I stacks from reference genome against predicted coding DNA sequence using a significant threshold of <1e−6.
3. The higher proportion for *Hin*P1I fragments indicates enrichment in transcribed genomic regions, likely due to the methylation sensitivity of this enzyme hence it is used to identify gene-rich regions.

Table 16
Parameter settings used in the stacks program for SNP discovery

Parameter	Description
-n 3	Number of mismatches allowed between loci during catalog building
-m 3	Minimum number of identical, raw reads required to create a stack
-M 2	Number of mismatches allowed between loci when processing a single individual
-t	Remove, or break up, highly repetitive RAD-tags during unstacks

3.7.17 Detection of Repetitive DNA

Examine GBS libraries for the presence of repetitive DNA using BLAST searches of all *Bsr* GI and *Hin* P1I against the TIGR Brassicaceae Repeat Database ver2_0_0.

3.7.18 Detection of Organellar DNA Contamination

1. To identify contamination from organellar genomes, RAD fragments should be compared against plastid genome and mitochondrial genome of *Gossypium hirsutum* L. This is performed using BLASTN (https://shorturl.at/hY4r3).
2. The proportion of plastid and mitochondrial sequences detected should be very low, indicating minimal contamination.

3.7.19 SNP Identification Between Taxa

For loci shared between any two cotton taxa:

1. Sequence polymorphisms should be identified.
2. Classify SNP using marker predictions generated by Stacks ver. 0. 998. Total SNP markers detected and should be submitted to the CottonGen database (http://www.cottongen.org).

3.7.20 Comparison with Existing SNP Databases

To evaluate novelty of the SNP markers, fragments should be compared with 183,035 SNPs available in CottonGen using the batch BLAST tool (https://shorturl.at/SkrpG).

3.7.21 In-Silico SNP Validation

1. A subset of SNP markers from reference genome should be validated computationally using BLASTN (https://shorturl.at/hY4r3) at a significance threshold of <1e−6.
2. Markers should be compared with genome sequences of desired samples. Alignments should be constructed using Geneious bioinformatics package (https://www.geneious.com/).

3.7.22 Experimental SNP Validation Using Cleaved Amplified Polymorphic Sequences (CAPS) Markers

Selected SNPs should be experimentally validated using CAPS (Cleaved Amplified Polymorphic Sequence) markers.
Marker selection criteria: SNPs should be selected if:

1. SNP must be near the middle of the 60-bp sequence to allow for design of flanking primers.
2. To design primer, flanking sequences should be suitable with G+C content for primer design (30–60%) and lack simple sequence repeats.
3. Specific primers should be designed (using the alignments) to avoid amplification of known paralogs and homeologs using Primer3 (https://primer3.ut.ee/software).
4. SNP should be occurred within the recognition site of a commercially available restriction enzyme.

3.7.23 PCR Amplification

PCR amplification should be performed using the KAPA3G Plant PCR Kit Kapa (Biosystems, Roche, Basel, Switzerland) and Analyze PCR products using 4% agarose gel electrophoresis.

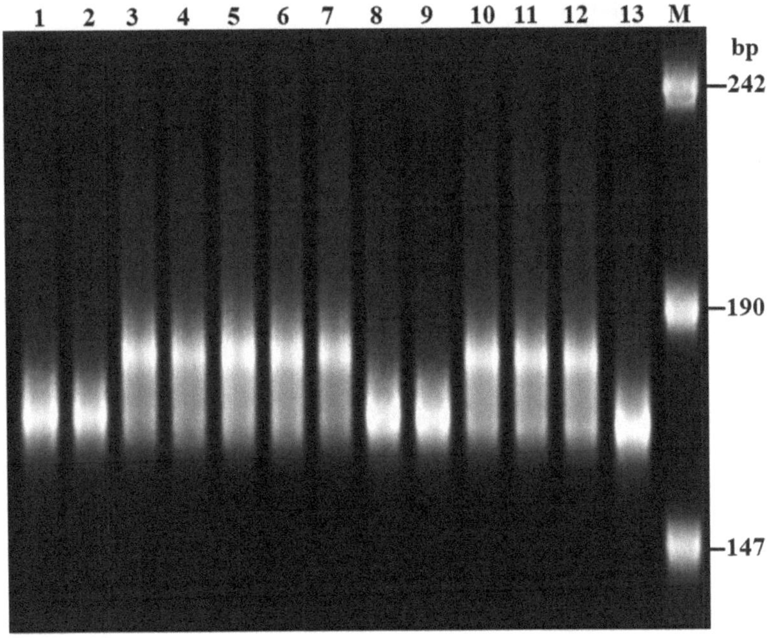

Fig. 2 SNP validation by genotyping method. The representative image shows that template DNA of various genotypes was analyzed with multiple specific alleles called PAMSA (PCR Amplification of Multiple Specific Alleles). The SNP PCR products were separated by 3% high-resolution agarose gel and stained. For PAMSA markers, allele-specific amplification produces single amplicons corresponding to each allele. Homozygous genotypes (AA or TT) show a single distinct band, while heterozygous genotypes (AT) display two bands, representing amplification of both alleles [17]

3.7.24 Restriction Digestion and Marker Scoring

1. PCR amplicons should be digested with the appropriate restriction enzymes corresponding to CAPS markers. Digested products should be resolved on agarose gels to detect allelic polymorphism (Fig. 2).
2. Restriction digestion of PCR amplicons produced distinct banding patterns corresponding to different SNP alleles. The observed polymorphism confirmed successful conversion of the SNP into a CAPS marker and subsequently used for genotyping cotton genotypes and validating SNP.

3.8 Method 3: SNP Identification Based on Restriction-Site Associated DNA Sequencing (RAD-Seq) [18]

3.8.1 Plant Material and DNA Extraction

To prepare a good-quality genomic library, high-quality genomic DNA should be used.

Table 17
Preparation of restriction enzyme digestion

Requirements	Volume (50 µl)
Genomic DNA	5 µl
10× restriction buffer	5 µl
*Eco*RI	2 µl
Nuclease-free water to the final volume of	50 µl

Table 18
Preparation of P1 adapter ligation

Requirements	Volume (10.5 µl)
P1 adapter (100 nM)	2.5 µl
ATP (10 mM)	1 µl
10× NEB buffer 4	1 µl
T4 DNA ligase (1000U)	1 µl
Nucleases-free water	5 µl

3.8.2 Restriction Enzyme Digestion

1. The isolated genomic DNA samples should be digested with the restriction enzyme *Eco*RI (*see* **Note 20**). Mix all components mentioned (see Table 17) gently in a PCR tube and incubate at 37 °C for 60 min.
2. It should be heated 65 °C for 20 min to stop enzyme activity (*see* **Notes 14 and 15**).

3.8.3 Adapter Ligation (P1 Adapter)

1. Once the restriction digestion over, the P1 adapter should be ligated by adding the following components mentioned (see Table 18).
2. Incubate at room temperature for 20 min and heat-inactivate for 20 min at 65 °C.

3.8.4 DNA Fragmentation

Here, the adapter-ligated DNA fragments should be randomly sheared using Bioruptor sonicator (Diagenode, Seraing, Belgium), which produces an average size of 500 bp suitable for sequencing.

3.8.5 Size Selection

The sheared fragments should be for size selection using 1.5% agarose gel electrophoresis, 0.5 TBE gel and DNA fragments from 300 to 700 bp should be isolated using MiniElute gel extraction kit (Qiagen).

Table 19
Preparation of reaction mixture for PCR

Requirements	Volume 100 μl
RAD library DNA	20 ng
Phusion master mix	20 μl
Solexa amplification primers	5 μl
Nuclease-free water to the final volume of	100 μl

Table 20
PCR reaction conditions

Steps	Conditions
Initial denaturation	98 °C for 30 s
Denaturation	98 °C for 12 s
Annealing	65 °C for 30 s
Extension	72 °C for 30 s
Fina extension	72 °C for 1 min
No. of cycles	18

3.8.6 End Repair (End Blunting)

Here, the overhangs should be converted to blunt ends using End-blunting enzymes (Enzymatics, USA). The DNA should be purified again using MinElute columns (Qiagen).

3.8.7 Addition of A-Overhangs

Exo-Klenow enzyme (Enzymatics, USA) should be added and incubated at 37°C to generate 3′ adenine overhangs and purificd using MinElute columns (Qiagen).

3.8.8 P2 Adapter Ligation

The samples should be purified, and 1 μl of 10 μM P2 adapter, a divergent modified Solexa adapter (Illumina, Inc.), must be ligated to these DNA fragments at 18 °C. Samples should be again purified using MinElute columns (Qiagen) and eluted in 50 μl (*see* **Notes 22 and 23**).

3.8.9 Library Quantification

Concentration RAD libraries should be measured using Qubit fluorometer and about 20 ng DNA should be used for PCR amplification.

3.8.10 PCR Amplification

PCR amplification should be carried by using the following components (see Table 19) under the temperature conditions mentioned (see Table 20).

3.8.11 Final Size Selection

The PCR products should be run on agarose gel. The fragments that have 300–700 bp should be selected.

3.8.12 Library Normalization

Final RAD libraries should be diluted to 1 nM concentration for sequencing platforms.

3.8.13 Restriction Site-Associated DNA Sequencing

The normalized samples should be sequenced using paired-end sequencing on an Illumina genome analyzer II (https://www.illumina.com).

3.8.14 Raw Sequence Data Processing

Raw sequencing reads should be subjected to quality filtering to remove low-quality or contaminated reads. Reads should be discarded if they contain:

1. More than 2% N calls.
2. Poly-A structures.
3. Adapter contamination.
4. Base (quality scores ≤5) number accounts for 50% of the reads in PE libraries. Additional filtering should be performed using the FASTX Toolkit Updated (http://hannonlab.cshl.edu/fastx_toolkit/) to obtain high-quality paired-end reads for downstream analysis.

3.8.15 De-Novo Assembly of RAD Reads

Filtered reads from each parental genotype should be assembled de novo using the Velvet sequence assembler (version 1.0.18).
Assembly parameters included:

1. Hash length: 31 bp.
2. Minimum contig length: 200 bp.
3. Other parameters should be kept at default settings.
4. The assembled contigs represented genomic fragments derived from RAD loci.

3.8.16 Identification of Shared Sequences

To identify common sequences between the cotton genotypes, contigs should be compared using CAP3 software (https://doua.prabi.fr/software/cap3).
Assembly parameters should include the following:

1. Minimum overlap length: 80 bp.
2. Minimum overlap similarity: 95%.

This analysis should be generated for a non-redundant dataset consisting of

1. Singleton sequences from each parental genotype.
2. Contigs shared between the parental genotypes.

These assembled sequences should be subsequently used for molecular marker discovery and polymorphism analysis.

3.8.17 Functional Annotation of SNP-Containing Sequences

To determine potential gene functions, SNP-containing sequences should be subjected to BLASTX (https://blast.ncbi.nlm.nih.gov/Blast.cgi) analysis against the TAIR10 protein database (http://www.arabidopsis.org/) (*see* **Note 13**).
Search parameters included:

1. E-value threshold: 1×10^{-15}.
2. The annotated sequences should be assigned functions based on Arabidopsis GO SLIM (ftp://ftp.arabidopsis.org/home/tair/Ontologies/Gene_Ontology/).
3. Then mapped to higher-level categories (plant GO slim) using GOSlim Viewer21 (https://usegalaxy.eu/root?tool_id=goslimmer).

3.8.18 SNP Discovery

1. RAD sequencing reads from each genotype should be aligned to the assembled contigs of the alternate genotype to identify sequence variations.
2. Alignment should be performed using SOAPaligner v2.21 (http://soap.genomics.org.cn/soapaligner.html) with a maximum of two mismatches allowed per read.
3. The alignment results should be used to identify single-nucleotide polymorphisms (SNPs) using SOAPsnp software.

3.8.19 SNP Genotype Classification

Identified SNP sites should be classified based on minor allele frequency (MAF) mentioned (see Table 21). This filtering ensured that only reliable SNP polymorphisms should be retained for marker development.

3.8.20 Primer Design for SNP Marker Validation

Contigs containing SNP loci should be used for primer design using Primer3 program (http://bioinfo.ut.ee/primer3/). Primer design parameters mentioned (see Table 22).
If multiple markers should be located within 500 bp, primers must be designed to amplify the entire region. However, the final PCR product size should be maintained below 800 bp.

3.8.21 SNP Marker Development

1. New primers should be designed from RAD sequencing.

Table 21
SNP genotype classification

Minor allele frequency	SNP genotype classification
<0.10	Homozygous
>0.25	Heterozygous
0.10–0.25	Unknown genotype

Table 22
Primer design parameters

Parameter	Value
Primer length	18–24 bp
Optimal primer size	20 bp
GC content	40–60%
Optimal GC content	50%
Annealing temperature	57 °C
PCR product size	100–500 bp

Table 23
Analysis of PCR products by polyacrylamide gel systems

Gel type	Purpose
6% denaturing PAGE	Used for SSR analysis
8% native PAGE	Used for SSCP analysis

2. Previously reported markers should be used from earlier cotton studies.

These markers should be used to screen polymorphism between parental cotton lines.

3.8.22 PCR Amplification of Marker Regions

For each marker:

1. Primers should be used to amplify genomic DNA from the parental lines (*see* **Note 28**).
2. PCR products should be generated for each marker locus.

After PCR, the amplified fragments should be analyzed to detect DNA sequence variation.

3.8.23 Separation of PCR Products

PCR products should be separated using polyacrylamide gel electrophoresis (PAGE). Two types of gels used are mentioned (see Table 23). Polyacrylamide gels provide high resolution, allowing detection of very small DNA differences.

3.8.24 SSCP Analysis for SNP Detection

A single nucleotide change (SNP) can change the secondary structure of single-stranded DNA. So, the following steps are used for SSCP Analysis for SNP Detection.

Table 24
Parameters of SSCP analysis

Parameters	Conditions
Gel	8% native polyacrylamide
Acrylamide ratio	1:29 (bis-acrylamide: acrylamide)
Electrophoresis power	15 W
Temperature	25 °C
Run time	3.5–4 h

Table 25
Banding pattern of the two parental lines

Parent	Band pattern
Parent A	Band position 1
Parent B	Band position 2

3.8.25 Electrophoresis Conditions

The SSCP analysis should be performed under controlled conditions mentioned (see Table 24). These conditions help separate DNA fragments based on small structural differences caused by SNPs.

3.8.26 Visualization of DNA Bands

1. After electrophoresis, DNA fragments should be detected using silver staining.
2. Silver staining is very sensitive and allows visualization of small quantities of DNA bands.
3. The gel should be fixed and stained with silver nitrate. Visualize the band pattern.

3.8.27 Identification of SNP Polymorphism

Researchers compare the banding pattern of the two parental lines. Example shown (see Table 25). If the band patterns are different, it indicates sequence variation may be often caused by SNPs; this means the marker is polymorphic.

3.8.28 Conversion into SNP Markers

Polymorphic band should be detected by the following

1. The PCR fragment contains a sequence difference.
2. That variation corresponds to a SNP locus discovered by RAD sequencing.
3. The primer pair flanking that SNP becomes a usable SNP marker.

4 Notes

1. Young, healthy cotton samples should be used as starting material for DNA extraction.
2. Efficient grinding of tissue in liquid nitrogen is critical (see "DNA extraction CTAB method"). Incomplete grinding leads to poor cell lysis and low DNA recovery.
3. A modified CTAB method containing PVP and β-mercaptoethanol is strongly recommended for cotton tissues. These components neutralize phenolic compounds that otherwise bind and degrade DNA.
4. All reagents used for DNA extraction should be fresh and of molecular biology grade (see Sect. 2). Oxidized or contaminated reagents often result in brownish DNA pellets and low A260/A280 ratios.
5. Gentle mixing during lysis and extraction steps should be maintained to avoid mechanical shearing of high molecular weight DNA.
6. Samples should be processed immediately or stored at −80°C to prevent DNA degradation
7. *Gossypium hirsutum* L. has ~2.5 Gb; random SNP discovery without targeting is inefficient.
8. Select SNPs located in or near fiber-related genes or QTL regions for further validation.
9. For validation, design primers flanking the SNP (PCR amplicon ~50–200 bp).
10. Amplify candidate SNP regions before sequencing or marker development.
11. Confirms SNP identity and detects heterozygosity; reduces false positives.
12. Correct for population structure (Q) and kinship (K) in MLM models to avoid false associations.
13. Identify genes within ±500 Kb of significant SNPs for functional annotation (cell wall, cellulose synthase, and hormone signaling).
14. Use ≤1 μg DNA; more DNA may need scaling of enzyme volumes.
15. Buffers are usually provided with the enzyme mix, so follow the kit's instructions.
16. Many labs combine end repair and A-tailing sequentially in one tube to reduce DNA loss.
17. Adapter: insert molar ratio

- Typically, 10: 1 adapter: insert molar ratio is used.
- Too much adapter → more adapter dimers.
- Too little adapter → low ligation efficiency.
- Number of cycles should be minimized to prevent over-amplification and PCR bias.
- Typically, 8–12 cycles are sufficient for 250–800 bp libraries.

18. Short-insert libraries: PCR is used to enrich and amplify ligated DNA.
19. Long-insert libraries (>5 kb): PCR is avoided to preserve fragment length and reduce bias.
20. Do not over-incubate, as active restriction enzymes may start cutting desired ligated fragments.
21. Pooling before purification ensures uniform removal of contaminants across samples.
22. MinElute columns efficiently remove small adapter dimers while retaining DNA fragments of desired sizes.
23. Elution volume (10–30 μl) affects concentration—use lower volume for higher concentration if needed for sequencing.
24. Clean-up after chimera removal ensures only high-quality, correctly ligated DNA is retained.
25. This step reduces chimeric reads in sequencing, improving SNP calling accuracy.
26. In silico digestion is a computational simulation, not a wet-lab digestion. It helps identify potential plastid-derived reads to remove.
27. NEBNext Fill-In and ssDNA isolation kits provide all necessary reagents and a detailed protocol, but incubation times and volumes may vary slightly depending on input DNA quantity.
28. Use high-quality, intact genomic DNA; degraded DNA can cause PCR and genotyping failures.

References

1. Majeed S, Rana IA, Atif RM, Ali Z, Hinze L, Azhar MT (2019) Role of SNPs in determining QTLs for major traits in cotton. J Cotton Res 2(1):5
2. Wang J, Zhang Z, Gong Z, Liang Y, Ai X, Sang Z, Zheng J (2022) Analysis of the genetic structure and diversity of upland cotton groups in different planting areas based on SNP markers. Gene 809:146042
3. Dwiningsih Y, Rahmaningsih M, Alkahtani J (2020) Development of single nucleotide polymorphism (SNP) markers in tropical crops. Adv Sustain Sci Eng Technol 2(2):343558
4. Kushanov FN, Pepper AE, Yu JZ, Buriev ZT, Shermatov SE, Saha S, Abdurakhmonov IY (2016) Development, genetic mapping and QTL association of cotton PHYA, PHYB, and HY5-specific CAPS and dCAPS markers. BMC Genet 17(1):141

5. Gowda SA, Fang H, Tyagi P, Bourland F, Dever J, Campbell BT, Kuraparthy V (2024) Genome-wide association study of fiber quality traits in US upland cotton (*Gossypium hirsutum* L.). Theor Appl Genet 137(9):214
6. Joshi B, Singh S, Tiwari GJ, Kumar H, Boopathi NM, Jaiswal S, Jena SN (2023) Genome-wide association study of fiber yield-related traits uncovers the novel genomic regions and candidate genes in Indian upland cotton (*Gossypium hirsutum* L.). Front Plant Sci 14:1252746
7. Liu W, Song C, Ren Z, Zhang Z, Pei X, Liu Y, Li W (2020) Genome-wide association study reveals the genetic basis of fiber quality traits in upland cotton (*Gossypium hirsutum* L.). BMC Plant Biol 20(1):395
8. Sun Z, Wang X, Liu Z, Gu Q, Zhang Y, Li Z, Ma Z (2017) Genome-wide association study discovered genetic variation and candidate genes of fibre quality traits in *Gossypium hirsutum* L. Plant Biotechnol J 15(8):982–996
9. Iqbal MA, Rahman MU (2017) Identification of marker-trait associations for lint traits in cotton. Front Plant Sci 8:86
10. Rogers SO, Bendich AJ (1994) Extraction of total cellular DNA from plants, algae and fungi. In: Gelvin SB, Schilperoort RA (eds.) Plant molecular biology manual. 2nd edn. Kluwer Academic Publishers/Springer, Dordrecht, Vol. D1, pp 183–190
11. Porebski S, Bailey LG, Baum BR (1997) Modification of a CTAB DNA extraction protocol for plants containing high polysaccharide and polyphenol components. Plant Mol Biol 15(1):8–15
12. Gurusamy D (2014) Transcriptome analysis of cotton (*Gossypium hirsutum* L.) during boll development under drought condition and functional characterization of drought tolerance genes. PhD. thesis, https://hdl.handle.net/10603/215568
13. Sivakumar JS, Gurusamy D, Chellamuthu S, Rathnasamy SA, Krishnan V (2015) Characterization of transgenic cotton plants by Southern blotting. In: Transgenic plants. Springer US, New York, NY, pp 171–181
14. Sivakumar JS, Gurusamy D, Chellamuthu S, Rathnasamy SA, Krishnan V (2025) Characterization of transgenic cotton plants by PCR. In: Transgenic plants. Springer US, New York, NY, pp 157–170
15. Li F, Fan G, Lu C, Xiao G, Zou C, Kohel RJ, Yu S (2015) Genome sequence of cultivated Upland cotton (*Gossypium hirsutum* L., TM-1) provides insights into genome evolution. Nat Biotechnol 33(5):524–530
16. Logan-Young CJ, Yu JZ, Verma SK, Percy RG, Pepper AE (2015) SNP discovery in complex allotetraploid genomes (*Gossypium* spp., Malvaceae) using genotyping by sequencing. Appl Plant Sci 3(3):1400077
17. Gaudet M, Fara AG, Sabatti M, Kuzminsky E, Mugnozza GS (2007) Single-reaction for SNP genotyping on agarose gel by allele-specific PCR in black poplar (*Populus nigra* L.). Plant Mol Biol Rep 25(1):1–9
18. Wang H, Jin X, Zhang B, Shen C, Lin Z (2015) Enrichment of an intraspecific genetic map of upland cotton by developing markers using parental RAD sequencing. DNA Res 22(2):147–160

Chapter 26

SSR Marker-Assisted Selection and Monitoring of High-Metabolite Cell Lines in Plant Tissue Culture: Protocols and Somaclonal Variation Assessment

W.M.A.C. Premasiri, R.M.Y.I. Rathnyake, C.M.N.T. Chandrasekara and D.L.C.K. Fonseka

Abstract

Simple Sequence Repeat (SSR) marker-assisted selection has proven to be the most powerful tool for genetic monitoring and selection in in vitro culture systems. It is widely used in detecting somaclonal variation and producing true-to-type in vitro plantlets with superior quality. Genetic fidelity is the critical point in plant tissue culture, and SSR markers act as a supporter to pass this critical point through locus specificity, codominancy and high polymorphism. SSR markers allow the precise detection of genetic changes during callus formation, organogenesis, somatic embryogenesis, and long-term subculturing. Secondary metabolite production can be enhanced through an in vitro culture system, and SSR markers enable the identification and continuous monitoring of genetic stability and high metabolite production in cell lines. Their co-dominant nature allows differentiation between homozygous and heterozygous alleles with deeper insights into allelic stability across successive culture cycles. Moreover, SSR markers exhibit high reproducibility and genome-wide distribution, making them reliable for long-term genetic monitoring. Overall, SSR marker–based analysis strengthens the reliability, efficiency, and commercial applicability of plant tissue culture by ensuring genetic uniformity and supporting the selection of elite in vitro materials. This chapter aims to review all the aspects related to the overview of SSR markers, real-world applications, somaclonal variations, phytochemical production, genetic fidelity, challenges, and future prospects.

Keywords In vitro, DNA markers, Phytochemical production, Somaclonal variation, SSR marker

1 Introduction

In modern biotechnology, plant tissue culture plays a big role. It acts as the most powerful tool for rapid propagation, conservation of plant genetic resources, and secondary metabolites production under in vitro conditions [1]. Plant-derived secondary metabolites are widely used in the pharmaceutical industry. Therefore, it is essential to identify and maintain high metabolite-producing cell lines for efficient metabolite production. Conventional methods have many constraints

Nitish Kumar (ed.), *Molecular Markers in Plants*, Methods in Molecular Biology, vol. 3097,
https://doi.org/10.1007/978-1-0716-5663-1_26,

and are mainly based on morphological traits and growth characteristics [2]. Even though these methods provide direct information on metabolic content, they are time-consuming, labor-intensive, costly, and destructive. In addition, somaclonal variation may result in undesirable metabolite production due to unpredictable changes in biosynthetic pathways [3]. Therefore, continuous monitoring of genetic stability is crucial for the development of reliable high-metabolite cell lines.

Integration of molecular markers with the plant tissue culture system is the best way to overcome these limitations. Molecular markers provide direct DNA-level information on genetic variation. The key advantage of using these molecular markers is that they are not influenced by environmental factors or developmental stages [4]. Among the available marker systems, Simple Sequence Repeat (SSR) markers are more suitable due to their high polymorphism, codominant inheritance, wide genome distribution, and high reproducibility [5–8]. Simple Sequence Repeat (SSR) markers are short tandem repeats of one to six base pairs distributed throughout nuclear and organellar genomes. SSR markers are highly polymorphic, co-dominant, reproducible, and PCR-based, making them suitable for genetic analysis in tissue culture systems [9]. SSR markers are widely applied for the detection of somaclonal variation, assessment of genetic fidelity, identification and maintenance of elite cell lines, and marker-assisted selection linked to metabolite traits [10]. Their ability to detect subtle genetic changes at early culture stages makes SSR markers particularly valuable in in vitro metabolite production systems.

2 Materials

2.1 Plant Materials

1. Explants from genetically stable, healthy, and disease-free mother plants for in vitro culture [11].
2. Early-stage in vitro cultures, callus and suspension cultures [12] and regenerated plantlets [13] (Fig. 1).

Fig. 1 Stages of in vitro regeneration of medicinal plants. **a** Early-stage in vitro cultures, **b** callus culture, **c** suspension culture, and **d** regenerated plantlet stage

2.2 DNA Extraction

1. CTAB extraction buffer or commercial plant DNA extraction kit (*see* **Note 1**).
2. 5–10 ml of liquid nitrogen (at −196 °C) is needed to completely cover and freeze the 50–100 mg of plant tissue before DNA extraction (*see* **Note 2**).
3. Chloroform/isoamyl alcohol at a ratio of 24: 1 and stored at 4 °C.
4. 500 μl of chilled isopropanol: incubated at −20 °C for 20–30 min and centrifuged to pellet the DNA.
5. 70% ethanol (*see* **Note 3**).
6. TE buffer: 10 mM Tris-HCl, 1 mM EDTA, pH 8.0 (*see* **Note 4**).
7. RNase A: 1–2 μl (10 mg/ml stock). Incubated at 37 °C for 15–30 min (*see* **Note 5**).

2.3 SSR Marker Selection Materials

1. Published SSR primer sequences for species-specific or transferable markers.
2. NCBI database access for primer information (if needed).
3. Synthesized forward and reverse primers.

2.4 Polymerase Chain Reaction (PCR) Amplification Materials

1. PCR machine for programmed temperature cycling.
2. Taq DNA polymerase (DNA amplification enzyme).
3. 10× PCR buffer.
4. $MgCl_2$: required for Taq polymerase activity, final concentration at 1.5–2.5 mM final.
5. dNTP mix: 200 μM of each nucleotide (dATP, dTTP, dCTP, dGTP).
6. Forward and reverse SSR primers: 0.2–0.5 μM from each.
7. Template genomic DNA: 20–50 ng per reaction.
8. Nuclease-free water**:** usually 25 μl volume.

2.5 Electrophoresis

1. Agarose gel electrophoresis: agarose gel (0.8–2%), TBE or TAE buffer (50–100 ml), DNA ladder (100 bp or appropriate size range), Ethidium bromide or safe DNA stain, Gel casting tray, and comb, power supply, UV transilluminator or gel documentation system.
2. Capillary electrophoresis: fluorescently labeled primers and a capillary electrophoresis system.

3 Methods

3.1 In Vitro Propagation

1. Collected explants from mother plants are inoculated on suitable culture media under in vitro conditions.

2. In vitro established explants are subcultured to produce callus, suspension cultures, and regenerated plants.
3. Sampling at regular subculture intervals is needed.

3.2 Genomic DNA Extraction

1. Genomic DNA is extracted from in vitro cultured plant tissues using the CTAB method or a commercial plant DNA extraction kit.
2. High-quality, pure DNA is ensured for SSR amplification by checking DNA quality on agarose gel and quantifying using a spectrophotometer.
3. RNase A is used to treat DNA to remove RNA contamination and ensure accurate PCR results.

3.3 Assemble Polymerase Chain Reaction (PCR) with SSR Primers

1. Selected informative SSR markers are used to ensure genome-wide coverage.
2. PCR reactions are prepared under standard thermal cycling conditions with template DNA, primers, Taq polymerase, buffer, $MgCl_2$, dNTPs, and nuclease-free water.
3. Amplified DNA is run on electrophoresis (agarose gel or capillary) to resolve PCR products. TBE or TAE buffer can be used with a DNA ladder for sizing.

3.4 Analyze SSR Banding Patterns and Assessment of Genetic Fidelity

1. Banding patterns are needed to compare with in vitro cultures, including early and late subculture stages and the mother plant.
2. Identification of any genetic deviations compared to the mother plant is required to detect stable lines and somaclonal variants (based on the presence or absence of bands, allele size variation, and appearance of novel alleles).
3. Cultures showing SSR profiles identical to the mother plant are considered genetically stable.
4. Genetically stable lines are selected for further subculturing, regeneration, or metabolite production.
5. Novel alleles, missing bands, or allele size variations of cultures are classified as somaclonal variants.
6. These somaclonal variants can be discarded or studied further to understand genetic variation induced during tissue culture.
7. All SSR data, including gel images, allele sizes, and profiles, are recorded and stored.
8. This documentation allows tracking of genetic stability over time and verifies reproducibility in long-term cultures

3.5 Identification and Monitoring of High-Metabolite Cell Lines

1. Genetically stable cultures confirmed by SSR analysis are further screened to identify high-metabolite-producing cell lines.
2. Selected high-metabolite lines are maintained under controlled culture conditions.

3. SSR markers are periodically applied to these lines to ensure continued genetic fidelity while maintaining consistent metabolite production over successive subcultures.

3.6 Data Analysis and Interpretation

1. SSR banding patterns from all stages (mother plant, initial stages of in vitro cultures, subcultures, and high-metabolite lines) are compared to assess genetic stability.
2. Genetically stable lines and somaclonal variants are identified, confirmed, and documented.
3. Statistical analyses can be done using GenAlEx, PowerMarker, and NTSYS, which are applied to calculate genetic similarity, diversity indices, or dendrograms.
4. Data are interpreted to select genetically stable, high-metabolite cell lines for further propagation or application.
5. This step ensures reproducibility, reliability, and practical application of SSR-assisted selection in tissue culture.

4 Notes

1. 100 mg of fresh leaf or in vitro-derived tissue is enough to be frozen in liquid nitrogen and homogenized in 700 μl of preheated (65 °C) 2% CTAB extraction buffer containing 2% CTAB, 1.4 M NaCl, 100 mM Tris-HCl (pH 8.0), and 20 mM EDTA, with freshly added 0.2% β-mercaptoethanol to prevent oxidation of phenolic compounds. The mixture will be incubated at 65 °C for 30 min to facilitate cell lysis and release of nucleic acid.
2. Liquid nitrogen will be used to freeze the plant tissue before DNA extraction. About 5–10 ml of liquid nitrogen is sufficient to completely cover and freeze 50–100 mg of tissue in a mortar. The tissue will be ground into a fine powder while keeping it frozen. This step prevents DNA degradation and helps break the cell walls effectively, ensuring good-quality DNA for SSR marker analysis.
3. For SSR marker analysis, after DNA precipitation with isopropanol, the DNA pellet is washed with about 500 μl of 70% ethanol to remove impurities
4. The TE buffer is used to dissolve and store the extracted genomic DNA. Tris maintains a stable pH, while EDTA chelates divalent ions like Mg^{2+} to protect DNA from degradation by nucleases. For SSR analysis, DNA is typically dissolved in 50–100 μl of TE buffer to achieve a good concentration for PCR.

5. RNase A is an enzyme added to the DNA solution to remove RNA contamination after extraction. This ensures that only pure DNA is present, which is important because RNA can interfere with accurate quantification and PCR amplification of SSR markers.

References

1. George EF, Hall MA, De Klerk G-. (2008) Plant propagation by tissue culture. Springer
2. Verpoorte R, van der Heijden R, Memelink J (2002) Engineering the plant cell factory for secondary metabolite production. Transgenic Res 9:323–343
3. Bairu MW, Aremu AO, Van Staden J (2011) Somaclonal variation in plants: causes and detection methods. Plant Growth Regul 63(2):147–173
4. Agarwal M, Shrivastava N, Padh H (2008) Advances in molecular marker techniques and their applications in plant sciences. Plant Cell Rep 27(4):617–631
5. Powell W, Machray GC, Provan J (1996) Polymorphism revealed by simple sequence repeats. Trends Plant Sci 1:215–222
6. Gupta PK, Varshney RK (2000) The development and use of microsatellite markers for genetic analysis and plant breeding. Euphytica 113:163–185
7. Rani V, Raina SN (2000) Genetic fidelity of organized meristem-derived micropropagated plants: a critical reappraisal. In Vitro Cell Dev Biol Plant 36(5):319–330
8. Shi Z, Zhao W, Li Z, Kang D, Ai P, Ding H, Wang Z (2022) Development and validation of SSR markers related to flower color based on full-length transcriptome sequencing in *Chrysanthemum*. Sci Rep 12(1):22310–22310
9. Lebedev VG, Subbotina NM, Maluchenko OP, Lebedeva TN, Krutovsky KV, Shestibratov KA (2019) Transferability and polymorphism of SSR markers located in flavonoid pathway genes in *Fragaria* and *Rubus* species. Genes 11(1):11
10. Maurya RL, Kumar M, Sirohi U, Priya, Chaudhary V, Sharma VR, Yadav MK (2022) An effective micropropagation protocol and determination of the clonal fidelity of in vitro developed microshoots of carnation (*Dianthus caryophyllus* L.) using SSR markers. Nucleus 65(1):49–55
11. Bhardwaj S, Kumari M, Chethan T, Hosamani M, Paul M, Moirangthem A, Verma S (2025) Advances in micropropagation and tissue culture for horticultural crops: a review. Plant Cell Biotechnol Mol Biol 26(11-12):166–182
12. Ioannidis K, Tomprou I, Mitsis V, Koropouli P (2022) Genetic evaluation of *in vitro* micropropagated and regenerated plants of *Cannabis sativa* L. using SSR molecular markers. Plants 11(19):2569
13. Kapoor B, Sharma M, Sharma R, Zadokar A, Thakur A, Sharma P, Kumar S, Rozar KP, Kumar KS, Hegde N, Pandey D (2023) De novo transcriptome profiling and development of novel secondary metabolites based on genic SSRs in medicinal plant *Phyllanthus emblica* L. (Aonla). Sci Rep, 13(1):17319

Chapter 27

ISSR Marker-Based Authentication of In Vitro-Derived Medicinal Plant Lines for Enhanced Phytochemical Production

W.M.A.C. Premasiri, W.G.I. Madushani and D.L.C.K. Fonseka

Abstract

Medicinal plants have been used to heal people with thousands of diseases. Phytochemicals are the main reason for a plant being a medicinal plant. Compared to field-grown plants, in vitro-derived plantlets have high phytochemical production. In vitro-derived plantlets are also the most reliable method for producing healthy, disease-free medicinal plants of superior quality, with enhanced phytochemical properties. Due to the presence of phytochemical compounds such as phenolic compounds, flavonoids, tannins, and terpenoids, medicinal plants possess therapeutic properties. These therapeutic properties open doors to making a significant contribution to the pharmaceutical industry. Genetic fidelity is a critical factor in in vitro-derived plantlets, ensuring true-to-type plants. The application of Inter Simple Sequence Repeat (ISSR) markers produces true-to-type in vitro plantlets by verifying genetic fidelity while supporting the scaling up of phytochemical production. The ISSR marker is an ideal marker for in vitro-based cultivation systems compared to other markers. ISSR, a DNA-based marker, is simple and fast. Even though ISSR markers are the most reliable markers for in vitro plantlets, the application process can be challenging. In this chapter, we discuss all aspects related to ISSR marker-based authentication of in vitro-derived medicinal plant lines for enhanced phytochemical production, including real-world applications, challenges and limitations, general procedures, and future prospects.

Keywords Genetic fidelity, In vitro, ISSR marker, Medicinal plants, Phytochemicals

1 Introduction

Medicinal plants are a priceless source of bioactive compounds for pharmaceuticals, nutraceuticals, and traditional medicines. Diverse global studies have shown that almost 70%–80% of the global population depends on plant-based medication for primary health care, while about 25% of modern drugs are derived directly or indirectly from plant secondary metabolites [1]. Recent studies have reported successful in vitro propagation methods for various medicinal plants, aiding in the large-scale production of disease-free and uniform plantlets [2, 3]. Cultured cells or regenerated plantlets with genetic

Nitish Kumar (ed.), *Molecular Markers in Plants*, Methods in Molecular Biology, vol. 3097,
https://doi.org/10.1007/978-1-0716-5663-1_27,

instability can adversely change the biosynthetic pathways leading to qualitative and quantitative changes in metabolites. Inter-simple sequence repeat (ISSR) is the most reliable and informative tool for evaluating the genetic stability of in vitro-derived plantlets [4].

ISSR analysis with microsatellite is mutually supportive, and this combination provides primers for the other [5]. It has been confirmed that ISSR-PCR analysis is identical to that of the mother plants, without any variation in major phytochemicals [6, 7]. ISSR-based approaches have been successfully applied in medicinal plants to correlate genetic diversity with biochemical variability, thereby supporting marker-assisted selection and conservation strategies aimed at improving phytochemical yield and quality. Although ISSR markers are dominant in nature, their efficiency, cost-effectiveness, and ability to detect polymorphism make them valuable tools for preliminary screening and enhancement of phytochemical traits in plant breeding and biotechnology.

2 Materials

Explants such as nodal segments, shoot tip, and young leaves from the healthy and disease-free mother plants (Fig. 1).

3 Methods

1. Explants are selected for in vitro cultivation and inoculated on a suitable culture medium following surface sterilization.
2. Shoot-initiated explants are subcultured for plant regeneration (Fig. 2) and callus induction (Fig. 3) in plant growth regulator-supplemented culture media.

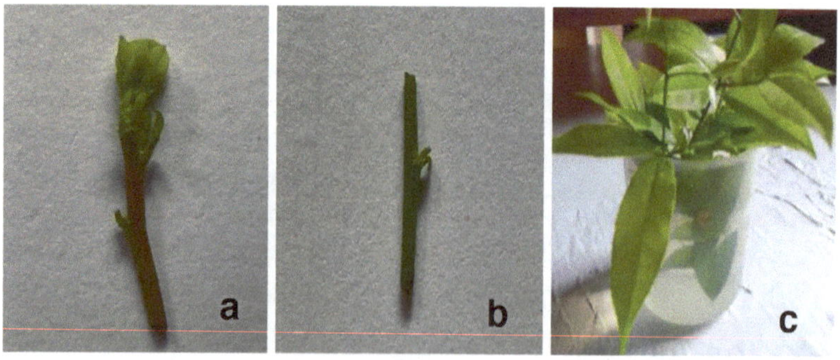

Fig. 1 Different types of explants. **a** Shoot tip, **b** nodal segment, and **c** young leaves

Fig. 2 Selected regenerated in vitro medicinal plants. **a** *Evolvulus alsinoides*, **b** *Salacia reticulata*, **c** *Zingiber officinale*, **d** *Curcuma longa*, and **e** *Flueggea leucopyrus*

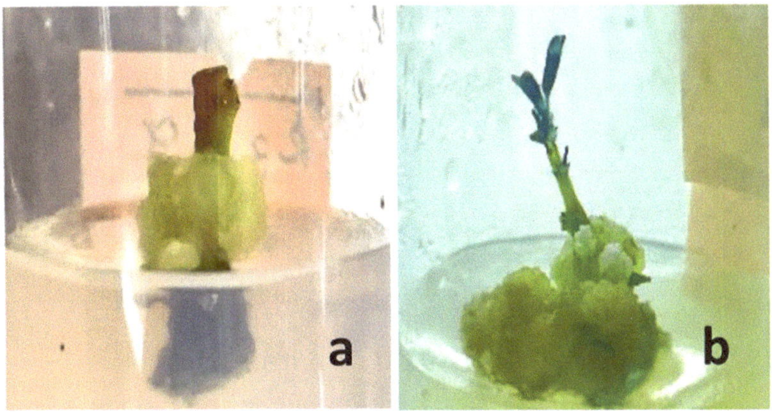

Fig. 3 Callus induction from different explants: **a** callus derived from nodal segment and **b** callus derived from shoot tip

3. All the cultures are maintained under sterilized and controlled conditions. Field-grown mother plants are used as reference controls for genetic authentication.
4. Leaf samples from the in vitro-derived plantlets are immediately frozen in liquid nitrogen and are stored at −80 °C before DNA extraction to prevent degradation [8, 9].
5. Genomic DNA is extracted using the cetyltrimethylammonium bromide (CTAB) method [10].
6. A set of 8–15 inter-simple sequence repeat (ISSR) primers containing di- and tri-nucleotide repeat motifs is selected based on reproducibility and clarity of amplification profiles (*see* **Note 1**).
7. Polymerase chain reaction (PCR) amplification is undertaken in a 25 μl reaction mixture consisting of template DNA, PCR buffer with $MgCl_2$, dNTPs, primer, Taq DNA polymerase, and nuclease-free water [11–13].

8. To ensure reproducibility, each amplification reaction must be repeated at least twice.
9. Amplified ISSR products are separated on 1.5%–2.0% agarose gels prepared in 1× TBE buffer and are stained with ethidium bromide or SYBR Safe (*see* **Note 2**).
10. Banding patterns are visualized under a UV transilluminator, and all the visualized patterns are documented using a gel imaging system [14].
11. A binary data matrix is created, where clear and reproducible ISSR bands are scored as 1 (present) and 0 (absent) for all samples [15].
12. Jaccard's similarity index is used to calculate the genetic similarity coefficients, and cluster analysis is performed using the unweighted pair group method with arithmetic mean (UPGMA) to generate dendrograms illustrating genetic relationships [16].
13. Comparison between ISSR banding profiles with those of the mother plant confirms the genetic fidelity of in vitro-derived plantlets.
14. The presence of monomorphic bands indicates genetic stability, whereas polymorphic bands suggest somaclonal variation [17].
15. ISSR-based authentication is employed to ensure that enhanced phytochemical production observed in selected plant lines is not associated with culture-induced genetic variation.
16. Genetic similarity coefficients are calculated using Jaccard's similarity index, and cluster analysis is conducted using the unweighted pair group method with arithmetic mean (UPGMA) to generate dendrograms illustrating genetic relationships among the plantlets.
17. Correlation analysis is performed to evaluate the association between ISSR-derived genetic variation and quantitative phytochemical traits.
18. All experiments are conducted in triplicate, and statistical analyses are performed to assess both phytochemical variation and genetic fidelity [18].

4 Notes

1. DNA quality is assessed by electrophoresis on 0.8% agarose gel, while DNA concentration and purity are determined spectrophotometrically using the A260/A280 ratio [19].

2. PCR cycling conditions are validated with an initial denaturation at 94 °C for 3 min, followed by 35–40 cycles of denaturation at 94 °C for 45 s, primer-specific annealing at 45–60 °C for 45 s, and extension at 72 °C for 1–2 min, with a final extension at 72 °C for 7 min [20].

References

1. Atanasov AG, Zotchev SB, Dirsch VM, Supuran CT (2021) Natural products in drug discovery: advances and opportunities. Nat Rev Drug Discov 20(3):200–216
2. Moraes RM, Cerdeira AL, Lourenço MV (2021) Using micropropagation to develop medicinal plants into crops. Molecules 26(6):1752
3. Anuruddi HIGK, Nakandalage N, Fonseka DK (2023) New insights for the production of medicinal plant materials: ex vitro and in vitro propagation. In: Biosynthesis of bioactive compounds in medicinal and aromatic plants: manipulation by conventional and biotechnological approaches. Springer, Cham, pp 181–212
4. Patel D, Sharma S, Reddy V (2023) ISSR marker selection for reproducible amplification in plant biotechnology studies. 3 Biotech 13(1):120
5. Rathore N, Sharma P, Singh V (2021) ISSR-based genetic stability assessment of micropropagated plants. J Appl Genet 62(3):311–322
6. Rani S, Verma R, Singh A (2020) Documentation of PCR-amplified ISSR products using gel imaging systems. Plant Cell Rep 39(8):1057–1065
7. Ahmad S, Khan MR, Sharma P (2024) Evaluation of somaclonal variation in in vitro-derived plants using ISSR markers. Plant Biotechnol J 20(5):821–835
8. Rani S, Singh A, Dhawan O (2021) Leaf sampling and storage for high-quality DNA extraction in medicinal plants. Plant Biotechnol Rep 15(4):455–466
9. Singh R, Verma R, Bairu MW (2023) Cluster analysis of ISSR data for genetic relationship assessment in in vitro-derived plants. Mol Biol Rep 50(7):6547–6559
10. Sharma A, Gupta R, Singh P (2023) CTAB-based DNA extraction for high secondary metabolite medicinal plants. Plant Mol Biol Report 41(2):123–135
11. Zietkiewicz E, Rafalski A, Labuda D (1994) Genome fingerprinting by SSR-anchored polymerase chain reaction amplification. Genomics 20(2):176–183
12. Patel D, Rani S, Bairu MW (2024) Binary data analysis and statistical assessment of ISSR polymorphisms in plant populations. Plant Mol Biol Report 42(1):15–28
13. Rathore N, Ahmad S, Verma R (2022) Evaluation of somaclonal variation in in vitro-derived plants using ISSR markers. Plant Biotechnol J 20(5):821–835
14. Verma R, Singh A, Bairu MW (2023) Reproducibility and reliability of ISSR markers in medicinal plant biotechnology. Plant Cell Tissue Organ Cult 144(2):415–429
15. Bairu MW, Aremu AO, van Staden J (2011) The role of plant growth regulators in tissue culture of medicinal plants. S Afr J Botany 77(4):847–857
16. Sneath PHA, Sokal RR (1973) Numerical taxonomy: the principles and practice of numerical classification. W. H. Freeman and Company, San Francisco, CA
17. Biswas P, Nandy S, Dey A, Tikendra L, Nongdam P (2022) Molecular markers in assessing genetic clonal fidelity for in vitro propagated endangered medicinal plants. In: Molecular genetics and genomics tools in biodiversity conservation. Springer Nature, Singapore, pp 97–149
18. Ahmad S, Khan MR, Sharma P (2024) Genetic fidelity assessment of in vitro-propagated medicinal plants using ISSR markers. Plant Cell Tissue Organ Cult 137(1):75–88
19. Kumar S, Pandey V, Singh P (2022) Spectrophotometric assessment of DNA quality and purity in plant molecular studies. Mol Biol Rep 49(5):4567–4575
20. Gupta S, Singh A, Yadav K, Pandey N, Kumar S (2022) Micropropagation for multiplication of disease-free and genetically uniform sugarcane plantlets. In: Advances in plant tissue culture. Academic Press, pp 31–49

FSC
www.fsc.org
MIX
Papier | Fördert
gute Waldnutzung
FSC® C083411

Kolibri 360 GmbH
Ferdinand-Jühlke-Straße 7
99095 Erfurt, Deutschland
produktsicherheit@kolibri360.de